Lecture Notes in Mobility

The book series Lecture Notes in Mobility (LNMOB) reports on innovative, peer-reviewed research and developments in intelligent, connected and sustainable transportation systems of the future. It covers technological advances, research, developments and applications, as well as business models, management systems and policy implementation relating to: zero-emission, electric and energy-efficient vehicles; alternative and optimized powertrains; vehicle automation and cooperation; clean, user-centric and on-demand transport systems; shared mobility services and intermodal hubs; energy, data and communication infrastructure for transportation; and micromobility and soft urban modes, among other topics. The series gives a special emphasis to sustainable, seamless and inclusive transformation strategies and covers both traditional and any new transportation modes for passengers and goods. Cutting-edge findings from public research funding programs in Europe, America and Asia do represent an important source of content for this series. PhD thesis of exceptional value may also be considered for publication. Supervised by a scientific advisory board of world-leading scholars and professionals, the Lecture Notes in Mobility are intended to offer an authoritative and comprehensive source of information on the latest transportation technology and mobility trends to an audience of researchers, practitioners, policymakers, and advanced-level students, and a multidisciplinary platform fostering the exchange of ideas and collaboration between the different groups.

Constantinos Antoniou · Fritz Busch · Andreas Rau · Mahesh Hariharan
Editors

Proceedings of the 12th International Scientific Conference on Mobility and Transport

Mobility Innovations for Growing Megacities

Springer

Editors
Constantinos Antoniou
TUM School of Engineering and Design
Technical University of Munich
Munich, Germany

Fritz Busch
TUM School of Engineering and Design
Technical University of Munich
Munich, Germany

Andreas Rau
TUM Asia Pte Ltd.
Singapore, Singapore

Mahesh Hariharan
Railway Engineering
TUM Asia Pte Ltd.
Singapore, Singapore

ISSN 2196-5544 ISSN 2196-5552 (electronic)
Lecture Notes in Mobility
ISBN 978-981-19-8360-3 ISBN 978-981-19-8361-0 (eBook)
https://doi.org/10.1007/978-981-19-8361-0

This Springer imprint is published by the registered company Springer Nature Singapore Pte Ltd.
The registered company address is: 152 Beach Road, #21-01/04 Gateway East, Singapore 189721, Singapore

Contents

Editorial

Constantinos Antoniou, Fritz Busch, Andreas Rau, and Mahesh Hariharan

"Mobility Innovations for Growing MegaCities"—was the topic of mobil.TUM 2022, International Scientific Conference on Mobility and Transport. Based on a historical review on the developments in transportation engineering, we wish to contribute to the ongoing transformation processes in this field. More than 60 peers joined this event held online in April 2022.

This twelfth edition of the mobil.TUM was co-organized by TUM Asia in Singapore and the TUM professorships of the Mobility Systems Engineering Department to organize a truly interdisciplinary event. In particular, the conference aimed to invoke debates among keynote speakers from diverse and multi-faceted research backgrounds, enabled through strong bonds between the wide range of topics and methods such as data analytics, active mobility solutions, transportation demand management, cloud computing and internet of things applied in ITS, transport impacts on climate change, traffic microsimulation etc.

Selected papers of the conference are published in the book series, Lecture Notes in Mobility, in the form of full papers. We are grateful to all participants for three inspiring and conversation-stimulating days. Given the times that we live in, the decision to host mobil.TUM online was an extremely difficult, but necessary one. We have observed intense in-depth discussions about highly specialized methodologies,

C. Antoniou (✉) · F. Busch
TUM School of Engineering and Design, Technical University of Munich, Munich, Germany
e-mail: c.antoniou@tum.de

F. Busch
e-mail: fritz.busch@tum.de

A. Rau · M. Hariharan
TUM Asia Pte Ltd, Singapore, Singapore
e-mail: andreas.rau@tum-asia.edu.sg

M. Hariharan
e-mail: mahesh@tum-asia.edu.sg

C. Antoniou et al. (eds.), *Proceedings of the 12th International Scientific Conference on Mobility and Transport*, Lecture Notes in Mobility,
https://doi.org/10.1007/978-981-19-8361-0_1

exchange between the disciplines in cross-cutting sessions, where a common research subject (e.g. V2V, V2X, TOD etc.) was analyzed from different viewpoints (planning, modelling, traffic management), and also interactions between different generations of researchers.

We would like to express our deepest gratitude to all participants of mobil.TUM 2022. We are grateful that you contributed to making this conference a very special one, especially in the year that marks TUM Asia's 20th Anniversary. Finally, we would like to thank our entire team who made this digital conference possible, including the Scientific Committee who supported the review process to ensure the quality as well as the organizational team, who did all the hard work in the background.

To a better future.

The Impact of Autonomous Vehicles and Their Driving Parameters on Urban Road Traffic

Bernd Kaltenhäuser, Sascha Hamzehi, and Klaus Bogenberger

Abstract Traffic congestion might be partly solved by using autonomously driving vehicles which are expected to enter the market at a significant rate within the next years (Kaltenhäuser et al. in Transp Res Part A: Policy Pract 132:882–910, 2020, [1]; Bansal and Kockelman KM in Transp Res Part A: Policy Pract 95:49–63, 2017, [2]; Nieuwenhuijsen et al. in Transp Res Part C: Emerg Techno 86:300–327, 2018, [3]). Several studies have been undertaken to examine the impact of autonomous vehicles (AVs) on road traffic. Also, autonomous vehicles and connected autonomous vehicles (CAVs) have been simulated in the literature with different operational parameters, leading to different results. Hence, in our study we examine how different parameters for the operation of AVs and CAVs influence urban traffic in the case of Munich, Germany. Furthermore, the impact of different percentages of AVs and CAVs on urban traffic is studied. For this, the traffic will be studied for the whole city, as well as for certain travel routes, e.g. in the main travel direction (into the city in the morning), in opposite direction or along the highway surrounding Munich. Last but not least, future scenarios with an enhanced travel behaviour will be studied. The results show that the headway and reaction times of the vehicles have the largest impact on urban traffic. Here, vehicles with large reaction times have a negative impact on urban traffic while short reaction times have a positive one. The results can be used to configure future AVs such that they reduce congestions and optimize urban traffic flow.

B. Kaltenhäuser (✉)
Department of Technical Management, Baden-Wuerttemberg Cooperative State University, Villingen-Schwenningen, Germany
e-mail: bernd.kaltenhaeuser@dhbw-vs.de

B. Kaltenhäuser · S. Hamzehi · K. Bogenberger
Technical University of Munich, Chair of Traffic Engineering and Control, Munich, Germany
e-mail: sascha.hamzehi@bmw.de

K. Bogenberger
e-mail: klaus.bogenberger@tum.de

S. Hamzehi
BMW Group, Munich, Germany

C. Antoniou et al. (eds.), *Proceedings of the 12th International Scientific Conference on Mobility and Transport*, Lecture Notes in Mobility,
https://doi.org/10.1007/978-981-19-8361-0_2

Keywords Autonomous Vehicles (AVs) · Urban traffic · Driving parameters · Road network capacity

1 Problem Statement

At present, the automotive industry is undergoing a significant change, mainly due to the large-scale introduction of electric and hybrid drives, shared mobility and individual mobility offers. The biggest social changes, however, will arise from the development of autonomous driving. Initially, it will be just an optional feature where the driver is still in control of the vehicle, but, on the contrary, with the introduction of autonomous taxis without a steering wheel, as proposed e.g. by Waymo, the technology will soon have a key influence on daily life. Besides the usage prediction of all these types of mobility, the impact of autonomously driven vehicles on urban and intercity traffic is a major field of investigation. Most authors focus on single and double lane traffic as it is found on highways, where to our knowledge, only three publications have studied urban traffic: Bailey [4], Yeola et al. [5] and Mavromatis et al. [6].

2 Research Objectives and Motivation

Hence, in our study we examine how different parameters for the operation of AVs and CAVs influence urban traffic in the case of Munich, Germany. These parameters are as different as the headway time, the acceptance of speed limits, or the acceptance of small (risky) gaps for overtaking manoeuvers. The influence of the parameters will be studied for the traffic in total as well as for certain travel routes, e.g. in the main travel direction (into the city in the morning), in opposite direction or along the highway surrounding Munich. The used key performance indicators are the travel time, the harmonic average speed and the delay time. Furthermore, the impact of different percentages of AVs and CAVs on urban road traffic will be studied. The results might be used to configure future AVs such that they reduce congestions and optimize urban traffic flow.

3 Literature Review

Several studies that examine the impact of autonomous vehicles on traffic have been undertaken. Table 1 shows a summery of the literature. As most studies focus on highways or straight road strips, these are shown first, followed by the three studies examining urban road traffic and then two studies examining platoons in urban traffic. As the headway time (HWT] chosen for the AVs is the most important parameter in

Table 1 Survey of the relevant literature

Topic	Parameters and results	Literature
Single and double lane traffic	Longer HWT leading to reduced traffic flow or capacity	[8–11, 23, 24]
	Lower HWT or higher HWT variance leading to increased traffic flow or capacity	[4, 8–11, 13, 14, 17, 19]
Urban traffic	Lower HWT leading to increased traffic flow or capacity	[4–6]
	Platoons in urban traffic	[34, 35]
The parameters of autonomous driving	Using single values single values for the HWT or reaction time	[4, 8, 13, 14, 20, 23, 25]
	Using a range of values for the HWT or reaction time	[7–11, 19, 26–28]
	Further parameters defining the behaviour of AVs	[7, 8, 13, 14, 29]
Literature on KPIs		[4, 5, 7 – 9, 13, 20, 30–33]

most publications, the studies are sorted according to this parameter first. Afterwards, studies using different parameters for the simulation of AVs are shown, followed by studies using different KPIs.

It can be seen that a lot of studies simulated AVs and CAVs on single and double lane roads. Also, lots of parameters have been used for the definition of AVs and CAVs and the parameters could be used in this study. Furthermore, the literature on KPIs is thorough and the KPIs can be used for this study. However, as the literature on the impact of autonomous driving vehicles on urban road traffic is quite sparse, this will be the main topic of the traffic simulation.

In the following, a more detailed survey of the literature is given, following the order of Table 1.

Studies showing increasing and decreasing traffic flow according to the headway time

Krause et al. [8] and Hartmann et al. [9] examined the impact of AVs and CAVs on highway capacity using VISSIM. Compared to AVs, human drivers were simulated with a higher variance in the HWT, centered for both around 1.1 s. In contrast, CAVs used a lower HWT (0.5 and 0.9 s), but only to other CAVs, while AVs used a longer HWT (1.8 s). The simulations have shown that AVs reduce highway capacity while CAVs increase it, mainly due to the modified HWT. The effects increased with a higher percentage of the respective vehicles, with CAVs increasing the road capacity by up to 30%.

Also, Papamichail et al. [10] have shown that the HWT (0.8–2 s) has an impact on road capacity on a single lane road. Here, a shorter HWT resulted in a higher capacity and vice versa. Ntousakis et al. [11] found that highway capacity increases linearly with the penetration rate of adaptive cruise control (ACC) vehicles with desired time gaps less than 1.1 s. Likewise, the capacity decreased for desired time gaps exceeding 1.5 s. Here, the car-following model proposed by Wang and Rajamani [12] was used.

Studies showing an increasing traffic flow, most of them due to short headway times

Aria et al. [13] have used VISSIM to simulate a 3 km long highway including drive-ups. With the usage of CAVs with a HWT of 0.3 s, the average speed increased and in return the travel time was reduced. Due to the higher velocity, less vehicles were using the road at the same time. Motamedidehkordi et al. [14] have simulated a 8.5 km strip on the German highway A5. Here, the road capacity increased with a higher percentage of AVs using a HWT of 0.5 s. Arnaout and Arnaout [15] have simulated a multi-lane highway with a mix of cooperative cruise control (CACC) vehicles and manually driven ones. At low magnitudes of the vehicle flow, which reflect moderate traffic conditions, there was not found any significant statistical difference between cases with varying amounts of CACC vehicles. In contrast, in heavy traffic conditions an increase in CACC vehicles was found to increase the traffic flow significantly. However, for CACC penetration rates of less than 40%, the effect was minimal.

Bailey [4] has simulated AVs with a reaction time of 0.1 s in Aimsun. The scope was one road with a crossing using different traffic light parameters. Here, an increasing number of AVs leads to a higher traffic flow and shorter travel times. Vanderwerf et al. [16] have used an own car-following model to study the impact of ACC- and CACC-equipped vehicles on freeway operations. They have concluded that at 100% penetration, ACCs increased highway capacity and CACCs-equipped increased it even further. Kesting et al. [17] have used the Intelligent Driver Model (IDM) by Treiber et al. [18] to examine a 13 km long three-lane freeway section. The parameters used are a safe time of 1.6 s (instead of 1.5 s) and higher accelaration values. Already with 25% AVs, traffic jams were non-existent anymore when using AVs with low time-gaps, which was found to be the most important factor. Shladover et al. [19] have used ACC parameters from Nissan and CACC and manual vehicle parameters from Chinese literature. For the HWT they used 31.1% of all vehicles with 2.2 s, 18.5% with 1.6 s and 50.4% with 1.1 s for ACCs; and 12% with 1.1 s, 7% with 0.9 s, 24% with 0.7 s and 57% with 0.6 s for CACCs. Here, ACCs did not have any impact on the road capacity as the headway times were similar to human drivers. In contrast, the road capacity increased steadily with an increasing number of CACCs.

Studies showing a reduced traffic flow, mainly due to larger headway times

Fountoulakis et al. [20] have used Aimsun with the modified Gipps driving model [21, 22] to study mixed traffic under congested as well as free-flow conditions. They have simulated a 10 km long 3-lane strip and the reaction time of all vehicles was set to 1 s. Here, the ACCs lead to more congestion. Van Arems et al. [23] have used a HWT of 1.5 s which resulted in a reduced traffic flow when the percentage of ACCs exceeded 40%. Also, Bierstedt et al. [24] have concluded that cautiously programmed AVs with conservative headways could reduce flows, densities and capacities on highways.

Publications showing an increased traffic flow or capacity in urban traffic due to lower HWTs

Bailey [4] has simulated the city of Lausanne, where more AVs with a reaction time of 0.1 s lead to a higher traffic flow (reaction time of 0.1 s). Yeola et al. [5] used Vissim to simulate the impact of CAVs in the Urban area of Singapore. The used key performance indicators (KPIs) were travel time, average speed and the queue lengths at junctions. The latter ones are expected to be decreased by 47% with respect to their future base scenario of the year 2030 (0% CAVs). Similarly, the average delay at the junctions is likely to decrease by 30%, resulting in an increasing average speed of the CAVs by 36%. Likewise, the link flow capacities on the Singapore expressways are expected to increase from 2,500 up to 3600 vehicles per hour with 100% CAVs. However, as the paper was just submitted, the used parameters are not known to us yet.

Mavromatis et al. [6] simulated the road networks of Manhattan, Paris, Berlin, Rome and London. Here, AVs with a headway time of 0.5 s (instead of 1.69 for human driven vehicles) and CAVs with a HWT of 0.1 s had a positive effect on traffic flow. However, solely randomly inserted vehicles were used instead of calibrated OD matrices.

Literature on platooning in urban traffic

The reduced headway or reaction time is often used to form platoons of CAVs. With a lower distance between the vehicles, they can cross traffic intersections (the bottlenecks in urban traffic) at a higher rate [34]. Thus, with platooning, urban traffic might be optimized as well. This has been shown by Lioris et al. [35] who simulated a road network with 16 intersections and 73 links. Here, the throughput doubled when forming CAV platoons.

The parameters headway and reaction time

In the studies shown above, mainly the HWT and the reaction time have been used to define the behaviour of AVs. A longer HWT as well as a longer reaction time both correspond to a larger distance to the vehicle in front and it can be easily shown that in a simple breaking model with a constant deceleration, equal headway and reaction times lead to exactly the same results, although differently defined. However, they are not always comparable as they are differently used in different car following models.

A huge range of headway times has been used throughout the literature:

- 0.3 s [13, 25]
- 0.5 s [14]
- 1.1 s [8], where human drivers use the same value, but with a higher variance, and
- 1.5 s [23].

Other papers study a range of values here:

- Papamichail et al. [10]: 0.8–2 s (instead of 1.1 s for human drivers).
- Hartmann et al. [9]: 0.5, 0.9 and 1.8 s (instead of 1.1 s for human driven vehicles (HVs)).
- Ntousakis et al. [11]: less than 1.1 and more than 1.5 s.
- Stogios [7]: 0.5–2.1 s.
- Van Arem et al. [26]: 0.5 s when following another CACC equipped vehicle and 1.4 s when following a non-CACC equipped vehicle.
- Gouy et al. [27]: 0.3 s and 1.4 s.
- Kesting et al. [28]: 0.9 – 2.5 s.
- Mavromatis et al. [6]: 0.5 s for AVs and 0.1 s for CAVs.

In contrast, Shladover et al. [19] used a distribution of parameters: 31.1% of all vehicles with 2.2 s, 18.5% with 1.6 s and 50.

Studies using the reaction time instead of the HWT have been conducted by

- Bailey [4] with 0.1 s and
- Fountoulakis et al. [20] with 1 s for all vehicles.

Both use the Gipps driving model.

Further parameters defining the behaviour of AVs

Aria et al. [13] have used the speed limit ±2 km/h as target speed. AVs change lanes at crossings and turns earlier than human drivers. CAVs were cooperating with each other when changing lanes. Krause et al. [8] have defined AVs to move faster back to the right driving lane. When changing lanes, AVs were defined to accept less delays for the following vehicle and smaller gaps between the cars on the target lane. And for CAVs, smaller gaps to other CAVs were allowed. Krause et al. [29] have defined AVs to keep a higher standstill distance (2.0 instead of 1.5 m; the distance the vehicles keep to front vehicle when stopping at crossings), have added more safety time (reacting 10 s instead of 8 s before reaching the safety distance), did not vary the distance to the vehicle in front, have used the same accelerations as human driven vehicles, but monitored one vehicle in front (instead of two). Furthermore, their acceleration when changing lanes was equal to HVs, except the vehicle was not able to break fast enough; also, they have used a higher minimum distance when changing lanes (1 m instead of 0.5), and kept a constant velocity when driving idle (HVs with varying speed). Motamedidehkordi et al. [14] have used a standstill distance of 1 m.

Stogios [7] has varied the standstill distance, the threshold for entering 'Following' (controls the start of the deceleration process when a driver identifies/senses a preceding slower vehicle), the negative and positive 'Following' thresholds (control the speed differences during the following state: smaller values lead to a more sensitive reaction to accelerations or decelerations of preceding vehicles), the oscillation acceleration (the fluctuations in acceleration during the acceleration process), the standstill acceleration (the desired acceleration when starting from standstill), the acceleration at 80 km/h, the min. headway (front/rear) and the safety distance reduction factor (this factor reduces the safety distance during a lane change). Calvert et al.

[36] have used the desired gap of the vehicle to distinguish between human driven vehicles, AVs and CAVs. The values were statistically distributed.

Literature on KPIs

Throughout the literature, several KPIs, mainly depending on the methodology and the research design, have been used to study the impact on traffic. The most important KPIs are the

- average travel speed [5, 7, 8, 13, 20, 30, 32]
- the (average) travel time [4, 5, 13, 30, 36],
- the average delay [7],
- the total network delay [30],
- the queue lengths at junctions [5, 30],
- the average vehicle density [13, 30, 31],
- the road capacity [8, 9] and
- the traffic flow [30, 33, 36].

Other study designs used more specific KPIs, e.g.

- the inflow on highway segments [20],
- emissions [7] and
- platoon sizes [33].

4 Methodological Approach

A road network model of Munich including the „outer ring "-highway is used with calibrated origin-demand matrices, all implemented in Aimsun. As a warm-up, the network is filled with vehicles from 5.30 to 7 am and the simulation is then run from 7 to 10 am Fig. 1.

For the evaluation of the total traffic, the whole time window of three hours is used. Additionally, 150 tracked vehicles are placed into the simulation at 7 am and another 150 at 8 am. Of these, respectively 30 are driving inside the inner city, 30 around the middle part of the city, 30 into the city, 30 out of the city and 30 along the „outer ring " (a highway surrounding Munich). Each of the 30 vehicles is split into 10 human driven cars, 10 AVs and 10 trucks. Each simulation is run with 10 replications and the base scenario (no AVs) with 30 replications for statistical significance. 10 of these routes are exemplary shown in the Figure.

However, today it is still unclear how the AVs driving algorithms will be designed in detail as only few AVs are available on the market and driving details are not public. To give the users a safe feeling, it is likely that they will simulate the behaviour of human drivers. For this, we use the modified version of the Gipps driver model (1981, 1986), which is implemented in Aimsun.

The general rule for changing parameters is that they are usually distributed for human drivers, while for AVs they are set to fixed values.

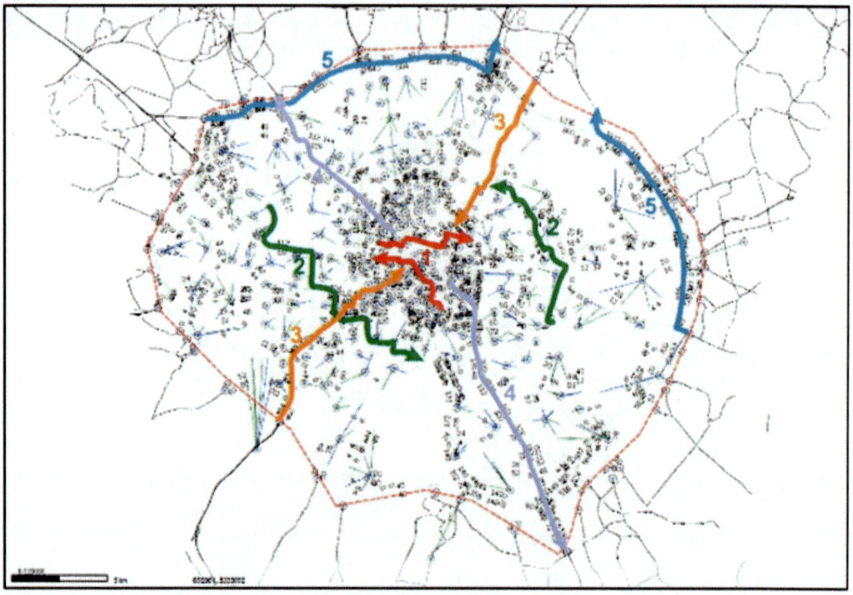

Fig. 1 The scope of the road network and its main OD entry and exit points

- The headway time is the most often discussed parameter in the literature. In the Gipps model it is represented by the reaction time following. Furthermore, it determines the minimum of the reaction time at stop and the reaction time front vehicle (reaction to green traffic lights). For human drivers they are set to 0.8, 1.2 and 1.6 s, respectively. For AVs, several values have been discussed in the literature, ranging from 0.1 to 2.2 s. Here, we use a value of 1.2 as this results in a headway to the front vehicle that equals German driving laws (distance in meters = half value of the tachometer in km/h) in the Gipps model. For CAVs, it's also likely that these will keep a distance according to traditional laws to ensure safety. However, it might also be possible that they will be able to reduce their distance to the front vehicle significantly due to the inter-vehicle communication, so their parameter is set to 0.1. Thus, they will differ here from AVs. However, this value can only be applied if all cars are CAVs.

  The following parameters are applied to AVs and CAVs likewise:

- Aggressiveness level: this parameter determines if the vehicle accepts small gaps between vehicles when changing lanes (0 for normal gaps, 1 for smaller gaps). For human drivers it is distributed between 0 and 1, for AVs it's set to 0, as AVs will likely drive safely [11].
- Margin for Overtaking Manoeuvre: this adds an additional safety factor when estimating if an overtaking manoeuvre can be finished in time due to oncoming traffic. Human drivers: 3–7, for AVs it's set to 7 to give the users a safe feeling [11].

- Distance zone factor: this determines if lanes are changed in advance when lining up at red traffic lights. For human drivers it's distributed 0.8–1.2, for AVs it's set to 1.
- Speed acceptance: this determines if the vehicle accepts the speed limit. For human drivers it's centred around 1.1, i.e. human drivers usually drive faster than allowed. For AVs it's set to 1.
- Clearance: this determines the distance to the front vehicle when stopping, e.g. at traffic lights. For human drivers: 0.5–1.5, for AVs: 1.
- Maximum give way time: if a vehicle waits longer than this time at an intersection, it accepts shorter gaps for crossing. Human drivers 4–14 s, AVs 14 s to give the user a safe feeling [11].

Other vehicle parameters like the length of the vehicle or its maximum acceleration and deceleration are left unchanged, as it is still unclear if AVs will be designed differently from traditional cars. Also, further parameters where AVs are unlikely to differ from human driven vehicles are left unchanged.

Furthermore, the future of urban traffic was studied, using the scenario described by Kaltenhäuser et al. [1]. They predicted the total vehicle miles travelled (VMT) in Germany to increase from 100% in 2015 to a maximum of approx. 123% in 2036. Afterwards, the total vehicle miles are expected to decrease slightly. In 2036, about 55% of the VMTs are driven by HVs and the remaining 45% by AVs, while trucks remain unchanged. Here, the main KPIs are studied again: the mean queue (representing the delay times), the density, the total travel time and the harmonic speed.

Beyond the parameters used here, the main features of CAVs are the communication with each other and with infrastructure. The first one is applied here indirectly with the usage of short reaction times which can only be realised with communication. The latter one is beyond this study as the communication can be simulated solely with single intersections and not yet for a whole city so far. For this topic, see e.g. [37, 38].

5 Results

First, the impact of the driving parameters on the total traffic was studied, using the KPIs described above. The results are summarized in Table 2.

Here, it can be seen that both the driving aggressiveness and the margin for overtaking manoeuvre have only a low and not significant impact on the KPIs. In contrast, the distance zone factor, speed acceptance, clearance and maximum give way time all have a negative impact on urban traffic flow. When modifying the reaction times, it must be generally distinguished between AVs and CAVs. Using a reaction time of 1.2 s, associated with AVs, the traffic flow is decreased significantly, whereas using quick reaction times of 0.1 s, associated with CAVs, the traffic flow

Table 2 Results of the parameter study. Numbers marked with * are not significant at a 95% level. *Note* He "reaction time at stop" and the "reaction time front vehicle" require the basic reaction time to be lower or equal (here: 0.1 s). Thus, the results shown for these two KPIs are relative to a reaction time of 0.1 s

Parameter	Value (s)	Mean queue (%)	Stop time (%)	Delay time (%)	Density (%)	Travel time (%)	Harmonic speed (%)
Aggressiveness	0	−0.7*	−0.8*	−0.6*	−0.2*	−0.2*	0.3*
Margin for overt. manoeuvre	7	0.8*	1.2*	1.0*	0.4*	0.2*	0.5*
Distance zone factor	1	15.8	6.9	6.8	7.9	2.2	−3.7
Speed acceptance	1	8.8	6.5	2.2*	5.3	1.9	−4.9
Clearance	1	10.7	6.5	5.5	4.6	1.0	−2.9
Maximum give way time	14	13.8	9.9	8.4	5.7	2.4	−4.4
All reaction times	1.2	45.1	39.5	38.7	24.7	9.5	−19.8
Reaction time	0.1	−21.5	−21.0	−27.2	−15.8	−5.9	19.1
Reaction time at stop	0.1	−65.4	−57.6	−52.4	−31.2	−23.0	37.1
Reaction time front vehicle	0.1	−21.0	−17.4	−15.8	−9.8	−4.9	8.8

gets enhanced. Thus, these parameters have a similar effect on urban traffic as it is studied here and intercity traffic as it is studied in the literature.

Then, the impact of autonomous vehicles on urban traffic was studied, using the parameters described above.

Total traffic

The impact of an increasing percentage of AVs on urban traffic is shown in Fig. 2.

It can be seen that all KPIs (except the harmonic speed) are steadily increasing with an increasing percentage of AVs. Here, especially the waiting times (queue, stop time, delay) are increasing significantly by 72, 60 and 54%, respectively, while the density and travel time are increasing by just 36% and 12%, respectively. This corresponds to a decreasing harmonic speed (−22%).

Traffic in specified travel directions

Here, 300 additional vehicles were placed in the simulation as described above. Of these, 20 HVs, 20 AVs and 20 trucks travelled around the inner city, and the same amount respectively around the middle part of the city, along the highway surrounding Munich, into the city and out of the city. Their travel behaviour was

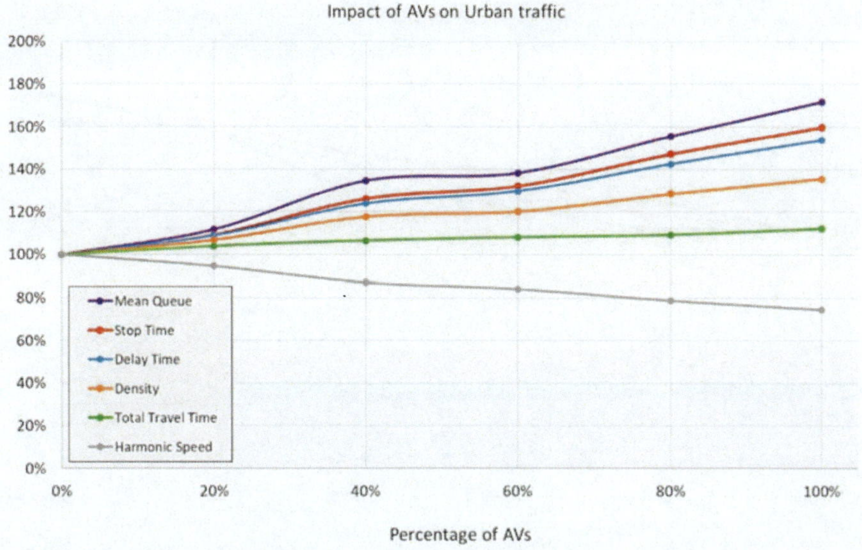

Fig. 2 Relative KPI change with an increasing percentage of AVs

studied depending on the percentage of AVs. First, Fig. 3 shows that the relative travel time increases almost linearly and similarly for all types of vehicles with an increasing percentage of AVs (exemplary shown for the traffic in the inner city, but equally for the other travel directions). For 100% AVs, it increases by around 20%.

Thus, the relative travel time is averaged for the three types of vehicles and shown in Fig. 4 for the five specified travel directions.

It can be seen that the travel time increases also almost linearly for all directions, except the traffic around the middle city. Here, the travel time decreases first, before the linear increase starts at an AV percentage of 40%. Furthermore, the time increase in the main travel directions (such as into the city a along the outer ring in the morning) is with 28 and 47% significantly larger than the increase in the sparsely travelled directions (out of the city and around the middle city) with 6 and 15%.

These results reflect the fact that all AV driving parameters have a negative impact on urban road traffic, as shown in Table 2. Combining all effects together must then necessarily have an increased negative impact on urban road traffic. In return, human drivers "optimize" the traffic flow by not following regulations and recommendations.

The impact of connected autonomous vehicles

In contrast to AVs, CAVs were designed with very short reaction times of 0.1 s. First, their impact on the total traffic was studied by comparing the penetration rates of 0 and 100%. Table 3 shows the relative changes.

The table shows that especially the waiting times (mean queue, stop and delay) get significantly reduced by about 70%. Also, the vehicle density and the travel time get reduced by 42 and 28%, respectively. This corresponds to an increased harmonic

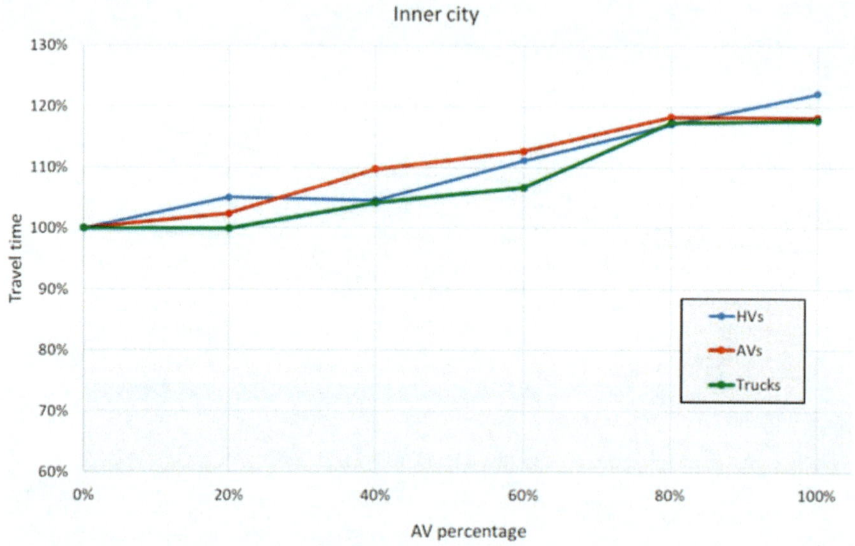

Fig. 3 Impact of an increasing AV percentage on the travel time of HVs, AVs and trucks around the inner city

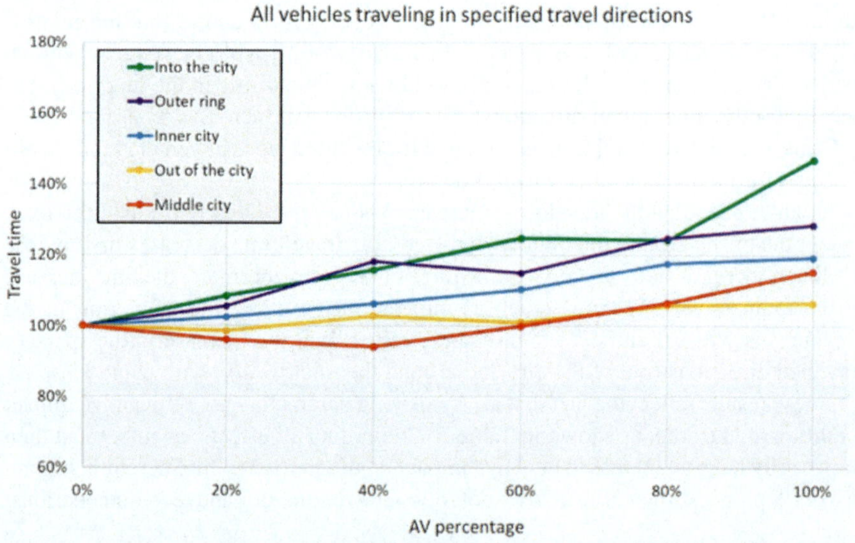

Fig. 4 Impact of an increasing AV percentage on the travel time along specified routes. As the travel time of HVs, AVs and trucks increases similarly (see Fig. 3), the relative travel time is shown as an average of these vehicles

Table 3 Relative KPI change due to 100% CAVs (compared to 100% HVs)

KPI	Relative change (%)
Mean queue	−76
Stop time	−70
Delay time	−70
Vehicle density	−42
Total travel time	−28
Harmonic speed	+62

Table 4 Relative change in the travel time with 0 and 100% CAVs. All vehicle types feature almost the same changes

Vehicle type	Inner city (%)	Middle city (%)	Into the city (%)	Out of the city (%)	Outer ring (%)
HVs	−26	−28	−31	−22	−53
CAVs	−25	−23	−32	−21	−52
Trucks	−24	−24	−29	−21	−52

speed (+62%). Thus, the reduced reaction times (increasing the flow) have a much higher impact than the other driving parameters (decreasing the flow). This is in accordance with Table 2 where the impact of the single parameters is shown and the reaction times have a far higher impact.

Also, the directional traffic was studied using the 300 additional HVs, CAVs and trucks. The relative change in the travel time between 0 and 100% CAVs is shown in Table 4.

It can be seen that most travel times are being reduced around 21–32%. Here, especially the less travelled direction out of the city (in the morning) shows a lower reduction, while the reduction along the outer ring with high velocities shows a relatively large reduction of about 52%. These results reflect the fact that the reaction time dominates all other studied AV driving parameters, as shown in Table 2. The traffic might be further optimized with CAVs communicating with infrastructure like traffic lights.

Predicted future traffic with AVs

Furthermore, the simulation was run for the scenarios described by Kaltenhäuser et al. [1]. Their predicted AV and HV miles travelled and the resulting KPIs are shown in Fig. 5.

It can be seen that the mean queue, the traffic density and the total travel time will increase significantly in the future. The total travel time, which might be regarded as the most important variable, increases to a maximum of about 124% (compared to 2015) in 2036 and will slightly decrease afterwards due to a declining population. The increased travel time corresponds to a lower speed, which is being reduced by approx. 39%.

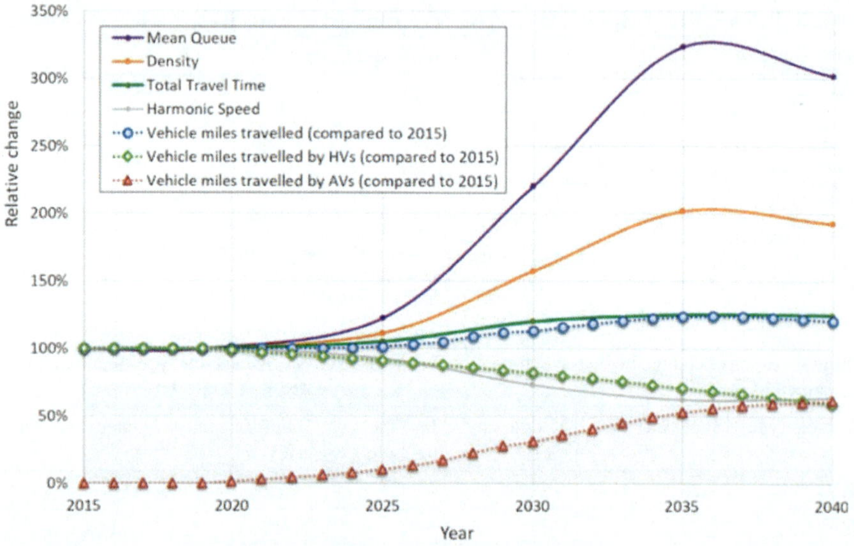

Fig. 5 Predicted vehicle miles travelled by HVs and AVs and the resulting KPIs mean queue, density, total travel time and harmonic speed. For a better clarity, the stop and delay time are not shown

However, due to the increased usage of autonomous driving vehicles, the extended travel time might be used for activities like reading, working or sleeping, which could compensate it at least in parts.

6 Conclusions

In previous studies, the impact of autonomous driving vehicles and their driving parameters on urban road traffic were examined. However, these studies focussed mainly on single and double lane traffic as it is found e. g. on highways, while the impact on urban road traffic is still unclear.

For this, a traffic simulation using a calibrated city model of Munich was run with several features of AVs. With this, the AV driving parameters influencing urban road traffic were examined. Here, especially the reaction times had an impact, where larger values used for AVs had a negative impact on traffic flow, while shorter values used for CAVs had a positive effect. This is in accordance with the literature, although the results are barely comparable as mainly single and double lane traffic was used, while in our study a complete city network was used and results could have been different for example due to vehicle behaviour at crossings or traffic lights. In our study, the impact is highest when traveling along the main travel directions (into the city in the morning) and it is higher for highways than for (slow) urban traffic.

A similar result was produced when the impact of AVs and CAVs, simulated with their respective driving parameters, was simulated. Here, AVs, associated with longer headway times lead to reduced traffic flow while CAVs, associated with shorter headway times, lead to an increased traffic flow. Thus, it could be seen that the headway times dominated the other features associated with autonomous driving that were identical for both types of vehicles.

Hence, it is important to reduce the reaction and headway times of AVs and CAVs to reduce travel times, traffic and thus emissions. This will not just be a technically difficult task, but also a policy issue, as road safety must be ensured at all times. This topic will gain in importance when looking at future scenarios that predict an increased traffic.

As the effects are higher in the main travel directions, an intelligent and collaborative routing of AVs could as well to reduce traffic.

AuthorContribution The authors confirm contribution to the paper as follows: study conception and design: B. Kaltenhäuser, K. Bogenberger; data collection: B. Kaltenhäuser; analysis and interpretation of results: B. Kaltenhäuser; draft manuscript preparation: B. Kaltenhäuser, S. Hamzehi, K. Bogenberger. All authors reviewed the results and approved the final version of the manuscript.

References

1. Kaltenhäuser B, Werdich K, Dandl F, Bogenberger K (2020) Market development of autonomous driving in Germany. Transp Res Part A: Policy Pract 132:882–910
2. Bansal P, Kockelman KM (2017) Forecasting Americans long-term adoption of connected and autonomous vehicle technologies. Transp Res Part A: Policy Pract 95: 49–63. https://doi.org/10.1016/j.tra.2016.10.013
3. Nieuwenhuijsen J, de Almeida Correia GH, Dimitris M, van Arem B, van Daalen E (2018) Towards a quantitative method to analyze the long-term innovation diffusion of automated vehicles technology using system dynamics. Transp Res Part C: Emerg Technol 86:300–327. https://doi.org/10.1016/j.trc.2017.11.016
4. Bailey NK (2016) Simulation and queueing network model formulation of mixed automated and non-automated traffic in urban settings. Master thesis, Massachusetts Institute of Technology
5. Yeola P, Margreiter M, Abeyweera B (2020) Impact of connected and autonomous vehicles (CAVs) on the urban road network performance of Singapore CBD. Submitted to MobilTUM 2020. Singapore
6. Mavromatis I, Tassi A, Piechocki RJ, Sooriyabandara M (2020) On urban traffic flow benefits of connected and automated vehicles. *arXiv preprint* arXiv:2004.00706
7. Stogios C (2018) Investigating the effects of automated vehicle driving operations on road emissions and traffic performance. Master Thesis, University of Toronto
8. Krause S, Motamedidehkordi N, Hoffmann S, Busch F (2017a) Mikroskopische Simulation von automatisierten Fahrzeugen zur Ermittlung der Wirkungen auf die Kapazität von Autobahnen. Straßenverkehrstechnik: 831–838
9. Hartmann M, Motamedidehkordi N, Krause S, Hoffmann S, Vortisch P, Busch F (2017) Impact of automated vehicles on capacity of the German freeway network. ITS World Congress
10. Papamichail I, Bekiaris-Liberis N, Delis AI, Manolis D, Mountakis KS, Nikolos IK, Roncoli C, Papageorgiou M (2019) Motorway traffic flow modelling, estimation and control with vehicle automation and communication systems. Annu Rev Control 48:325–346

11. Ntousakis IA, Nikolos IK, Papageorgiou M (2015) On microscopic modelling of adaptive cruise control systems. Transp Res Procedia 6:111–127
12. Wang J, Rajamani R (2002) Adaptive cruise control system design and its impact on highway traffic flow. IEEE.In: Proceedings of the 2002 American Control Conference, vol 5. pp 3690–3695
13. Aria E, Olstam J, Schwietering C (2016) Investigation of automated vehicle effects on driver's behavior and traffic performance. Transp Res Procedia 15:761–770
14. Motamedidehkordi N, Margreiter M, Benz T (2016). Effects of connected highly automated vehicles on the propagation of congested patterns on freeways. In: 95th Annual Meeting of the Transportation Research Board
15. Arnaout GM, Arnaout J-P (2014) Exploring the effects of cooperative adaptive cruise control on highway traffic flow using microscopic traffic simulation. Transp Plan Technol 37(2):186–199
16. Vanderwerf J, Shladover S, Kourjanskaia N, Miller M, Krishnan H (2001) Modeling effects of driver control assistance systems on traffic. Transp Res Rec: J Transp Res Board 1748:167–174. https://doi.org/10.3141/1748-21
17. Kesting A, Treiber M, Schönhof M, Helbing D (2008) Adaptive cruise control design for active congestion avoidance. Transp Res Part C: Emerg Technol 16(6):668–683
18. Treiber M, Hennecke A, Helbing D (2000) Congested traffic states in empirical observations and microscopic simulations. Phys Rev E 62(2):1805
19. Shladover SE, Su D, Lu XY (2012) Impacts of cooperative adaptive cruise control on freeway traffic flow. Transp Res Rec 2324(1):63–70
20. Fountoulakis M, Bekiaris-Liberis N, Roncoli C, Papamichail I, Papageorgiou M (2017) Highway traffic state estimation with mixed connected and conventional vehicles: Microscopic simulation-based testing. Transp Res Part C: Emerg Technol 78:13–33
21. Gipps PG (1981) A behavioural car-following model for computer simulation. Transp Res Part B: Methodol 15(2):105–111
22. Gipps PG (1986) A model for the structure of lane-changing decisions. Transp Res Part B: Methodol 20(5):403–414
23. van Arem B, Hogema J, Smulders S (1996) The impact of autonomous intelligent cruise control on traffic flow. In: Proceedings of the 3rd world congress on intelligent transport systems. Orlando
24. Bierstedt J, Gooze A, Gray C, Peterman J, Raykin L, Walters J (2014) Effects of next-generation vehicles on travel demand and highway capacity. FP Think Working Group 8:10–11
25. Laquai F, Duschl M, Rigoll G (2011) Impact and modeling of driver behavior due to cooperative assistance systems. International Conference on Digital Human Modeling, Springer, 6777:473–482
26. van Arem B, van Driel CJ, Visser R (2006) The impact of cooperative adaptive cruise control on traffic-flow characteristics. IEEE Trans Intell Transp Syst 7(4):429–436
27. Gouy M, Wiedemann K, Stevens A, Brunett G, Reed N (2014) Driving next to 1 automated vehicle platoons: How do short time headways influence non-platoon 2 drivers' longitudinal control? Transp Res Part F 27:264–273
28. Kesting A, Treiber M, Schönhof M, Helbing D (2007) Extending adaptive cruise control to adaptive driving strategies. Transp. Res. Rec.: J. Transp. Res. Board 2000(1):16–24
29. Krause S, Motamedidehkordi N, Hoffmann S, Busch F, Hartmann M, Vortisch P (2017b) Auswirkungen des teil-und hochautomatisierten Fahrens auf die Kapazitaet der Fernstrassen-infrastruktur. FAT-Schriftenreihe 296
30. Elmorshedy L, Mostafa TS, Taha I, Abdulhai B. Quantifying the impact of driving automation on Ontario's freeways. University of Toronto. https://uttri.utoronto.ca/files/2019/07/3-Elmorshedy-et-al_Quantifying-the-imapcts-of-driving-automation-on-ontario-freways_June-2019.pdf. Accessed 20 June 2020
31. Delis AI, Nikolos IK, Papageorgiou M (2015) Macroscopic traffic flow modeling with adaptive cruise control: Development and numerical solution. Comput Math Appl 70(8):1921–1947
32. Roncoli C, Papamichail I, Papageorgiou M (2014) Model predictive control for multi-lane motorways in presence of VACS. In: 17th International IEEE Conference on Intelligent Transportation Systems (ITSC). pp 501–507

33. Mahmassani HS (2016) Autonomous vehicles and connected vehicle systems: Flow and operations considerations. Transp Sci 50(4):1140–1162. https://doi.org/10.1287/trsc.2016.0712
34. Calvert S, Mahmassani H, Meier JN, Varaiya P, Hamdar S, Chen D, Li X, Talebpour A, Mattingly SP (2018) Traffic flow of connected and automated vehicles: Challenges and opportunities. Road Veh Autom 4: 235–245. Springer
35. Lioris J, Pedarsani R, Tascikaraoglu FY, Varaiya P (2017) Platoons of connected vehicles can double throughput in urban roads. Transp Res Part C: Emerg Technol 77:292–305
36. Calvert SC, Schakel WJ, Van Lint JWC (2017) Will automated vehicles negatively impact traffic flow? J Adv Transp
37. Guo Q, Li L, Ban XJ (2019) Urban traffic signal control with connected and automated vehicles: A survey. Transp Res Part C: Emerg Technol 101:313–334
38. Yang K, Guler SI, Menendez M (2016) Isolated intersection control for various levels of vehicle technology: Conventional, connected, and automated vehicles. Transp Res Part C: Emerg Technol 72:109–129

Enhancing Robustness Against Component Failures in Intelligent Transportation Systems Through Self-diagnosis Functionality

Christian Creß, Lukas Rabe, and Alois Knoll

Abstract The intelligent transportation systems (ITS) are part of possible solutions to the problems in transportation. Current systems generate digital twins of traffic participants. The traffic can be interpreted, and control signals can be sent to vehicles. Malfunctions could have disastrous consequences. Therefore, we present a self-diagnosis functionality for ITS which enhancing robustness against component failures. First, we identified sources of failures. Then, we compared existing failure detection approaches in use case of ITS. Based on this, we developed the methods Heartbeat, Sensor Metadata Checking, Process Pipeline Checking and Measurement Point Cross-Checking. To react to malfunctions, we introduce a remediation component. For testing, we used the real environment test bed Providentia++. The unique setup enables novel approaches for enhancement of robustness. In particular, Measurement Point Cross-Checking was tailored to our unique sensor setup. During the experiments, we verified the effectiveness of our methods. In future work, we suggest more plausibility checks.

Keywords C-ITS · Intelligent transportation systems · Intelligent infrastructure systems · Robustness · Self-diagnosis · Failure detection

1 Introduction

The problems in transportation segment are omnipresent. These include, for example, traffic accidents due to human error, air pollution and traffic congestion. Intelligent traffic concepts can make an effective contribution to solving the mentioned issues. Therefore, the intelligent transportation systems (ITS) has become a very relevant research topic in the previous years. The current systems are increasingly using high precise sensors, such as cameras, radars and LiDARs. With sufficient computing

C. Creß (✉) · L. Rabe · A. Knoll
Chair of Robotics, Artificial Intelligence and Real-time Systems, Technical University of Munich, Boltzmannstr. 3, 85748 Garching, Germany
e-mail: christian.cress@tum.de

C. Antoniou et al. (eds.), *Proceedings of the 12th International Scientific Conference on Mobility and Transport*, Lecture Notes in Mobility,
https://doi.org/10.1007/978-981-19-8361-0_3

power, these systems allows the generation of a digital twin of each individual traffic participant in real-time. Therefore, common value-added services, such as warning of slippery road, construction sites and traffic accidents, can be extended to more safety-critical value-added services, which needs more information density [1]. The ITS of today are able to interpret the traffic flow as well as specific traffic situations. Based on this, the systems can send control signals to the vehicles, which are connected via V2X technology.

Alternatively, electronic gantry bridges can also send instructions to road users who have no connectivity functions. This development allows higher optimization of the road traffic in terms of safety and efficiency, especially when autonomous vehicles become part of the automotive landscape [2, 3]. For sure, the malfunctions in the ITS could have disastrous consequences for the road traffic. For this reason, these systems also count as critical infrastructure and need protection.

This drives the following research question: How to protect and secure ITS from malfunctions and external attacks? First, we need to analyze against which specific failures or threads we have to protect an ITS. Therefore, we create a list of possible failure scenarios which could produce wrong output, and rate their impact factor in terms of stability:

- **Breakdown of a sensor:** An easily identifiable failure scenario is the breakdown of a sensor. When a sensor does not generate data, in the best-case scenario, we lose accuracy. In the worst case, a digital twin of a road user can no longer be generated.
- **Breakdown of a measurement point:** A complete measurement point has a breakdown or the connection to this node is broken. The ITS cannot generate digital twins in a specific road section anymore. Furthermore, in case of distributed edge-computing, the overall system loses computing power.
- **Misbehavior of a software component:** A software component, such as the object recognition module or tracking software, can misbehave. In the best-case failure scenario, the component produces no output. In the worst-case, the software produces wrong data output, which leads to faulty and contradicted data streams in the processing pipeline. These anomalies can be assigned to the more complex failure scenarios and may only be recognizable in the global context over several measuring points.

The reasons for the failure scenarios can be defective software or hardware components, as well as external attacks. In any case, the system has to recognize the mentioned scenarios and generate a corresponding response. This is the only way to avoid potentially fatal decisions based on incorrect data.

In this paper, we present a self-diagnosis functionality using methods of Heartbeat, Sensor Metadata Checking, Process Pipeline Checking, and Measurement Point Cross-Checking. These methods detect malfunctions and therefore, they can increase the security and robustness of the overall system. For testing in real environment, we implement these methods on the Providentia++ test bed. The unique structure of the test bed is the overlapping field of views with different kind of sensors, and it enables new approaches for enhancement of the robustness. In particular, the method of Measurement Point Cross-Checking is a novel approach which is tailored to our unique sensor setup.

We organized the paper as follows. First, Sect. 2 explains well-known concepts for increasing the robustness of safety-critical software. The Sect. 3 describes our methods for self-diagnosis functionality in ITS. The Sect. 4 discusses the results which we achieved with our presented methods on the Providentia++ test bed. Last, a conclusion and outlook are given in Sect. 5.

2 Related Work

The software of safety-critical systems have to be robust. The survey [4] describes failsafe techniques in safety-critical automotive applications. These techniques can be summarized into the groups 1.) redundancy, 2.) hardware integrity checks, 3.) program execution checks and 4.) output monitoring. For example, the autonomous underwater vehicle described in [5] improves the safety with an additional hardware module that detects critical situations. According to the authors from [6], robustness with failure detection systems can be achieved with two approaches, which are usually used in software testing: 1.) An external component observes the output of another software component and handles it as a black box. 2.) A diagnosis component monitors internal states of the observed component. Thus, this approach treats the software module as a white box. A deviation from the expected output or in internal state can be recognized as failure.

The author from [7] presents a white box method for monitoring the internal state of a software module including the concept of self-healing. Here, each module has a service layer and a healing layer. If an unexpected state transition occurs or expected messages have not arrived in reasonable time intervals, the healing layer presumes a failure. In detail, aliveness-indication-routines are added to the code which notify an external observer about the current internal state of the process. If the process changes to an unexpected state, the observer becomes aware about a malfunction. The mentioned aliveness-indication-routines are used in automotive software applications [8] and can be implemented with a watchdog timer. Here, an external device resets the monitored system, when a specific amount of time has passed. The system can only avoid the reset by requesting a restart of the timer. In case of malfunction, this request is not possible. Therefore, the watchdog performs a system reset. This white box method has two disadvantages. First, the implementation of existing software modules has to be modified for submission of their internal state to a monitoring system. Second, duplicating a logic into the monitoring system has a risk of duplicating undetected errors too. According to Parnas et al. [9], the preferred approach for evaluation of safety-critical software is to treat the whole system as a black box.

Black box methods detect failures without the knowledge of the internal state of the software. Here, key performance indicators can be analyzed. For example, the frame rate of a video stream module can be an indicator. When the actual value is out of range in comparison to the expected value, the observer method detects a failure. The name of this approach is reasonableness test [4]. For implementation, the author [10] suggests checking a component with defined test data which leads to known associated output. This check can be executed, for example, in the start routine. Unfortunately, when a component has no external inputs, this procedure is not possible. The analysis of the network traffic is another black box method. Here, the monitoring of software modules is possible without modification of existing components. This approach is a watchdog [11]. The authors in [12] offer an example of such a network-centric watchdog approach. For example, when a node generates, less traffic than expected, a failure is assumed. The monitoring focuses on data quantity. Therefore, incorrect content cannot be detected. A watchdog, with focus on the data content, is presented in [13]. Here, the algorithm extracts statistical features from consecutive packets. A fuzzy logic system decides about any malfunction.

As mentioned before, the ITS are safety-critical systems and require protection against malfunction and cyberattacks. Therefore, we need to ensure their functionalities with the mentioned methods. The ITS developed in [14] optimizes the traffic flow with methods of data analytics automatically. Unfortunately, the generated data is not used for anomaly and failure detection. The authors [15] have proposed a fully decentralized multiagent system. From a technical point of view, each sensor or measurement point of the ITS, is an agent. The low-level agents, such as sensors and acturators, are architecturally separated from high-level agents, such as object classification and tracking. The focus lies on the architecture and the communication between the agents. Furthermore, the system does not address robustness.

The system described in [16] contains elements for enhancing robustness. Even though the main focus of the video surveillance system is not the optimization of the traffic, the technical challenges are similar to an ITS. Distributed nodes with cameras as sensors send data over the network to a central cloud unit. Even in case of malfunction, a computing node or the central cloud goes offline, the system tolerates the failure and the other nodes continue operating. Interestingly, resource and image processing monitoring recognize overloaded or stopped computing nodes. Therefore, the system can restart the elements. Although, the system checks if nodes are running, it does not check the data content quality.

3 Methodology

Using the example of the ITS Providentia++ [17–19], we present in this work methods for self-diagnosis functionality, which detect malfunctions and thus increase the robustness and security of the overall system. First, background knowledge of the test bed Providentia++ is necessary to understand the developed methods: On a length of 3.5 *km*, the Providentia++ system is located at the autobahn A9 and highway B471 near Munich. The test bed consists of seven measurement points having overlapping field of views distributed over the entire test bed. The purpose is creating digital twins of the road traffic. For this, each measurement point has multi-modal sensors. These

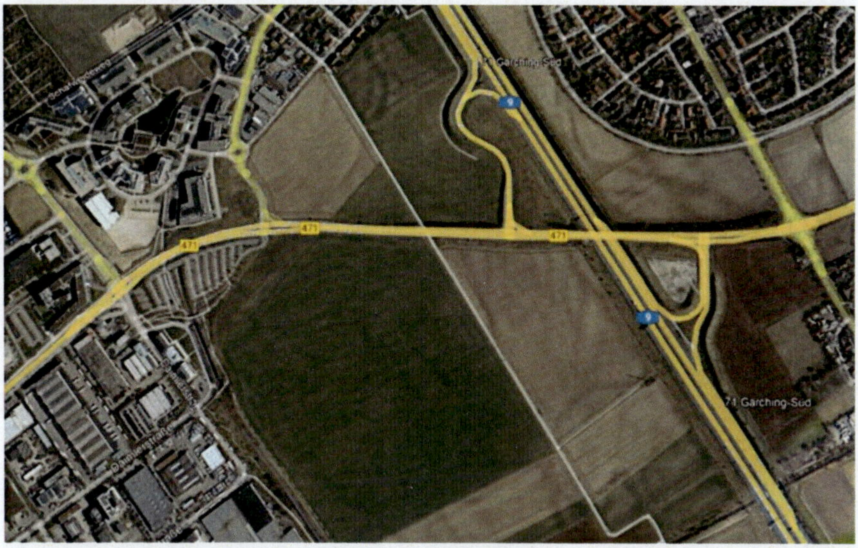

Fig. 1 Overview of the test bed Providentia++ (Graphics made with Google)

include cameras, event-based cameras, radars and LiDARs. Each measurement point creates a local digital twin. These local partial results are merged into a global digital twin, which could be imported into vehicles via V2X technology or made accessible to third parties via the internet. However, external applications such as traffic control or autonomous driving currently do not use this digital twin. Figure 1 shows all measurement points. The software on the Providentia++ system, which creates digital twins, is based on the ROS-framework. This framework enables a publish-subscribe mechanism between the individual hardware and software components. Because of this fact, the necessary software components of the ITS, such as device drivers, components for object detection and tracking, as well as data fusion, are ROS nodes. They are written in object orientated languages, such as C++ and Python.

At the time of writing, the mentioned software was used at the measurement points S40 and S50. These two measurement points are equipped with cameras and radars. The Fig. 2 shows the network between the backend component and the measurement points, which is realized using optical fiber. Based on this system, we implement conventional methods Heartbeat, Process Pipeline Checking and Sensor Metadata Checking for self-diagnosis functionality. Furthermore, we propose Measurement Point Cross-Checking, which detects complex failure scenarios in the program flow. It should be highlighted that Measurement Point Cross-Checking is a novel approach

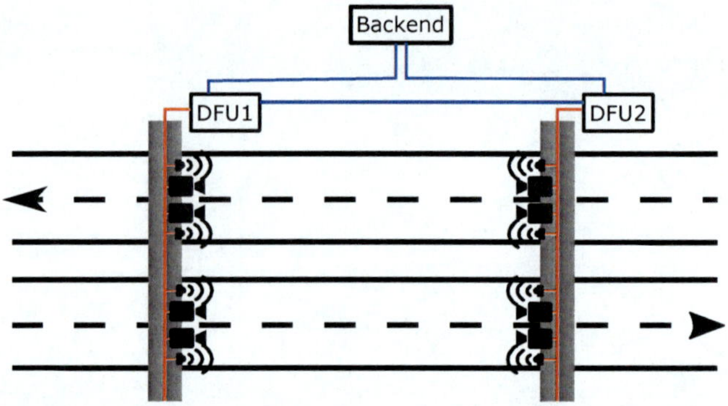

Fig. 2 Top-down view to the backend, DFU1 (S40) and DFU2 (S50). Each measurement point consists of a data fusion unit (DFU) and sensors which are installed on a gantry bridge

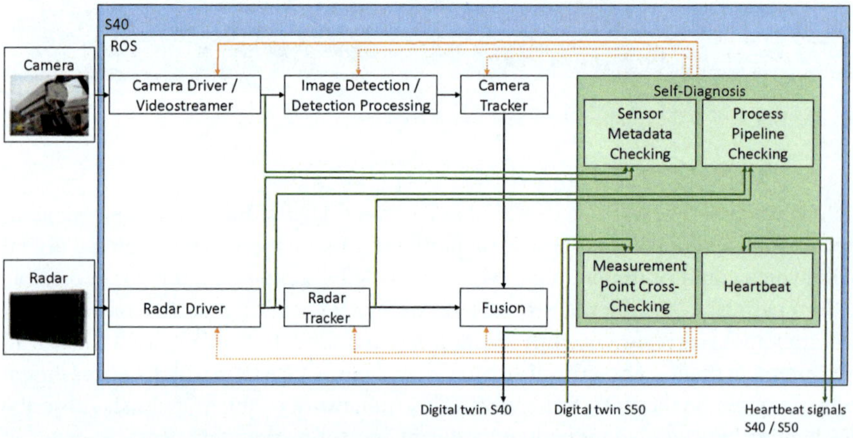

Fig. 3 This diagram gives an overview about the links between the existing components in the ITS Providentia++ and the self-diagnosis functionality for the measurement point S40. The outputs of the component are the digital twin and its heartbeat signal. The inputs include the digital twin and the heartbeat signal of the adjacent measurement point S50. The existing data flow for generating digital twins is shown by the black arrows. The self-diagnosis methods receive the data streams (green arrows) and check the components for wrong behavior. In the event of an error, a remediation mechanism can be triggered. This was implemented as proof of concept for the camera driver. Theoretically, all other components could also trigger a reaction (orange arrows). The measurement point S50 has the same structure

that is tailored to our unique sensor setup. The Fig. 3 illustrates the links between the presented methods of the self-diagnosis functionality and the existing components of the ITS Providentia++. Here, the outputs of the various software components are evaluated by the self-diagnosis component.

According to our mentioned methods, we record each detected malfunction of the system into a database. In the event of malfunction, this SQLite database enables better analysis of the reasons. Furthermore, we can trigger a reaction in the overall system. It is to be emphasized that the remediation logic must be specifically designed per component. Hence, we provide an interface which offers access to the broadcasted signals from the self-diagnosis functionality and implement interfaces for remediation. In this way, the specific components can define the customized remediation logic. However, the failure remediation is not the main focus of this paper, we implement a rudimentary reaction in case of failure as proof of concept. We describe this reaction as follows:

1. The camera sends images to the process chain at irregular intervals. The method Sensor Metadata Checking detects this misbehavior.
2. The software component, which implements Sensor Metadata Checking, reports this downtime into a database and publish a remediation request as a broadcast into the complete ROS-Network.
3. All components, which implement the mentioned remediation interface, receive the remediation request. In case of our implementation, the camera device driver receives the request and can reset its internal state.
4. The method Sensor Metadata Checking recognizes that the camera is back in normal operation. It stops sending remediation requests.

After the description of the remediation mechanism, we would like to present in detail the methods, which detect malfunctions and misbehavior in the safety-critical ITS.

3.1 *Heartbeat*

The method Heartbeat is able to detect downtimes of complete measurement points or the backend. When the instance of the self-diagnosis functionality goes offline, this method detects the malfunction. For implementation, each measurement point sends a signal about its own internal status to the network periodically. All others measurement points receive and evaluate the signals. As soon as a measurement point has a downtime, this can be recognized by other measurement points. The overall system is thus able to restart a complete measurement point. Therefore, the Heartbeat method serves the error scenario of total breakdown.

3.2 *Sensor Metadata Checking*

To deal with the scenario of a sensor breakdown, the method analyses the metadata of sensor outputs. Therefore, we measure the metadata, such as the frequency of data, of the radars and cameras and calculate the average data frame rate. Then, we evaluate the values in terms of plausibility. It is known from the Providentia++ system that the cameras and radars feed their data into the processing pipeline usually at 25 Hz and 13 Hz respectively. Ideally, the latest message should have arrived just a few milliseconds ago. As soon as a significant deviation is registered, the subsequent components in the process pipeline, such as object detection and tracking, are informed about the anomaly. However, scheduling buffers at the threshold, such as 20 % from the target of 25 *FPS*, avoids false positives.

3.3 Process Pipeline Checking

The method Process Pipeline Checking analyses the data flow in the data processing components of the overall system. Here, we compare the data outputs of a specific component with the data inputs of the subsequent component. Therefore, we realize a verification of the correct data consumption in the various components in real-time, so a possible malfunction in the process chain can be detected. In this work, we implement such a plausibility check with the pipeline of the radar sensor. Here, a software component is responsible for tracking the objects, which are detected by the radar. For this reason, we compare the number of detected objects reported by the sensor with the number of actively tracked objects. Also, the algorithm compares the input of the pipeline with its output. A detected failure could indicate the modification of data in an unintended way. Furthermore, large delays in the processing pipeline would be interpreted as malfunction too.

3.4 Measurement Point Cross-Checking

Another complex failure scenario is the output of faulty data streams from a complete measuring point. Because all traffic participants pass through the field of views of both measurement points, a comparison with surrounding measuring points can be helpful. For this reason, the method of Measurement Point Cross-Checking was developed. In this algorithm, the movement of each detected traffic object in one direction is simulated in real time based on the recorded speed. Since the positions of the measurements points are known, the arrival timestamps of the simulated twins at the neighboring measurement point can be calculated. Therefore, we introduce the formula

$$t_{\text{target}} = t_{\text{detection}} - \frac{d}{v_{\text{avg}}} \times \alpha, \tag{1}$$

where d is the distance between the measurement points, v_{avg} the average speed and $t_{\text{detection}}$ the current timestamp of the measured digital twin. The sensors do not look perpendicularly to the ground, instead they look into the distance. Therefore the required distance for a detected object between the measurement points is slightly less than d. This fact can be taken into account using the parameter α, which was determined empirically.

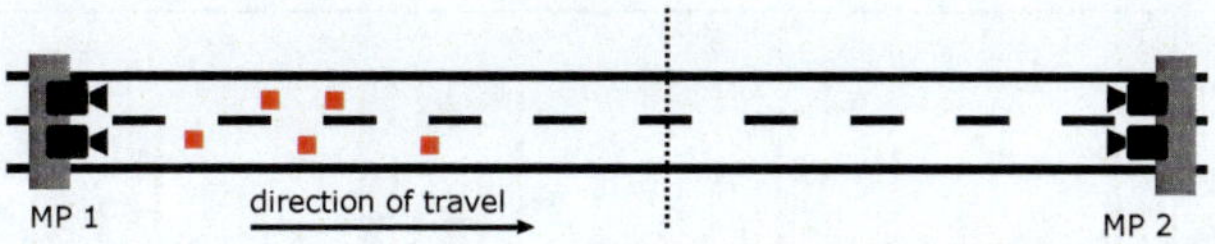

(a) First, the system stores the number of detected vehicles on the left measurement point in a buffer. This information will later be accessed for verification.

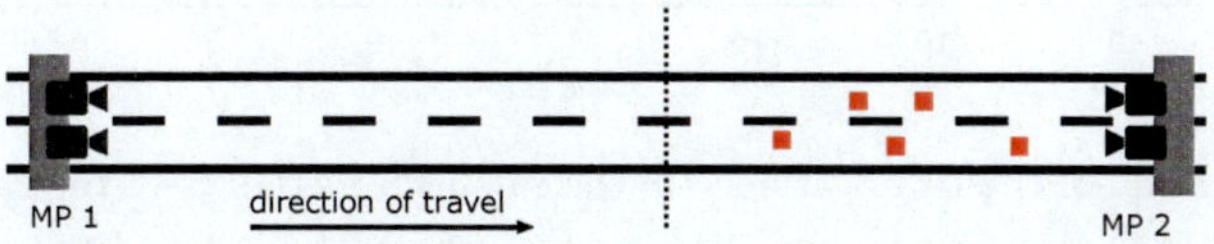

(b) Then, the system records the number of currently detected vehicles on the measurement point on the right and calculates their average speed. Then, the system estimates the timestamp of the traffic at the previous measurement point. This estimation can be compared with the measured values. Ideally, those indicators are similar.

Fig. 4 Principle of the measurement point cross-checking

Then, the simulated twins can be compared with the measured real digital twins at the respective measurement point. Because objects cannot suddenly appear or disappear on a closed route, a failure can be assumed, when there is a relevant deviation between the simulated and real traffic flow. The Fig. 4 gives an overview about the method. It should be noted, that the method makes a statement about plausibility only of the entire traffic flow. The method does not require personal data. If the vehicle speeds diverge significantly, the prediction will not be highly accurate.

4 Results

For evaluating the results, we have simulated downtimes on the Providentia++ system. Here, we have recognized that the presented self-diagnosis methods work effectively. The method Heartbeat has detected the downtime of a measurement point, and Sensor Metadata Checking as well as Process Pipeline Checking have detected malfunctions in the camera respectively the radar. Lastly, the Measurement Point Cross-Checking method has identified complex failures in the content of collected digital twins. Because the methods can initiate corrective actions, such as restart of the specific component, an increase in the stability and the robustness of the overall system has been achieved. This section describes the detailed findings, thresholds and limitations of the individual methods.

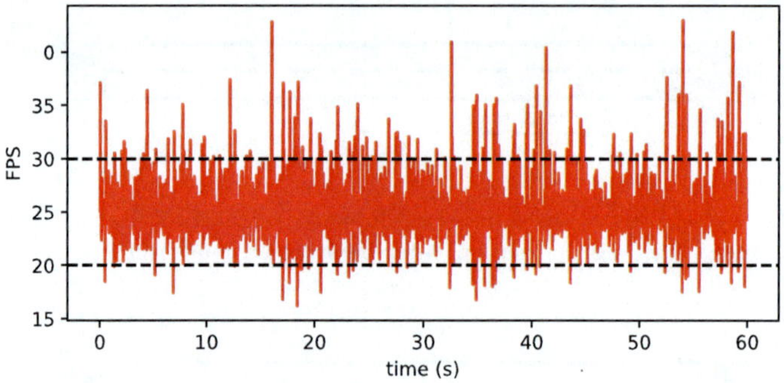

(a) An illustration of the large variance in the raw FPS metadata of the camera video stream.

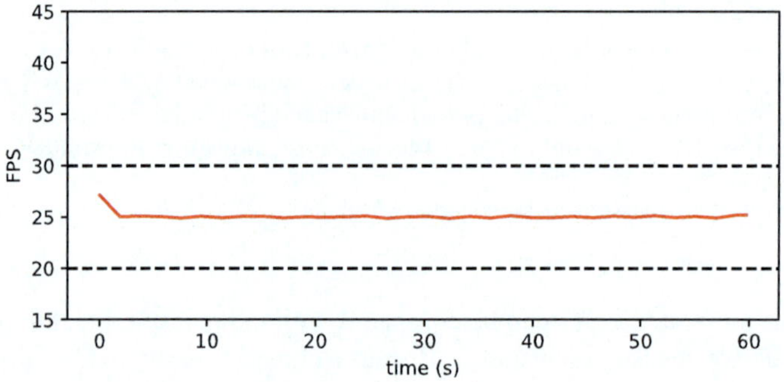

(b) By averaging the FPS of the last 50 video frames, outliers no longer erroneously trigger the downtime detection.

Fig. 5 Graphical representation of the sensor metadata checking, performed on the video stream of the camera during a 60-s interval. The dashed lines represent the 20% threshold tolerance around the target FPS value of 25

- **Heartbeat:** The method Heartbeat is a fundamental sanity check. Any failures detected by this module would indicate a downtime of the complete measurement point. During normal operation, all measurement points broadcast a heartbeat signal every second. We found three seconds to be an a propriate threshold value in order to trigger a failure event without false positives. Since this method only checks the responsiveness of a component, there is no connection between the described threshold and traffic situations. The most important limitation is the event of a total downtime of all system components. In this case, there are no opportunities to initiate corrective measures to restore the system.
- **Sensor Metadata Checking:** The method Sensor Metadata Checking monitors the camera and the radar. First, a threshold defines the maximum allowed time to wait for a data packet. After this time, an error event will be triggered. We have used 5 s for the camera and radar. In addition, we have considered the average frame rate over the last 50 received camera images, shown in Fig. 5. When a value

is 20% above or below of this target, the method triggers an error. Since this method only checks the performance parameters of a component, there is no connection between the threshold value and the traffic situation. The limitation of the method is the evaluation of metadata itself. If a sensor delivers wrong content, such as the camera always sends the same image, this could not be detected.

- **Process Pipeline Checking:** We have implemented this method to verify the data flows in the ITS, demonstrated on the radar process chain. Since the algorithm compares the input of the pipeline with the output, a threshold for an erroneous deviation must be defined. Normally, the data processing for the radar takes a few milliseconds, so that the deviation between input and output should be zero. Due to non-guaranteed time synchronization, we set the threshold to one. With this threshold, we should prevent false-positives. There is no relation between this parameter and specific traffic situations. The limitation of the current implementation is the comparison of metadata: We compare the number of tracked objects. A comparison of specific content, such as the speed and the exact position of traffic participants, would be better.
- **Measurement Point Cross-Checking:** The Measurement Point Cross-Checking reports a failure when the actual number of detected objects deviates highly from the predicted value. As mentioned before, the sensors do not look vertically to the ground. Thus, we have to determine a suitable threshold α. Figure 6 shows the results with several thresholds. With a suitable value, the two curves, which describe the number of detected objects and the number of predicted objects over the time, should look similar. The threshold value depends on the field of view of the sensors and not on the traffic situations. Therefore, according to our experience, the value $\alpha = 0.5$ was a suitable value under all tested traffic situations. The Fig. 7 shows the number of detected objects as well as the predicted values in several traffic situations. There were no disruption observered in the sequences. For our experiments, we set the threshold $\alpha = 0.5$. According to our experience, a deviation of less than 10 objects are in the tolerance range. Higher deviations can be interpreted as a failure. Our method works well in normal traffic as well as in dense traffic scenarios. The limits are in special traffic situations, such as traffic jams or accidents between the measurement points. In case of accident or traffic congestion, detected vehicles would be recorded at the first measurement point, but they would probably not arrive in time at the neighboring station. So, the actual number of objects would not match with the predicted number of objects. The system would assume an error falsely. Furthermore, there is logically a lack of traffic movements: No meaningful predictions can be generated. This can be observed very well in the fourth scenario.

In total, we have tested the methods in the productive system from 22.05.2021, 00:00 to 25.05.2021, 05:00 UTC. The results are summarized in Table 1 and they correspond to our observations on the real system. Interestingly, our self-diagnosis functionality uncovered a systematic weakness in the ITS Providentia++. The radar sensor has faced irregular failures during the night. Every few minutes between 9:00 pm and 4:00 am, the radar has stopped transmitting messages. According to

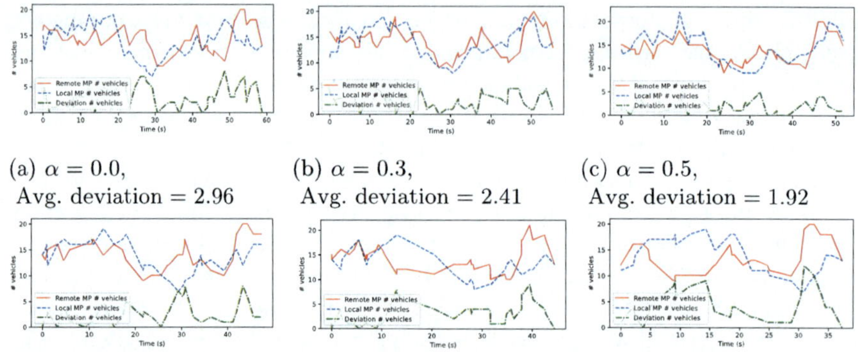

(a) $\alpha = 0.0$, Avg. deviation = 2.96

(b) $\alpha = 0.3$, Avg. deviation = 2.41

(c) $\alpha = 0.5$, Avg. deviation = 1.92

(d) $\alpha = 0.7$, Avg. deviation = 2.8

(e) $\alpha = 1.0$, Avg. deviation = 3.22

(f) $\alpha = 1.5$, Avg. deviation = 4.85

Fig. 6 The blue line represents the number of vehicles over the time, detected by the local measurement point. The red line is the corresponding prediction at the remote measurement point. The more similar the lines, the better the method can detect faulty sensor data. The value $\alpha = 0.5$ gives the smallest deviation and thus the best result. Higher values would represent a higher distance between the measuring points. Logically, with values >1.0 no meaningful similarity can be achieved

the collected data, this behavior was at least 500 times per night with a duration of around 10 s. According to the metric

$$\text{availability} = \frac{\text{total time} - \text{total downtime}}{\text{total time}}, \tag{2}$$

the radars achieve 95.03%. With support of the presented methods, we have identified this problem. Since the system is designed redundantly with other sensors, this availability cannot be classified as critical. A total breakdown of the system would have been detected by the heartbeat method and fortunately did not occur. So, the digital twin of road traffic could be reliably generated in the specified time. In summary, the methods provide an important contribution to the security and stability of the safety-critical ITS Providentia++.

5 Conclusion

In this paper, we have developed a self-diagnosis functionality which uses methods for failure detection. For this purpose, we have first identified potential sources of failures in the test bed Providentia++ and then, we have compared previous approaches to their detection.

Based on this knowledge, we have developed the methods Heartbeat, Sensor Metadata Checking, Process Pipeline Checking as well as Measurement Point Cross-Checking: 1.) The method Heartbeat detects a total breakdown of a measurement

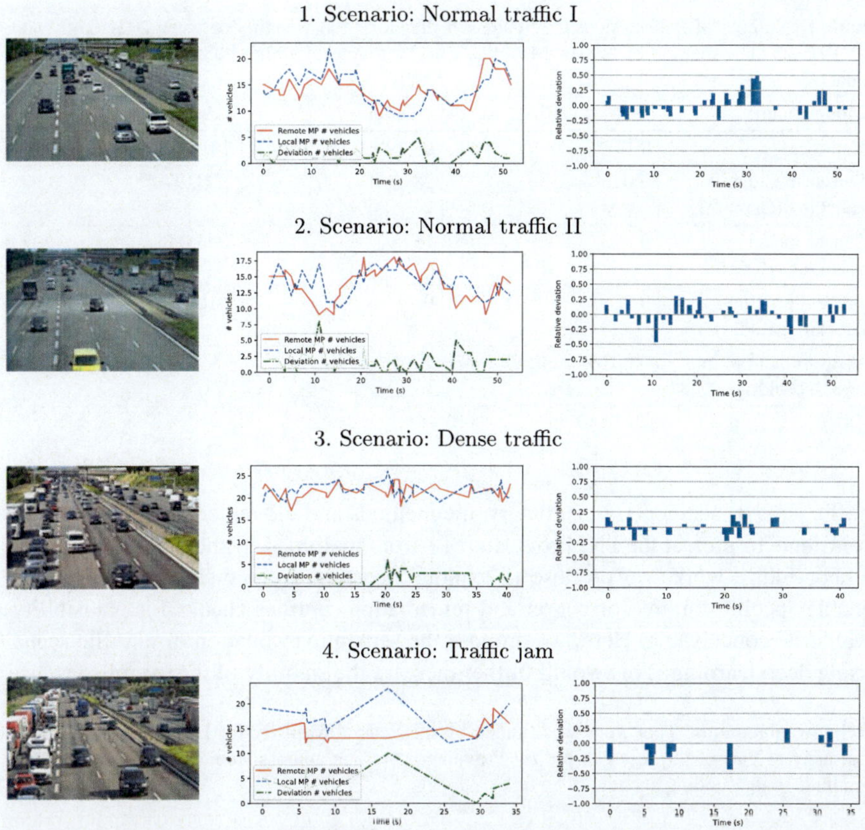

Fig. 7 Graphical representation of the analysis performed by the measurement point cross-checking module in several traffic situations. Although, strong correlation exists between the curves, we also notice minor deviations. The blue line represents the number of vehicles, detected by the local measurement point, the red line the corresponding prediction. Predictions cannot be performed correctly during traffic jams

point. 2.) The Sensor Metadata Checking detects failures based on the key performance indicators of the sensors without interpreting their data content. 3.) The Process Pipeline Checking analyses the data flow between the software components. 4.) The Measurement Point Cross-Checking interprets the data content between surrounding measurement stations. These methods follow a decentralized network-based monitoring approach. Furthermore, we have implemented a rudimentary remediation mechanism, which resets the camera driver in case of malfunction.

During our experiments in the real environment of the Providentia++ test bed, we were able to verify the effectiveness of our mentioned methods. For example, we uncovered a weakness in our radars. In addition, we also showed the limitations of our methods: For instance, Measurement Point Cross-Checking has difficulties during

Table 1 Number of failures detected by the self-diagnosis functionality between 2021-05-22 and 2021-05-25. The method of Sensor Metadata Checking detects a high number of failures in the radar

Module name	Downtime (h)	# Failures	Availability (%)
Heartbeat	0	0	100
Sensor metadata checking (Camera)	0.023	28	99, 97
Sensor metadata checking (Radar)	3.82	1375	95.03
Process pipeline checking (Radar)	0	0	100
Measurement point cross-checking	0.0016	1	99.99

traffic jams or accidents. In summary, the methods and the remediation mechanism contribute to protect the ITS Providentia++ from malfunction and external attacks.

For future work, we propose to explore smart recovery methods. This would enable specific failure correction and remediation. Further checks for plausibility would be conceivable. Here, we propose the semantic evaluation of a traffic scene using deep learning. This would further increase the stability of the overall system.

Acknowledgements This work was funded by the Federal Ministry for Digital and Transport, Germany as part of the research project Providentia++. The authors would like to express their gratitude to the funding agency.

References

1. Creß C, Knoll A (2021) Intelligent transportation systems using external infrastructure: a literature survey. In: ArXiv. https://arxiv.org/pdf/2112.05615
2. Wiegand G (2019) Benefits and challenges of smart highways for the user. In: IUIWorkshops'19
3. Buechel M, Schellmann M, Rosier H, Kessler T, Knoll A (2019) Fortuna: presenting the 5g-connected automated vehicle prototype of the project PROVIDENTIA. In: Researchgate. https://doi.org/10.13140/RG.2.2.24402.91842. https://www.researchgate.net/publication/337285135_Fortuna_Presenting_the_5G-Connected_Automated_Vehicle_Prototype_of_the_Project_PROVIDENTIA?channel=doi&linkId=5dcea3d3299bf1b74b43abcc&showFulltext=true
4. Leaphart EG, Czerny BJ, D'Ambrosio JG, Denlinger CL, Littlejohn D (2005) Survey of software failsafe techniques for safety-critical automotive applications. In: SAE transactions, vol 114. SAE International, pp 149–164. https://www.jstor.org/stable/44682414
5. Ortiz A, Proenza J, Bernat G, Oliver G (1999) Improving the safety of AUVs. In: Riding the crest into the 21st century. Marine Technology Society and IEEE Service Center, Washington, DC and Piscataway, pp 979–984. https://doi.org/10.1109/OCEANS.1999.805005
6. Mohan KK, Verma AK, Srividya A (2010) Software reliability estimation through black box and white box testing at prototype level. In: 2nd ICRESH-2010, pp 517–522. https://doi.org/10.1109/ICRESH.2010.5779604

7. Shin ME (2005) Self-healing components in robust software architecture for concurrent and distributed systems. Sci Comput Program 57(1):22–44. https://doi.org/10.1016/j.scico.2004.10.003. www.sciencedirect.com/science/article/pii/S0167642304001893
8. Chen X, Feng J, Hiller M. Lauer V (2007) Application of software watchdog as a dependability software service for automotive safety relevant systems. In: 37th Annual IEEE/IFIP international conference on dependable systems and networks, pp 618–624. https://doi.org/10.1109/DSN.2007.14
9. Parnas DL, van Schouwen AJ, Kwan SP (1990) Evaluation of safety-critical software. Commun ACM 33(6):636–648. https://doi.org/10.1145/78973.78974
10. Mori K (1993) Autonomous decentralized systems: concept, data field architecture and future trends. In: Proceedings ISAD 93: international symposium on autonomous decentralized systems, pp 28–34. https://doi.org/10.1109/ISADS.1993.262725
11. Beningo J (2010) A review of watchdog architectures and their application to Cubesats. https://www.beningo.com/wp-content/uploads/images/Papers/WatchdogArchitectureReview.pdf
12. Ramanathan N, Chang K, Kapur R, Girod L, Kohler E, Estrin D (2005) Sympathy for the sensor network debugger. In: Redi J (ed) Proceedings of the 3rd international conference on embedded networked sensor systems, ACM conferences. New York, p 255. https://doi.org/10.1145/1098918.1098946
13. Vollmer T, Manic M, Linda O (2014) Autonomic intelligent cyber-sensor to support industrial control network awareness. IEEE Trans Indust Inf 10(2):1647–1658. https://doi.org/10.1109/TII.2013.2270373
14. Morrissett A, Eini R, Zaman M, Zohrabi N, Abdelwahed S (2019) A physical testbed for intelligent transportation systems. In: Arxiv. https://arxiv.org/pdf/1907.12899
15. Abreu B, Botelho L, Cavallaro A, Douxchamps D, Ebrahimi T, Figueiredo P, Macq B, Mory B, Nunes L, Orri J, Trigueiros MJ, Violante A (2000) Video-based multi-agent traffic surveillance system. In: Proceedings of the IEEE intelligent vehicles symposium 2000. IEEE Service Center, Piscataway, pp 457–462. https://doi.org/10.1109/IVS.2000.898385
16. Limna T, Tandayya P (2016) A flexible and scalable component-based system architecture for video surveillance as a service, running on infrastructure as a service. Multimedia Tools Appl 75(4):1765–1791. https://link.springer.com/article/10.1007/s11042-014-2373-8. https://doi.org/10.1007/s11042-014-2373-8
17. Technical University of Munich: Project Providentia++ (2022). https://innovation-mobility.com/projekt-providentia/
18. Krämmer A, Schöller C, Gulati D, Lakshminarasimhan V, Kurz F, Rosenbaum D, Lenz C, Knoll A (2022) Providentia - a large-scale sensor system for the assistance of autonomous vehicles and its evaluation. J Field Robot. https://elib.dlr.de/135631/
19. Hinz GM, Buechel M, Diehl F, Chen G, Krämmer A, Kuhn J, Lakshminarasimhan V, Schellmann M. Baumgarten U, Knoll A (2017) Designing a far-reaching view for highway traffic scenarios with 5G-based intelligent infrastructure. In: 8. Tagung Fahrerassistenzsysteme, TÜV Süd, Munich, Germany. https://mediatum.ub.tum.de/doc/1421303/1421303.pdf

Can Carsharing Reduce Car Ownership and Emissions? An Analysis Based on an Intermediate Modelling Approach

Joren Vanherck, Santhanakrishnan Narayanan, Rodric Frederix, Athina Tympakianaki, Ferran Torrent, Constantinos Antoniou, and Georgia Ayfantopoulou

Abstract As cities grow larger, they often struggle in finding sustainable and liveable mobility solutions to accommodate this growth. Many alternative modes of transport—such as public transport, carsharing systems, bikesharing systems—exist next to private car travel. The effects of expanding those alternatives are often challenging to model. This is in particular the case for small and medium sized cities, which often use straightforward and easy-to-use four-step traffic models. The alternative modes of transport could be modelled using extensive agent-based traffic models. However, these are expensive to make and require a lot of data and expertise. In the context of the EU H2020 project "MOMENTUM", we developed an intermediate modelling approach that aims to reconcile the user-friendliness of four-step traffic models with the predictive power of agent-based models to investigate the effects of alternative modes of transport. In this paper, we demonstrate the modelling of a policy plan—away from private transport towards durable modes of transport such as shared mobility and public transport—in the city of Leuven, Belgium. We focus particularly on the developed disaggregate car-ownership model, induced demand model, and link-level emission model. It was found that an improved carsharing supply can significantly reduce the car ownership of a city's households. The largest reduction is seen in households that own several cars and decide they can do with one fewer. These households can use the carsharing system for the occasional trip they would make with the additional car. Moreover, policy measures for the promotion of alternative modes of transport—which might increase the travel times to reach the

J. Vanherck (✉) · R. Frederix
Transport & Mobility Leuven, Diestsesteenweg 57, 3010 Leuven, Belgium
e-mail: Joren.vanherck@tmleuven.be

S. Narayanan · C. Antoniou
Technical University of Munich, Arcisstrasse 21, 80333 Munich, Germany

A. Tympakianaki · F. Torrent
Aimsun SLU, Ronda Universitat 22 B, 08007 Barcelona, Spain

G. Ayfantopoulou
Centre for Research and Technology Hellas-Hellenic Institute of Transport, 6th km Charilao - Thermis Rd, Thessaloniki, Greece

C. Antoniou et al. (eds.), *Proceedings of the 12th International Scientific Conference on Mobility and Transport*, Lecture Notes in Mobility,
https://doi.org/10.1007/978-981-19-8361-0_4

city for privately owned cars—were found to be able to reduce the city's mobility-related emissions. In conclusion, we demonstrated that the developed intermediate modelling approach is versatile and applicable to the cities like Leuven, such that they can also account for new modes of transport. The developed models and concepts can help other small- and medium-sized cities to shape their mobility plans.

Keywords Shared mobility · Travel demand modelling · Transportation planning · Carsharing · Transport emissions modelling · Car ownership modelling

1 Introduction

Shared mobility services are slowly penetrating European cities. This includes services such as carsharing, bike-sharing, scooter-sharing, on-demand ride services and ride-sharing. These services can enable cost savings, provide convenience, and reduce vehicle ownership [20]. The growing interest in these new mobility solutions is a result of advancement in Information and Communications Technologies (ICTs), which has enabled personalised and convenient usage of these services through mobile applications. Introduction of such services in cities calls for proper evaluation of them, to avoid inefficiency and ineffectiveness. Hence, transport models that are capable of modelling them are a necessity to support the policymakers for making informed decisions.

Many European cities, especially small and medium sized ones, continue to use the traditional strategic four-step modelling approach, and there is an inertia to change due to several reasons, including but not limited to, insufficient data, deficit of technical expertise and the convenience of simpler models [10]. While cities continue to use the traditional strategic models, modelling of shared mobility calls for agent-based approaches. This can be observed in the existing pertinent literature, which are mostly based on agent and activity-based modelling approaches (e.g., [7, 14]). Considering the aforementioned reality, an intermediate modelling framework—adopting the disaggregate principle from the agent-based approaches and integrating it into the traditional aggregate four-step models—has been developed in the EU H2020 project "MOMENTUM" [15]. Besides the integration of the disaggregate principle, modules for estimation of emissions, car ownership and induced demand are also accommodated within this framework.

The intermediate modelling framework combines benefits of the four-step modelling approach with some advantages of agent based-models. The latter are based on a synthetic population of individuals, which is meant to represent the population living in a region by having socio-demographic attributes that agree statistically with the actual population. The synthetic population is believed to accurately mimic the behaviour of the actual population. The intermediate modelling framework consists of a subset of the existing steps from the four-step approach, such as trip generation and trip distribution. The OD (Origin-Destination) matrix from these steps is disaggregated using socio-demographic data to generate a synthetic population. A discrete mode choice model is used to assign modes of transport for the synthetic population. On the one hand, the disaggregate demand can be used in a set of fleet management algorithms that assign vehicles to the users and simulate the operations of the fleet operators. On the other hand, the car ownership of the synthetic population can be influenced by changes in carsharing supply. This has in turn effect on their mode choice, which can be aggregated again into OD matrices. Then, traffic assignment is performed and once an equilibrium is reached, post processing is carried out to calculate emissions and induced demand. For more details about the framework and the interactions between the aforementioned steps, the reader is referred to [15].

This paper elucidates the results of the application of this framework in the city of Leuven, after performing suitable adaptations required for the case study. The main contribution of this research is the evaluation of shared mobility services and the insights obtained, through the use of the (adapted) intermediate modelling approach and within the limits of available data. Besides the evaluation per se, this case study can also be an example to other cities, to adapt the intermediate modelling approach for their use case. The remainder of this paper is structured as follows: First the case study is detailed, including a discussion on the available data (Sect. 2). Next, the general methodological approach is discussed in Sect. 3 and the results of applying the methodology on the case study are consolidated in Sect. 4. Finally, the conclusions are summarized in Sect. 5.

2 Case Study Description

Leuven is a small city with approximately 100 000 inhabitants located in Flanders, Belgium, in the direct vicinity of Brussels. In cities like Leuven, carsharing systems can help reduce the demand for parking spaces, meanwhile freeing up public domain for non-motorised transport modes such as bikes and pedestrians.

Leuven has round-trip carsharing systems with stations all over the city that are operated by several companies, and is interested in analysing the effects of a further expansion of these carsharing systems on the cities inhabitant's car ownership. This expansion would be combined with supporting policy measures to reduce private car use and promote public transport. We will demonstrate in Sect. 4. how such policies to further promote shared mobility services can be evaluated more accurately using an induced demand model, which, apart from simple rerouting through traffic assignment, also estimates induced traffic flows. The combined measures will further affect the vehicle pollutant emissions [6]. The latter again contributes to evaluating the effect on living quality in the city.

Apart from the above points, the Leuven case study demonstrates how quantitative predictions and analysis on the effects of a growing shared-mobility ecosystem are possible, notwithstanding only having a limited amount of additional data on shared mobility services available. The carsharing data at our disposal is the carsharing supply in each district (Fig. 1 and Table 1) and a mobility survey conducted in the city [1].

Apart from carsharing-specific data, we made to a large extent use of open data sources, such as OpenStreetMap [8], traffic counts on main roads and Telraam [22]. The latter is a unique, open citizen science project that allows to fill blind spots in traffic count. It employs new technologies in innovative ways for measuring traffic in a continuous, multimodal, dense and efficient way. Finally, we make use of a full synthetic population of agents, performing trips that pass through Leuven. This synthetic population was obtained from the Flemish strategic traffic models [9] and

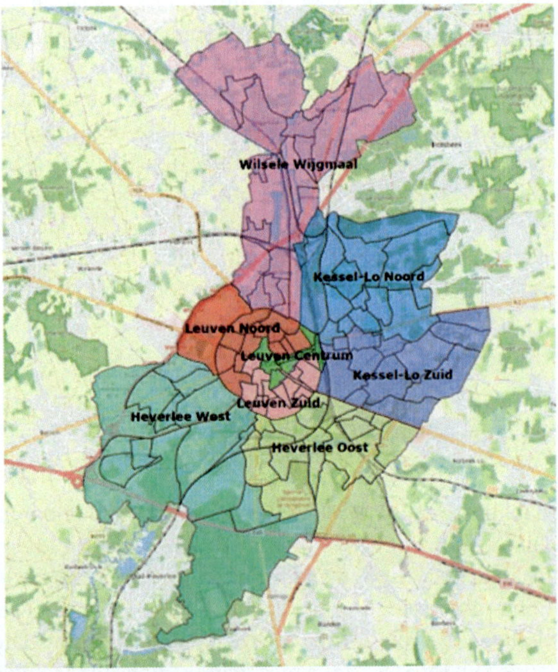

Fig. 1 Districts of the city of Leuven

Table 1 Carsharing supply (css) for each city district in Fig. 1. This is the number of carsharing cars available in 2017 (base case)

District	css
Heverlee Oost	18
Heverlee West	7
Kessel-Lo Noord	12
Kessel-Lo Zuid	18
Leuven Centrum	8
Leuven Noord	16
Leuven Zuid	15
Wilsele Wijgmaal	12

includes all trips of the agents, together with their mode shares (car, public transport, bike, foot).

3 Methodological Approach

The flowchart in Fig. 2 shows the main modules of the modelling approach applied to the city of Leuven. The synthetic population and its trips are here assumed to be inputs. If they are not available, they can be generated using the techniques described in [15]. The components in focus in this paper are the disaggregate car-ownership model, the emission model and the induced-demand model, which are all interlinked through a disaggregate mode-choice model and static stochastic traffic assignment. The light green parts indicate modules and data at disaggregate level, while the dark green parts indicate aggregated modules. The transition is made when aggregating trips with a mode choice into OD matrices that are used for the traffic assignment.

In scenarios, the carsharing supply can be changed to simulate and assess measures that are expected to promote the use of alternative modes of transport. The carsharing supply is one of the main inputs for the car-ownership model.

The following sections briefly describe the key modules of the modelling framework that is applied in this paper. A much more extended description, including information on the calibration process and parameters, as well as exact formulae, can be found in Deliverable 5.3 of the MOMENTUM project [9]. Henceforth, we will refer to this document as Momentum D5.3.

3.1 Synthetic Population and Mode Choice

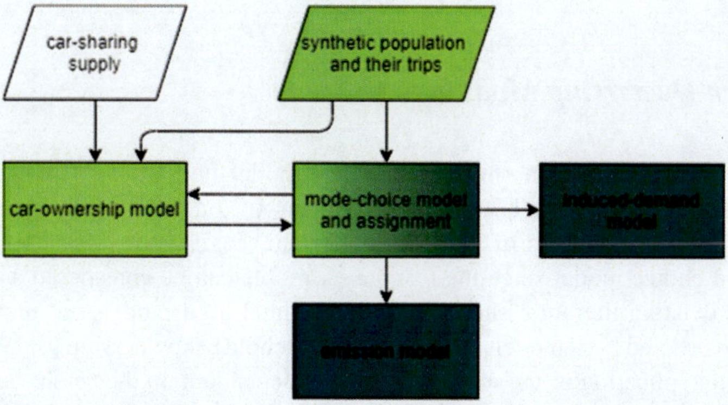

Fig. 2 Main model modules and their interactions in the model for Leuven. The light green colour indicates modules at disaggregated level, while dark green parts are aggregated models

The synthetic population and the individual trips they make are one of the main inputs to the model. It is used both for the car-ownership model and for the mode choice model. For the Leuven case study, a full synthetic population of agents, making trips that pass through the study area, is readily available from the Flemish government [9]. Nevertheless, if no synthetic population is available, it can be generated by the techniques of [15] or the open-source tool PopGen [17].

The disaggregate mode choice model uses the socio-demographic properties of the synthetic population together with trip-related properties (such as travel times) to predict mode shares for each of the trips. The possible modal choices are car, public transport, bike and foot. To propagate effects of a changing car ownership to other parts of the modelling framework, it is of key importance that car ownership is one of the parameters accounted for in the disaggregate mode choice model.

For Leuven, the synthetic population and its trips also include indicative mode shares for the base case. This provides a wealth of information to calibrate a custom disaggregate mode-choice model. As such, a multinomial logit mode choice model was iteratively calibrated using PandasBiogeme [4]. The variables that were found to be relevant for the modal choice are: mode-specific travel time (determined through traffic assignment), possession of a company car, the agent's occupation, e-bike availability and the number of cars in the household. The exact result for the model and its parameters can be found in Momentum D5.3. If there is not enough data in a city to calibrate a custom disaggregate mode-choice model on a reference scenario, it is advisable to use a mode-choice model that was calibrated on a city with similar characteristics.

The disaggregate mode-choice model allows to account for changes in the synthetic population, such as a reduced car ownership that results from an increased carsharing supply, in other parts of the model. It is difficult to have those kinds of dependencies in classical four-stage models. After applying the disaggregate mode-choice model, the trips and calculated mode shares are aggregated into OD matrices. From there on, the typical assignment step of four-stage models can be applied to estimate traffic flows on the network.

3.2 Car-Ownership Module

The car-ownership model is a multinomial logit model that estimates for each household whether it owns (i) no household private car, (ii) one car, (iii) two cars, or (iv) three or more cars. Instead of using a multinomial logit model, one could opt for an ordered choice model since the response variable can be considered as ordered. However, existing literature shows that multinomial logit models are more appropriate than ordered logit models for studying household car ownership [3, 19]. For the envisioned application of the model, it must be dependent on the carsharing supply in the living area of the synthetic population. The details of the car-ownership model are in [16] and Momentum D5.3.

The car-ownership model was estimated based on a local travel survey [1]. An analysis of the local travel survey showed that the most important variables in the car-ownership utility specification are: household size, age, income, cargo bike ownership, public transport pass availability, carsharing subscription, carsharing supply (number of vehicles per district) and commute travel distance. While other determinants of private car ownership may exist, they cannot be modelled here due to their absence in the local travel survey. One such determinant that is not included here, would be a sudden policy change on local pollution that could render people's currently owned private cars useless. In such a scenario, they would be immediately given the choice to buy a new car with electric traction or to use a shared alternative.

Along with the car-ownership model, a carsharing subscription model was calibrated using PandasBiogeme [4]. The details are once again given in Momentum D5.3. In this case, the local travel survey showed that the estimated model depends on: the carsharing supply in a region, the number of cars in a household, whether the household owns a cargo bike, the number of members of a household and a parameter measuring the willingness of households to use a carsharing system.

The car-ownership model depends on the availability of a carsharing subscription. Vice versa, the carsharing subscription model depends on the car ownership of a household. This means that the two models can be applied iteratively to reach self-consistency. We refer to the combination of these two models as the car-ownership module.

Not all utility variables in the car-ownership module are readily available in the existing synthetic population. The missing values are, of course, present in the local travel survey [1]—by definition, since the module used the survey as calibration dataset. Since the local travel survey contains other variables in common with the synthetic population, it can be used to enrich the latter. For the enrichment procedure, a random forest regression model [18] is developed to estimate the conditional probability of the unknown attributes with respect to a set of predefined common attributes. Random forests are ensembles of decision trees, which expand themselves using the explanatory variables (common attributes) to obtain leaves as simple as possible for the target variables (unknown attributes). Random forests are chosen instead of single decision trees as they reduce variance, by averaging weak decision trees (trees that are not fully expanded), and hence, reduce overfitting and improve generalization. The final random forest had a size of 1000 decision trees.

The car-ownership module can be applied to the part of the synthetic population that lives in one of the regions for which the module was calibrated. For these households, the effect of an increase in carsharing supply on the car ownership can be calculated. These changes propagate, through the disaggregate mode choice model, into the traffic assignment results. Since the car-ownership module operates at the disaggregate level, it allows cities to conduct a further detailed socio-demographic analysis of the impact of their policies. This helps to reach the targeted groups and, possibly, to install measures to make the policy more socially just.

3.3 Emission Model

While policies that reduce private car ownership lead, in the first place, to tactile improvements—such as reduced congestion and freeing up of public space that was formerly reserved as parking—they may also have secondary, less tactile benefits. One of the key invisible elements that make a city healthy to live in, is its air quality [6]. Our emission model helps cities to evaluate the contributions of mobility on this key metric. It is designed to be a one-stop-shop to calculate traffic-generated emissions, once assignment results are available.

In particular, the developed emission model merges the benefits of a detailed emission model with the spatial-detail benefits of a traffic model to produce a better estimate of traffic emissions. It allows to calculate emissions (CO, CO_2, NO_x, PM and VOC) on a link-specific basis using static assignment results. It accounts for the car flows for each link, the corresponding lengths and average speed for these links. Importantly, the model uses one-off fleet projection at EU-member state level for detailed fleet data [6]. The model was calibrated for 31 EU member countries for all years between 2016 and 2050. One must note that the future is inherently uncertain, and the model should be updated with the latest information to remain as precise as possible. Importantly, the emission model also accounts for speed-variations of emission factors at link level. Overall, the effect of a carsharing supply change on emissions can thus be evaluated by comparing the emission levels of the different scenarios.

The details of the model and its calibration can be found in Deliverable 4.1 [6].

3.4 Induced Demand

Cities that want to evaluate new policy measures—promoting a more prominent role for shared-mobility services, such as opening part of the regular car infrastructure to other transport modes—do not have the tools to do so at this moment. The traditionally used four-step transport assume demand to be constant, while this is unrealistic when, for instance, travel times increase heavily. On the contrary, fully activity-based models inherently account for such variability. However, they are very intensive on the computational side and require a lot of data. In contrast, our elasticity-calibrated induced demand model aims to partially correct for the effects of induced demand, without paying a large computational and complexity burden.

The induced demand model is based on a nested logit formulation, where the choice to travel or not to travel are balanced against each other. Doing so, and assuming that the utility of not travelling remains constant, new demand figures can be calculated as

$$D = \frac{P^t}{P_0^t} D_0 = \frac{e^U}{e^{U_0}} \frac{K + e^{U_0}}{K + e^U} D_0$$

In this formula, D_0 is the known demand in a reference scenario and P^t/P_0^t is the ratio of the new share of travellers as compared to the share of travellers in the reference scenario. In the second equality, this is expressed in terms of the logsum travel utility in the new scenario (U) and the reference scenario (U_0). These can be calculated using the scenario's travel times. The quantity K is the exponent of the unknown constant utility of not travelling and is the subject of the model's calibration. After calibration of the induced demand model, it only requires new travel times to calculate the new logsum travel utility U and the original demand D_0.

The induced demand model is calibrated by tuning the parameter K. The calibration is based on.

- The calibrated od demand matrix of the selected reference scenario;
- A representative car travel time-elasticity of car demand η_{car}. The short-term direct time–cost elasticity for car travel demand is set to $\eta_{car} = -0.6$ [13];
- The disaggregate mode choice calibration parameters. These are obtained during the calibration of the disaggregate mode choice model. Specifically, we use the utility coefficient for the mode travel time.

The detailed calibration procedure is described in Momentum D5.3.

4 Results

The framework of Sect. 3 was applied to the case study of the city of Leuven (Sect. 2). The city wants to investigate the effects of promoting alternative modes of transport. This promotion will be accomplished by increasing the carsharing supply and freeing up public space that can be used by the alternative modes of transport. The city wants to get insights in the changing traffic patterns, sociodemographic consequences and non-tactile effects on the traffic-induced emissions. We will refer to the reference situation without any changes as the 'base case'. The possible future situation, where alternative modes of transport are further promoted, will be referred to as the 'alternative case'.

The road network has 19,734 links and 8327 nodes. The results of an assignment of the base case between 7 and 8am for the cars is shown in Fig. 3.

Using the emission model, the emission of all pollutants and CO_2 can be calculated on a per-link basis. As an example, Fig. 4 shows the CO_2 emissions per link that corresponds with the above presented assignment. The calculation of pollutants yields similar results. The total cumulative emissions for the single modelling hour over the entire network are summarized in Table 2 for CO, CO_2, NO_x, PM and VOC.

The effects of increased carsharing supply and policy measures were studied by implementing two changes in the alternative case as compared to the base case:

- The carsharing supply was doubled in most city regions, and tripled in the Kessel-Lo regions (cf. Fig. 1 and Table 1);

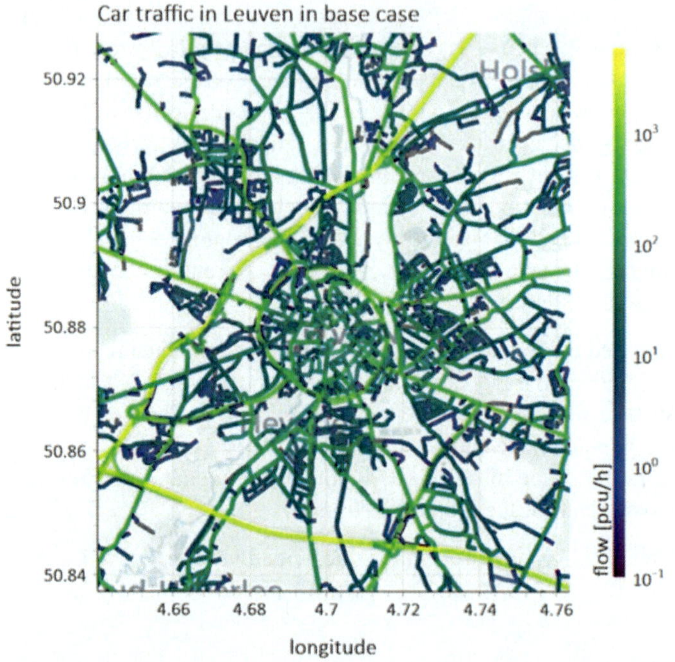

Fig. 3 Assignment results for car traffic on the central part of the Leuven network between 7 and 8 am in the base case. Note the logarithmic scale for the indicated flows

- New traffic light control schemes—that promote alternative modes of transport and that limit the inflow of private vehicles into the network—were implemented on the two main eastern entrance roads (Diestsesteenweg and Tiensesteenweg). This leads to a travel time increase for private vehicles entering the network through these roads.

Self-consistent application of the car-ownership module leads to changes in the number of cars of the households in the synthetic population (Fig. 5). Note that the modules can only be applied in the areas on which they are calibrated. The number of cars can be calculated as a function of sociodemographic variables. For the alternative case, the result of such calculations is shown as function of the net disposable monthly household income (Fig. 6) and household size (Fig. 7).

The total number of cars of the affected population went down from 42,788 to 38,880, while the total number of carsharing cars increased from 106 to 242. This means that every new shared car replaced, on average, 29 privately owned cars. This number has the correct order of magnitude, but is slightly too high according to general pertinent literature [5, 21]. Two possible reasons for this deviation arise:

- The context of the city of Leuven is rather unique as it has a very large carsharing uptake: 5% of its inhabitants have a roundtrip carsharing subscription, compared to only 3% for other cities in Flanders [2].

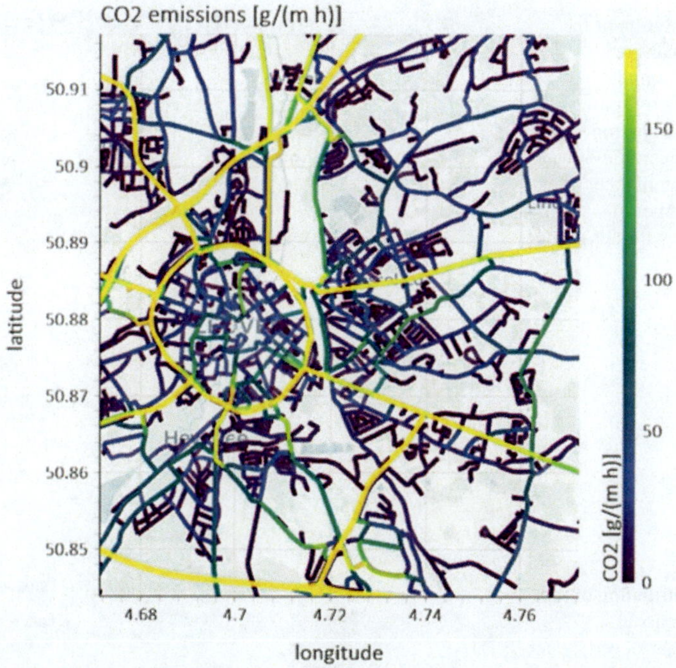

Fig. 4 CO_2 emissions on the central part of the Leuven network between 7 and 8 am in the base case. To normalize the emissions with respect to link lengths, they are expressed per meter of link length

Table 2 Total emissions over the entire Leuven network between 7 and 8 am. The emissions are given both for the base and the alternative case. For convenience, the realized reduction is calculated (positive when the alternative case has lower emissions)

Emission	Base	Alternative	Reduction
CO2 (ton/h)	131.13	130.36	0.77
CO (g/h)	241.43	240.01	1.42
NOX (g/h)	76.62	76.17	0.45
PM (mg/h)	3.644	3.623	0.021
VOC (mg/h)	15.222	15.133	0.089

- The number of households that start off with three or more cars is nearly halved in the alternative case, which is probably not realistic. We hypothesize that possessing such a large number of cars is in reality not directly related to a practical necessity, hinting at behaviour that is difficult to capture directly by the model. This hypothesis is corroborated by the results in Fig. 7. The car-ownership model should thus not be expected to yield realistic results for households owning three or more cars.

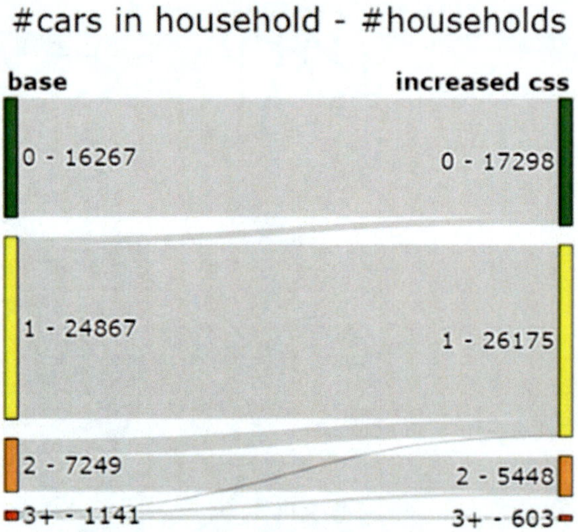

Fig. 5 Evolution of the number of cars per household from the base case to the alternative case with increased carsharing supply (css). The quantities always represent the number of cars and the number of households with this number of cars

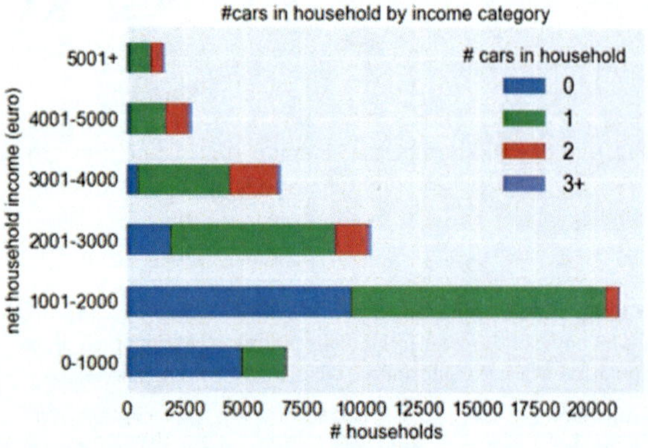

Fig. 6 Distribution of the number of cars in households, grouped by net disposable monthly household income. Households with lower incomes tend to have fewer cars

The model, including mode choice and induced demand, was further applied to the alternative case. The assignment differences are shown in Fig. 8. The main difference in the results is due to the induced demand model that works on the increased travel times for external travellers entering the network through the two main eastern roads. Notice that the effect of the reduced car ownership is negligible in the results. This can be explained by the fact that the shift in car ownership is mainly from households owning two cars to households owning a single car. Typically, the second car is used less frequently, such that it can be substituted by a shared car (which is, in turn, used even less frequently). As such, the reduced car ownership only has a small impact on the observed traffic flows. However, the reduced car ownership will have an impact on the number of parked cars. Reduced private car ownership will free up public space for other activities or mobility solutions.

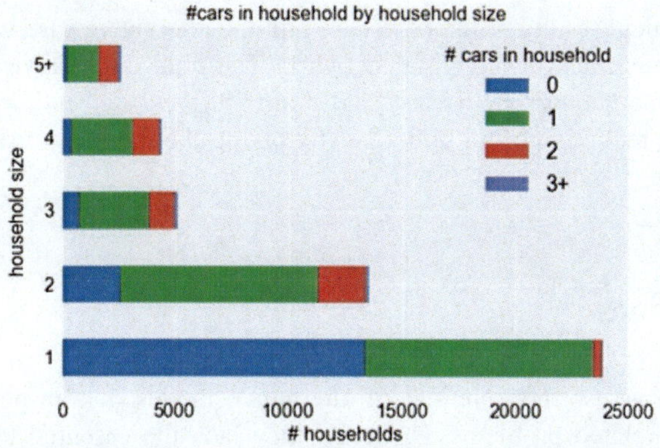

Fig. 7 Distribution of the number of cars in households, grouped by the household size. The single households are rather more likely not to have a car. In absolute terms, the number of households owning three or more cars is rather constant over all household sizes. It is unlikely that one-or two-person households need this many cars out of pure practical necessity, hinting at behaviour that is difficult to capture directly by the model

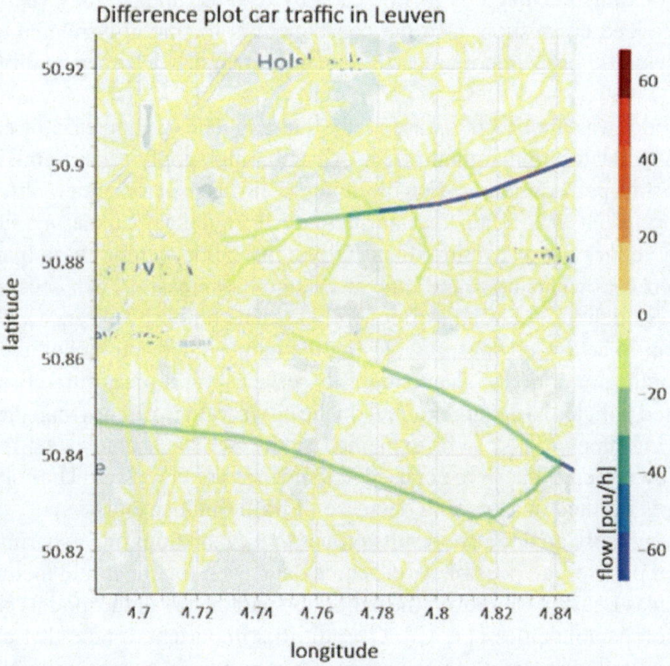

Fig. 8 Difference plot for the car traffic on the Leuven network between the base case and the alternative case between 7 and 8 am. The traffic volumes in the alternative case were lower than in the base case. The main effect is due to the travel time increase on the eastern entrance roads (which are clearly visible in the figure) and corresponding induced demand

The assignment results of the base case and the alternative case were both used to calculate emissions on the entire network. The total cumulative emissions for car traffic on the network between 7 and 8 am are presented in Table 2. There is an appreciable reduction of all pollutants by taking the measures that make up the alternative case.

5 Conclusions

Cities want to assess the potential of alternative modes of transport for solving their mobility-related problems. However, they lack the essential tools for doing so, without resorting to full activity-based models. We adapted an intermediate modelling approach, combining some of the benefits of agent-based modelling with a classical four-stage model. The adapted framework allows cities to model the effect of policy measures such as an increased shared mobility supply and the promotion of alternative modes of transport by transferring parts of the public domain that were previously mainly occupied by private cars. Next to the immediately visible effects such as reduced congestion, also the consequences for mobility-related emissions can be assessed. This is important, as it is an important driver for the quality of living in cities.

The model was applied in a case study that simulates an increase of carsharing supply, to evaluate the marginal effect of carsharing supply on car emissions and car ownership per household. For the latter, a further sociodemographic analysis was performed. It was found that an increase in roundtrip carsharing supply can effectively reduce car ownership. Especially households owning more than one car reduce their car ownership as the carsharing supply increases. Their second or third private cars might not be used as frequently as their first private car, which renders a shared car to be a good substitute. Using the induced demand model, the effect of policies for the promotion of shared mobility services was more accurately modelled. The adopted policies had a larger effect on the city's traffic volumes than the increase of carsharing supply. This can be explained by the fact that shared cars often replace privately-owned cars that were not used that much in the first place. The combination of measures resulted in an overall decrease of traffic-induce emissions.

Obtaining those quantitative results proves that evaluations of carsharing systems and related policies are possible with scarce data, even for small-and medium-sized cities, such as Leuven. Moreover, the calibrated parameters of the models are invaluable for comparable cities [11, 12]. Overall, it is important to note that carsharing is typically limited in terms of penetration, since many car owners will not abandon their private car (for instance, because they use it very intensively) and that people using carsharing services might still have a privately-owned car as well. In this sense, carsharing can also be complementary to private car use, for example within the context of Mobility as a Service (MaaS). On the other hand, some people might be forced into using carsharing services when policies on local pollution are applied—pushing towards the use of electric traction—and they cannot (yet) afford the substitution of their conventional non-electric car. This latter situation is not included in the current model and is suitable for further research.

Acknowledgements This research has been supported by European Union's Horizon 2020 research and innovation programme under grant agreement No. 815069 [project MOMENTUM (Modelling Emerging Transport Solutions for Urban Mobility)].

References

1. Agentschap Binnenlands Bestuur (2017) Stadsmonitor. Leuven. https://gemeente-stadsmonitor.vlaanderen.be/. Last Accessed 22 June 2022
2. Agentschap Binnenlands Bestuur (2020) Gemeente-stadsmonitor benchmark Vlaams Gewest. https://gemeente-stadsmonitor.vlaanderen.be/. Last Accessed 22 June 2022
3. Bhat CR, Pulugurta V (1998) A comparison of two alternative behavioral choice mechanisms for household auto ownership decisions. Transp Res Part B: Methodol 32(1):61–75. https://doi.org/10.1016/S0191-2615(97)00014-3
4. Bierlaire M (2020) A short introduction to PandasBiogeme. Technical Report TRANSP-OR 200605. Transport and Mobility Laboratory, ENAC, EPFL
5. Cambio (2022) Cambio is one of the main carsharing providers in the city of Leuven. www.cambio.be/nl-vla/onze-impact. Last Accessed 22 June 2022
6. Chapman DA, Eyckmans J, Van Acker K (2020) Does car-sharing reduce car-use? an impact evaluation of car-sharing in flanders Belgium. Sustainability 12(19):8155. https://doi.org/10.3390/su12198155
7. Ciari F, Balac M, Axhausen KW (2016) Modeling carsharing with the agent-based simulation matsim: State of the art, applications, and future developments. Transp Res Rec: J Transp Res Board 2564:14–20. https://doi.org/10.3141/2564-02
8. Contributors OpenStreetMap (2020) Leuven street network from Openstreetmap. www.openstreetmap.org. Last Accessed 22 June 2022
9. Departement Mobiliteit en Openbare Werken Vlaanderen. (2021). Strategische verkeersmodellen. www.vlaanderen.be/department-mobiliteit-en-openbare-werken/onderzoek/verkeersmodellen/strategische-verkeersmodellen. Last Accessed 22 June 2022
10. Givoni M, Beyazit E, Shiftan Y (2016) The use of state-of-the-art transport models by policymakers – beauty in simplicity? Plan Theory Pract 17:385–404. https://doi.org/10.1080/14649357.2016.1188975
11. H2020 MOMENTUM consortium (2021a) Deliverable 4.1: new transport modelling approaches for emerging mobility solutions. https://h2020-momentum.eu/wp-content/uploads/2021/06/MOMENTUM-D4.1-New-Transport-Modelling-Approaches-Issue-1-Draft-2.pdf. Last Accessed 22 June 2022
12. H2020 MOMENTUM consortium (2021b) Deliverable 5.3: Implementation of the Momentum decision support toolset in Madrid, Thessaloniki, Leuven and Regensburg. https://h2020-momentum.eu/wp-content/uploads/2022/02/D5.3-Implementation-of-the-MOMENTUM-Decision-Support-Toolset-in-Madrid-Thessaloniki-Leuven-and-Regensburg.pdf. Last Accessed 22 June 2022
13. de Jong G, Gunn H (2001) Recent evidence on car cost and time elasticities of travel demand in Europe. J Transp Econ Policy (JTEP) 35(2):137–160. www.ingentaconnect.com/content/lse/jtep/2001/00000035/00000002/art00001. Last Accessed 22 June 2022

14. Martinez LM, Correia GHA, Moura F, Mendes Lopes M (2017) Insights into carsharing demand dynamics: Outputs of an agent-based model application to Lisbon, Portugal. Int J Sustain Transp 11:148–159. https://doi.org/10.1080/15568318.2016
15. Narayanan S, Salanova Grau JM, Frederix R, Tympakianaki A, Antoniou C (2021) Modelling of shared mobility services—An approach in between traditional strategic models and agent-based models. In: 24th Euro Working Group on Transportation Meeting
16. Narayanan S, Antoniou C (2022) A multilevel analysis on what is common (and not) across the cities when it comes to private car-ownership: Behavioural, policy and modelling insights. Under Rev
17. MARG (2016) PopGen: Synthetic population generator [online]. Mobility analytics research group. Available at www.mobilityanalytics.org/popgen.html. Last Accessed 22 June 2022
18. Pedregosa F et al. (2011) Scikit-learn: Machine learning in Python. J Mach Learn Res (12):2825–2830
19. Potoglou D, Susilo YO (2008) Comparison of vehicle-ownership models. Transp Res Rec 2076(1):97–105. https://doi.org/10.3141/2076-11
20. Shaheen SA, Chan N (2015).Mobility and the sharing economy: Impacts synopsis. https://innovativemobility.org/88wp-content/uploads/Innovative-Mobility-Industry-Outlook_SM-Spring-2015.pdf. Last Accessed 22 June 2022
21. Shaheen SA, Cohen AP (2013) Carsharing and personal vehicle services: worldwide market developments and emerging trends. Int J Sustain Transp 7(1):5–34. https://doi.org/10.1080/15568318.2012.660103
22. Telraam (2020) Telraam. Leuven. https://www.telraam.net. Last Accessed 22 June 2022
23. TML (Transport & Mobility Leuven) (2021) Fleet model. Leuven. www.tmleuven.be/en/navigation/Fleet-Model. Last Accessed 22 June 2022
24. Vlaams Verkeerscentrum, Vlaamse Overheid (2020) Verkeersindicatoren. http://indicatoren.verkeerscentrum.be. Last Accessed 22 June 2022

Optimizing Passenger Flows in a Multimodal Personal Rapid Transit (PRT) Station Using Microscopic Traffic Simulation

Oytun Arslan, Max Reichert, Eftychios Papapanagiotou, and Silja Hoffmann

Abstract Personal rapid transit (PRT) systems are on-demand transport systems in which small autonomous vehicles travel on their own dedicated infrastructure. Passengers travel alone or in small groups in a vehicle without intermediate stops to their requested destination. Typically, they are considered as medium capacity systems of 5,000–20,000 passengers/h. This paper sets the focus on the optimal passenger guidance within PRT stations to maximize station capacity and in particular in the queue formation for the passengers within the PRT station. The queueing processes are modelled with a microscopic traffic simulation software. The examined scenario is a train station at which the passengers form queues before boarding to the PRT system. Two types of queues are examined: An S-shaped and a funnel-shaped queue. To determine the maximum number of the dispatched passengers of the PRT station, the percentage of the passengers alighting from the train is increased incrementally. The results of the simulation show that both queue types reach the same maximum number of dispatched passengers per hour. This indicates that there is a certain likelihood that station capacity is independent of the shape of queues within the station, but it depends rather on availability of vehicles within the station. We draw this conclusion from a visual observation of the model and a rough comparison with values from a previous study. Therefore, there is more potential to increase PRT system capacities in the optimization of the PRT vehicle-side processes in the station, rather than in the optimization of passenger-side queues in the station.

Keywords Personal rapid transit · PRT · Station design · VISSIM · Microscopic simulation · Microsimulation · Capacity analysis · Passenger flow optimisation

These authors contributed equally to this paper.

O. Arslan · M. Reichert (✉) · E. Papapanagiotou · S. Hoffmann
Fakultät für Bauingenieurwesen und Umweltwissenschaften, Professur für Intelligente, Universität der Bundeswehr München, Multimodale Verkehrssysteme, Werner-Heisenberg-Weg 39, 85577 Neubiberg, Germany
e-mail: max.reichert@unibw.de

C. Antoniou et al. (eds.), *Proceedings of the 12th International Scientific Conference on Mobility and Transport*, Lecture Notes in Mobility,
https://doi.org/10.1007/978-981-19-8361-0_5

1 Introduction

Assurance of mobility has been one of the most important issues in urban areas in recent years, where different measures such as traffic management, intelligent transportation systems, improvement of public transport, travel demand management or parking management are applied. Recently, on-demand systems have gained a lot of popularity and many private companies as well as municipal authorities offer this type of transport as an intermediate solution between individual and public transport.

Personal Rapid Transit (PRT) is a promising on-demand transport mode that often enables individual mobility, yet has many similarities to public transport. In a PRT system, fully autonomous electric vehicles (also known as pods) are deployed on their dedicated guideways, grade-separated from the rest of the traffic [1]. Vehicles are very small, with a capacity of 2–6 passengers, travelling with a maximum speed of around 60 km/h [2].

A PRT system essentially consists of a main guideway and stations. Advanced communication capabilities between vehicles can lead to very small headways, even as low as 0.5 s on the main guideway [1]. Passengers travel from one station to another as in public transport, though the main difference is that a trip does not have to include in-between stops which is common in public transport. Despite the dedicated guideways, this is possible thanks to two characteristics: The first feature is the off-line stationing of PRT stations.

The second characteristic is that PRT vehicles are occupied either individually or by a small group of people who travel to the same destination and usually know each other already (e.g. a family). This allows a direct service from A–B, but at the same time indicates the difference from popular ride-pooling services, where individuals with different destinations are "pooled" together in a ride. Traffic flow on the main guideway is not disrupted by stopping vehicles at stations, because stations are not located on the main guideway and rather off-line. This enables a non-stop travel from origin to destination as well as a high operational speed, since accelerations and decelerations for each and every stop are avoided. Another advantage of off-line stationing is that many stations with relatively short distances are possible especially in high-dense urban areas where the travel demand requires it, which in turn leads to short walking distances and a better accessibility [3].

From a passenger point of view, PRT systems allow a very limited to no waiting time at stations and act as an on-demand transport system [1, 2]. Vehicles travel only if there is a demand for it, there is no fixed timetable. Furthermore, to ensure an attractive system, the aim is to keep the waiting times as low as possible. Passengers are ensured to get a seat in the vehicle and are not obliged to rideshare, therefore an exclusive use of the vehicle during the trip is possible which can be seen as another positive aspect during pandemics. Thanks to fully autonomous electric vehicles, no direct vehicle emissions are exposed and operational costs (such as driver costs) are minimized, even though a 24/7 service can be provided at the same time [3].

Another essential aspect of PRT is the flexibility of the system. Different station sizes according to the local demand are possible, which differentiates it from rail-based public transport. Moreover, stations can be theoretically located in existing buildings which results in an effective use of urban space, less construction costs and better access of stations. Besides, narrow vehicles need narrow guideways that require less infrastructure space. The system is easily scalable since adding new stations or extending the main guideway with certain loops is mostly possible with a relative low complexity [3].

Unfortunately, not many PRT systems have been in operation around the world, yet. Some examples are the London Heathrow Airport Terminal 5 connection (UK), Morgantown at West Virginia University (USA), and Masdar City in Abu Dhabi (UAE) [1]. Yet, in these systems capacity was not taken as a crucial issue. Until now, the designs were made mostly for passenger comfort rather than high capacity offer. On the other hand, system capacity is an indispensable feature if a wide spread of PRT comes into question in the future.

Compared to other public transport modes, PRT systems offer a higher capacity than typical urban bus systems but a lower capacity than mass rail transit. This constraint positions it to a level of tram system or bus rapid transit (BRT), where a system capacity of 5,000–20,000 Passenger/h per direction is generally accepted. Typically, the capacity of a PRT system is affected by three factors: Capacity of the main guideway (constrained mostly by the allowed headways), number of vehicles in the system and station capacity [4]. Hereby, we focus on the station capacity and analyze the optimization of passenger flows in the station to maximize the number of dispatched passengers from a typical PRT station. Other factors as well as aspects, such as empty vehicle management or vehicle routing, is beyond the scope of this paper.

2 Related Work

The existing literature in the topic consists mainly of simulation studies of PRT networks as a whole [5–8] with limited focus on PRT stations. Modelling tools are developed specifically for PRT networks and sometimes even for a specific project. An analytical approach was carried out in [9], in order to calculate station capacities and compare different layouts. They investigate not only the capacity itself, but also length and area efficiency for serial, saw-tooth and saw-tooth high capacity stations. Moreover, different operation modes such as synchronous and asynchronous operation are investigated.

Some studies focus on developing alternative station layouts where a capacity gain is expected [1, 4, 9, 10]. Typical alternatives include avoidance of reversing of vehicles leaving berths, or suggest movement of vehicles in platoon for less vehicle interaction and hence less capacity loss. However, these alternatives have not yet been applied to real life PRT systems.

A recent study has investigated the impacts of different station layouts on the station capacities [9]. Five different layouts are modelled in a microscopic traffic simulation software for different numbers of berths and different boarding/alighting behaviors by using the autonomous vehicle car following models. The results show that the loosened design of a serial station with an additional station guideway offers the highest capacity among other layouts. The station layouts, where vehicles need to reverse or require time for using the 90° parking module, result in a considerable loss in capacity. Another finding is that increasing number of berths increases the capacity, but the relation is not directly proportional as the capacity increase. In particular, the capacity increase slows down per additional berth. However, in that study [3] the focus is set only on the vehicle side, and not on the passenger side. Only the effect of the number of vehicles served by the station is examined. Passenger interactions such as queueing in the station are not explored.

On the analogy of simulations for vehicles, also pedestrian simulations have macroscopic, mesoscopic or microscopic simulation approaches which vary in their level of detail [11]. Microscopic simulations have the highest level of detail and are meaningful when it comes to simulating individual behaviours of passengers, staff or other pedestrian groups which are of relevance [11, 12]. Applications of microscopic simulations for optimization tasks in airport can be found in literature. There are optimization tasks on passenger flows, passenger experience or minimizing passenger queue times in combination with non-passenger related indicators (mostly related to operational cost or revenue aspects) [11–13]. Although [11, 13] use mesoscopic approaches for their optimization tasks, microscopic models are applied to simulate the queueing processes in a detailed manner, which allows to utilize the detailed nature of microscopic models.

This paper aims to analyze the passenger interactions considering queue forms in a PRT station, as well as their impact on the station capacity. The Key Performance Indicator (KPI) used is the number of passengers that can be dispatched from a PRT station and is calculated with a microscopic traffic flow simulation tool.

3 Methodology

Taking also the passengers into consideration, we define PRT station capacity as the number of passengers that a station is able to handle in one hour. Based on the previous study [3], we selected the loosened serial station with an additional station guideway for our further investigations, since this has clearly given the highest capacity among different layouts. For the number of berths, 6 berths were chosen as the magnitude of capacity gain per additional berth decreased substantially after that. Taking the area efficiency into account as well, here we decided to use a layout with 6 berths, and not a larger one. Regarding the boarding/alighting behavior, we assumed a boarding type station, namely a station where passengers only board vehicles in these stations, for instance a station at a factory at the end of a work shift. Readers can refer to

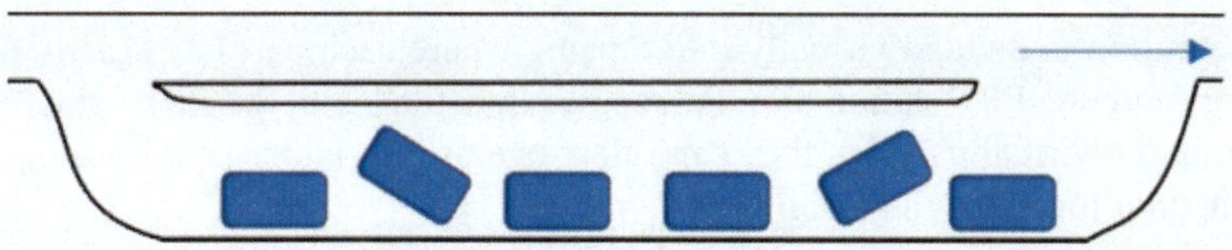

Fig. 1 PRT station layout selected for this study. Gaps between the stopped vehicles allow them to enter/leave their berths independently of other vehicles

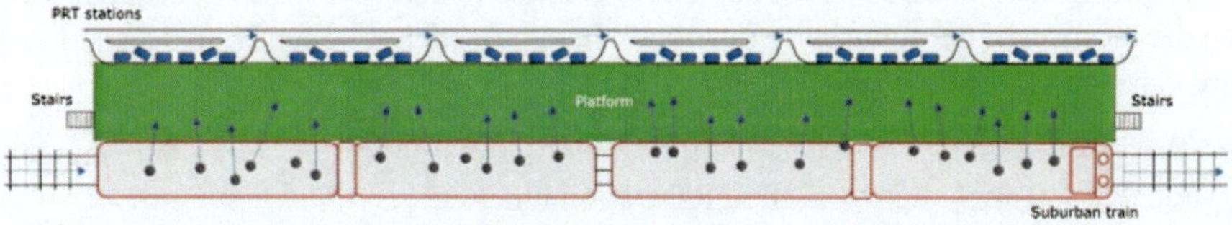

Fig. 2 Overview of the transfer station with suburban train stop and the PRT stations at the opposite side

the previous study [3] in order to get a deeper insight about these three parameters (station layouts, number of berths and boarding/alighting behavior). The selected station layout is shown in the following Fig. 1.

The scenario presumed for the simulation study is a suburban train station where passengers can transfer from the suburban train to the PRT vehicles located on the opposite side of the same platform. For example, employees working in a large commercial area in the vicinity of this station travel by suburban train in the morning peak and change here to the PRT vehicles to reach their final destinations. This scenario therefore depicts a situation of PRT integration into the existing public transport network. Following assumptions were made for the simulation model:

- The suburban train has a total length of approximately 195 m, which corresponds to the length of the station platform. For the whole length of the platform, 6 PRT station units (each having 6 berths) were designed (see Fig. 2).
- The train possesses 36 doors in total and passengers are distributed evenly in the train, which ensures a balanced usage of all train doors.
- Trains arrive at the station every 10 min, which means a peak demand is reached in every 10-min interval.
- PRT vehicles are used only by passengers that transfer from the suburban train, no other passengers start their journey from this station.
- Suburban trains are fully utilised, with 1.800 passengers per arrival. Number of actually alighting passengers was used as a parameter during the simulation study.
- 80% of the alighting passengers transfer to the PRT vehicles. 10% leave the station using the left-hand stairs as well as the other 10% leave using the right-hand stairs. Stairs are located at the very ends of the station. Passengers leaving the station are included in the model too, in order to observe their impact on the transferring passengers at the platform.

- The station is divided virtually into 6 parts, where each part has exactly 6 train doors and one PRT station unit. Passengers transferring to the PRT vehicles stay within the virtual part, but they can select one of the six berths according to the train door locations they alight.
- For the simplicity, only one virtual part of the suburban train station and the associated queueing situation of the PRT station was simulated since all other parts are identical.

As mentioned before, which passenger can use which PRT station berth depends on the location of the suburban train door where passenger alights. Here the six doors are classified in three types:

- Type 1 door is the one located at the left-most. Passengers alighting here can use the two left-most berths in the PRT station unit.
- Type 2 doors are located in the middle. Passengers alighting here can use the berth directly opposite as well as the neighboring berths (hence 3 alternatives in total).
- Type 3 door is the one located at the right-most. Passengers alighting here can use the two right-most berths in the PRT station unit.

Selection of a certain berth is carried out by comparing the waiting passengers at the queues. Each passenger reaching the decision area checks which queue has fewer passengers at the time of the decision and makes its decision accordingly. In this way, an even distribution of passengers to berths is ensured (see Fig. 3).

After selection of the PRT berth, the passenger moves forward and waits for his turn in the queue to board a PRT vehicle at that berth. As soon as a vehicle is present at the berth for boarding, the passenger is allowed to board. For the vehicle

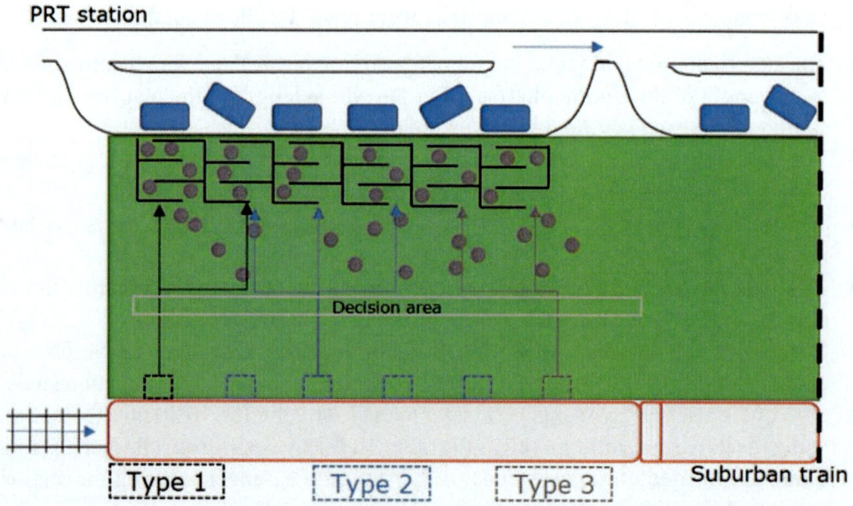

Fig. 3 Door-berth assignment of transferring passengers based on the train doors they alight

occupancies, it is assumed that 75% of the trips are made with 1 passenger and 25% with 2 passengers, as considered in the related study [3].

During the simulation it was observed that some queues dissolved faster than others due to certain stochasticity in the model. This sometimes led to waiting areas with no passengers, but still with an available PRT vehicle at the concerned berth. In order to avoid this inefficiency, an algorithm was developed that allows boarding from the longest queue to an available vehicle at another berth.

In case a vehicle is not available at the concerned berth, it is checked if the queue in which the passenger is waiting is the longest queue. If this is true, then it is checked if there is another berth where no passengers are present in its queue. If this is also fulfilled and there exists a PRT vehicle available at that berth, the passenger can board this vehicle, hence boarding a vehicle of another berth is enabled in those cases. The boarding process is explained in Fig. 4 by a flow diagram in more detail.

We select in that methodology the assumption that only passengers in the longest queue board the vehicle in which the waiting area of the concerned berths is empty.

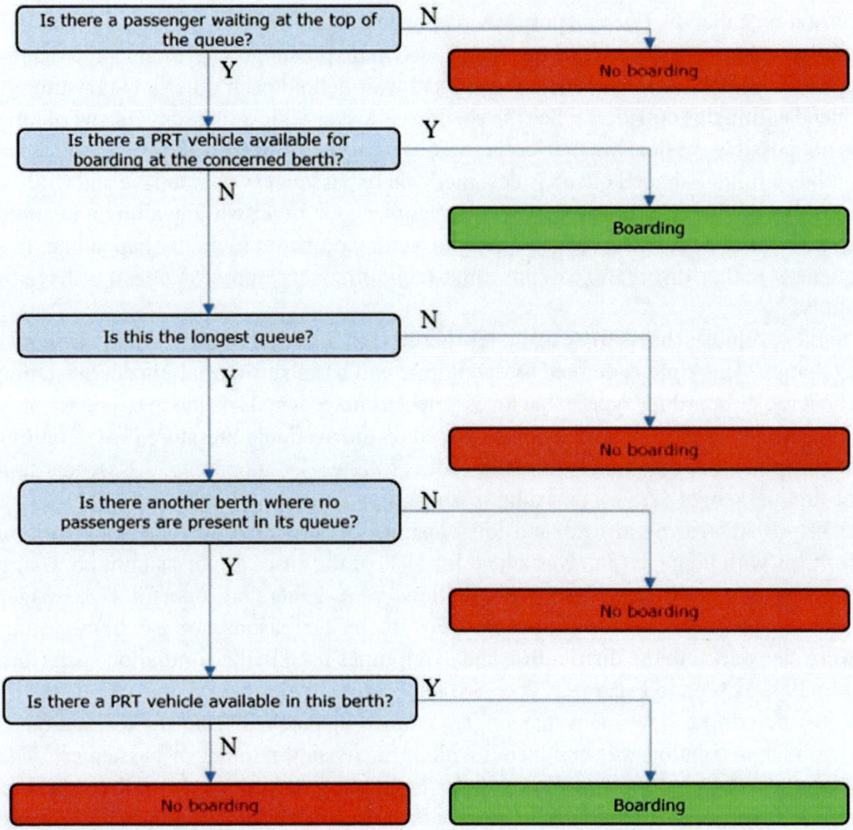

Fig. 4 Boarding process of transferring passengers at the PRT station

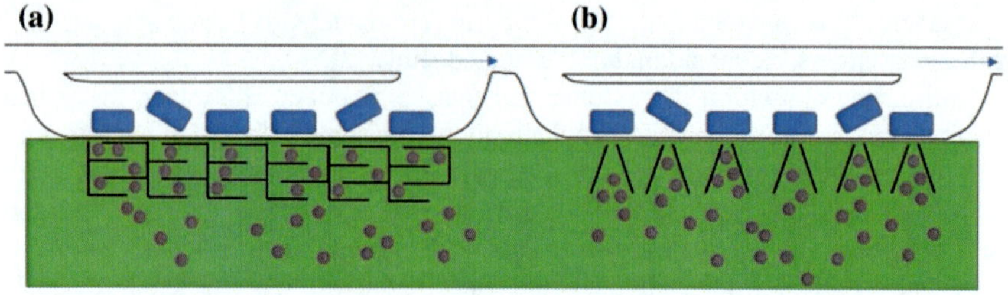

Fig. 5 Different passenger queue forms for the boarding in PRT stations: (**a**) S-shaped, (**b**) funnel-shaped

This procedure has the advantage that passengers in the longest queue are able to board a vehicle faster than passengers in the other queues due to the additional berth at which these passengers can board. In that way it is an element to reduce differences in the queue lengths. Nonetheless, other procedures are imaginable.

One aim of this study is the analysis of different queue forms and their impacts on station capacity. For this purpose, two queue forms are constructed as illustrated in Fig. 5. The first one is an S-shaped queue where passengers go along an S-shaped corridor to reach their vehicles. Passengers wait behind each other and constitute a queue within this corridor. Since the corridor is just as wide as passengers, overtaking is not possible, basically a first-come-first-serve process prevails. In the second alternative, a funnel-shaped queue is designed where passengers do not have a discipline to form a queue. Here, passengers can overtake each other while waiting and apply to each other a "social force" to move forward. Compared to the S-shaped one, this queue is rather disorganized and a first-come-first-serve process does not have to apply.

PRT vehicles have to wait at the berths for a certain time during passenger exchange. This time is defined as dwell time and has a substantial impact on station capacity. A boarding type behavior is selected here, which means passengers only board vehicles in these PRT stations. Based on the available literature [14] durations of sub-processes assumed in this study are as follows: Door opening (1.8 s), boarding of first passenger (1.5 s), boarding of second passenger—–if necessary–(1.125 s), ticket scan (2 s, own estimate) and door closing (1.8 s). Furthermore, an extra duration for trips with luggage (5 s) was added for 10% of the cases as an assumption. Using these values, two probabilistic distributions were generated, one for 1-passenger boarding and the other for 2-passenger. For stochastic reasons, some extreme values were also added to the distribution and dwell times used in the simulation span from 4 to 30 s, as seen in Fig. 6.

To determine how many passengers can be dispatched from a PRT station, a two-hour simulation was conducted with an increasing number of passengers. The procedure used here is as follows: A 60% alighting percentage from the suburban train is taken as starting point. Just before the next suburban train arrives (in 10 min), it is checked if there are still any passengers that wait for boarding a PRT vehicle, remaining from the previous train. If not, then the alighting percentage of the next

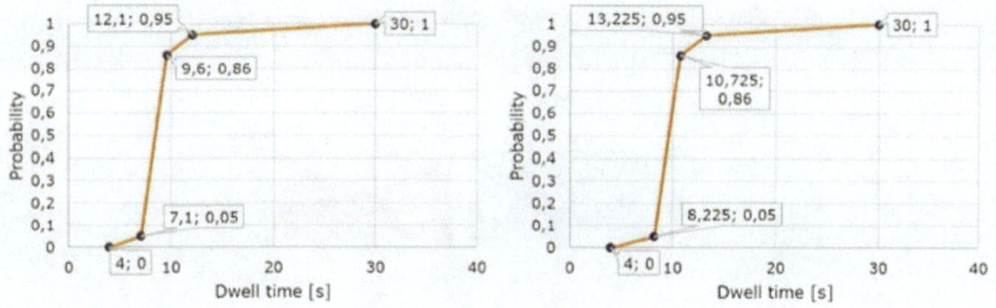

Fig. 6 Cumulative probability distributions of dwell times: **a** one passenger boarding, **b** two passengers boarding

train is increased by 2% and the simulation continues till the next train arrives. If on the contrary there are still remaining passengers, the alighting percentage is held constant. This process goes on until the end of the simulation. The alighting percentage value leading to no remaining passengers within a suburban train arrival cycle is noted down. By extrapolating this value, the number of passengers that can be dispatched from the PRT station is calculated as:

$$C_{PRT} = C_{Rail} * x * S_{PRT} * S_{Sim} * f_{Rail} \quad (1)$$

where:

- C_{PRT}—Number of dispatched passengers in one PRT station [passengers/h];
- C_{Rail}—Capacity of the railway vehicle [passengers/rail-vehicle];
- x—share of passengers which alight the train, based on the simulation results;
- S_{PRT}—Alighted passengers transferring to the PRT vehicles, here 0.8 (80%);
- S_{Sim}—Reduction factor due to the simulation of only one virtual PRT station, here 1/6;
- f_{Rail}—Frequency of rail arrivals per hour, here 6 railway-vehicles per hour.

On the vehicle side of the simulation, some rules were defined in order to ensure a seamless traffic flow in the PRT station, similar to the previous study [3]. The input traffic flow was regulated at the start of berth areas in order to avoid possible breakdowns in the station. A vehicle was able to enter the berth areas only if there existed a free berth to stop. An asynchronous mode was selected as the operation mode, which means that occupying the berths did not take place in a synchronous mode (such as odd berth numbers are filled together and then the even ones), rather free berths were occupied by the next available vehicle as soon as possible. Moreover, vehicles leaving the station without stopping at a berth were avoided, hence all vehicles were forced to stop at berths. Vehicles moved at berth areas with a maximum speed of 5 km/h due to safety reasons, whereas a maximum speed of 50 km/h was allowed elsewhere. Desired acceleration and decelerations rates were assumed to be 1.5 and −1.5 m/s2, respectively.

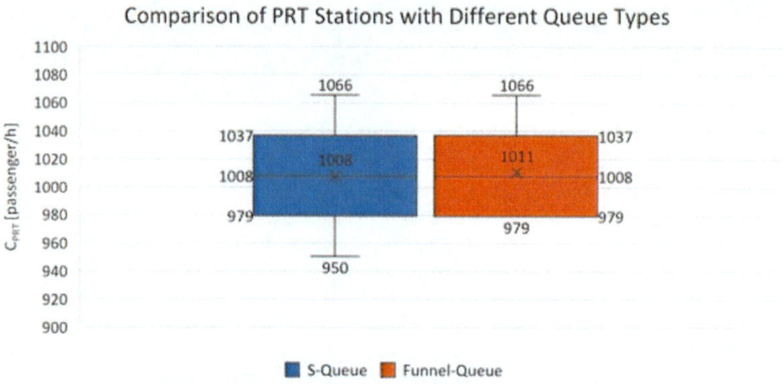

Fig. 7 Box plots to compare the number of passengers dispatched for PRT stations with an S-queue and a funnel queue

The simulations were performed in the microscopic traffic simulation software Vissim (Version 2020) from the company PTV AG [15]. For passenger movements and interactions, the pedestrian simulation add-on Viswalk was utilized. Passengers interactions were based on a social-force model developed by Helbing and Molnar [16]. Regarding the car following behavior which needed to represent the behavior of fully autonomous vehicles, the output parameters of the research project CoExist were used [17]. With the help of programming interfaces, characteristics such as PRT vehicle behavior and vehicle-passenger coupling mechanism were modelled.

To achieve statistically significant simulation results, the necessary sample size was estimated according to the German recommendations on microscopic traffic simulation studies [18]. With a confidence level of 0.95 and confidence interval of 2, the selected sample size was 20 for all simulation models. Mean and percentile values of these 20 runs were calculated and plotted in box-plots for both queue forms (see Fig. 7).

4 Results and Discussion

The results of the simulation are shown in Fig. 7. The figure displays the boxplots for the number of dispatched passengers in the PRT station with two different queue types. The results of a PRT station with an S-queue are shown in the first plot. The data of a PRT station with a funnel queue are visualized in the second plot. Interestingly, the results are nearly identical although the design of the queue types are very different (see Fig. 5). Most of the results are exactly the same or very similar: Data shown in the upper and the lower quartiles, the median values and the upper whiskers are exactly the same. A slight variation is existing in the mean values and the lower whiskers. But the difference between the queue types is within the selected confidence interval of 36 and is therefore negligible.

Considering the differences in queue designs, this gives a first impression that the queueing processes do not have a significant influence on the dispatching process. From a visual observation of our model, we see that the bottleneck is the dwell time of the vehicles in the station itself. The time which the passengers spend in the queue to wait for the vehicle is much longer than the time which passengers lose due to the processes in the queue themselves. We think that this effect masks the influence of the queues and therefore the dispatching process is rather limited by vehicle availability than by the queueing processes.

Further, the results from [3] support this conclusion. As already mentioned, in that paper we simulated three different boarding/alighting behaviors: Boarding Type Behavior, Alighting Type Behavior and 50/50 Type Behavior with the corresponding behavior of the passengers in the station. For the sake of brevity, [3] discussed only the 50/50 type behavior. Recapitulate, in this paper the boarding type behavior is relevant. But, unpublished simulation results from [3] for the vehicle capacity of a station with a boarding type behavior exist, too. The unpublished values for the capacity are in the order of magnitude of 800 vehicles per hour. These values are multiplied with the average number of passengers per vehicle of 1.25, the average value in the current study. From that we gain information that roughly 1000 passengers per PRT station and hour are processed. This is in the same order of magnitude as the results based on the number of alighting passengers from the railway vehicle (see Fig. 7). Nonetheless it is worth mentioning that the dwell times changed slightly in our current study. Furthermore, the current model has been neither calibrated nor validated due to a lack of empirical datasets. Therefore, we assess only a comparison in the order of magnitude as meaningful, and not a comparison of the exact numbers.

In our view, the visual observation and the comparison with the unpublished results from [3] strongly supports the conclusion that the limiting factor in PRT station capacity is the vehicle availability and not the passenger queuing processes themselves. We suggest therefore, for increasing PRT station capacity, focusing on optimizing the vehicle processes in the PRT station.

Literature on such detailed simulations of the processes in stations is not available. In addition, other researchers use different methodologies to model the PRT systems, as already discussed in the sections above. Therefore, we do not compare our current results with other research.

5 Summary and Conclusion

Continuously building on our previous research on an optimal PRT station layout, the focus of this study are pedestrian movements and queueing processes within the PRT station. Therefore, the current model considers two types of queues: S-shaped and funnel shaped queues. Applying microscopic traffic simulation for such purposes is a novel approach. The simulation setup models passenger interactions amongst each other as well as their behaviour within the station. Together with previous models [3], it considers on the vehicle side optimal operation processes and layouts of PRT stations. Combining detailed models on passenger and PRT vehicle issues with the focus on PRT stations is currently a unique characteristic of this approach.

The results of the simulation experiments with the queue types are shown in Fig. 7. With both queue types, the number of dispatched passengers per hour is on the same level. By multiplying the vehicle capacity of the PRT station from a previous study [3] with the average occupancy rate of the vehicles in the current study, a capacity value in the same order of magnitude as the results of our current simulations is received. This indicates with a certain likelihood, that queueing processes of the pedestrians within PRT stations are not decisive for the capacity of PRT stations. Instead, this denotes that the actual decisive factor is PRT vehicles and their interactions in stations. Nonetheless, for a concluding answer to that question an examination of other queue forms would be necessary.

We see in our simulation that a state-of-the-art microscopic traffic simulation is feasible to model the passenger flows in PRT stations. In doing so, the wide range of functionalities for different modes enable a detailed modelling process of the multimodal transport relations. A PRT system is in its behaviour neither a classical individual transport system nor a public transport system and a rare use case for microscopic simulations. Therefore, there are no dedicated functionalities to model PRT systems. However, there are a variety of processes which we could utilize to generate a similar functionality. In combination with an external programming interface, we were able to generate the expected behaviour. Last but not least, there is still no empirical data for calibration and validation of the model available. Hence, our results allow only qualitative statements and we advise against quantitative statements based on the current model.

This limitation also shows one field for further research: There are so far no empirical data for the calibration and validation of such microscopic models available. Data sets on actual vehicle dwell times in the station are necessary. Interesting would be to examine if parameters in the model fit to real-life PRT systems. Furthermore, our results emphasize the importance to optimize the operational processes of the vehicles in PRT stations. Such optimisations could be either implementation of already mentioned synchronised operation modes of vehicles in a station or a dynamic assignment of vehicles to their berths within a station, while they are already driving within the station. Another aspect which shows clear potential for further research would be the application of a sensitivity analysis on the existing models and their model parameters. This would enable to examine the influence of these parameters on station capacity. Generally, an examination of two classes of parameters is meaningful: On the one hand those which are related to the passenger side of the microsimulation, such as the arrival frequency of the suburban trains. On the other hand, there are those parameters which are related to the vehicle side, for instance vehicle car following model parameters or vehicle dwell times in the PRT station.

References

1. Mineta Transportation Institute (2014) Mineta Report. Automated transit networks (ATN): A review of the state of the industry and prospects for the future. In: Furman B, Fabian L, Ellis S, Muller P, Swenson R (eds). Mineta Transportation Institute, San José
2. Muller P (2015) Frequently asked questions about personal rapid transit, viewed 11 July 2021. https://prtconsulting.com/faq.html
3. Arslan O, Reichert M, Sellaouti A, Hoffmann S (2021) Investigation of personal rapid transit station capacities using microscopic traffic simulation. In: Proceedings of the 7th international conference on models and technologies for intelligent transportation systems (MT-ITS). IEEE
4. Lowson M, Hammersley J (2011) Maximization of PRT Station Capacity. In: Transportation Research Board (ed.) TRB 90th annual meeting compendium of papers DVD. Transportation research board 90th annual meeting. Washington DC, 1/23—1/27
5. Castangia M, Guala L (2011) Modelling and simulation of a PRT network. In: Pratelli A, Brebbia C, Brebbia CA (ed) Urban transport XVII in the 21st century, Bd. 116. WIT Press (WIT transactions on the built environment, 116), Southhampton, pp 459–472
6. Choromanski W, Daszczuk W, Grabski W, Dyduch J, Maciejewski M, Brach P (2013) Personal Rapid Transit (PRT) computer network simulation and analysis of flow capacity. In: Proceedings of the 14th international conference on automated people movers and transit systems. Arizona, Phoenix
7. Schweizer J, Danesi A, Rupi F, Traversi, E (2012) Comparison of static vehicle flow assignment methods and microsimulations for a personal rapid transit network. J Adv Transp 46 (4):340–350
8. Schweizer J, Parriani T, Traversi E, Rupi F (2016) Optimum vehicle flows in a fully automated vehicle network. In: Proceedings of the international conference on Vehicle Technology and Intelligent Transport Systems (VEHITS), Rome, Italy
9. Schweizer J, Mantecchini L, Greenwood J (2011) Analytical capacity limits of personal rapid transit stations. In: Proceedings of the 13th international conference on automated people movers and transit systems. Paris, France
10. Lowson M, Cook C, Hughes D (2010) Station design for personal rapid transport systems on the 13 April 2004. Patent Registration Nr 10/552,167. Publication Nr US 7,681,505 B2. WO2004/089679
11. Adacher L, Flamini M (2020) Optimizing airport land side operations: check-in, passengers migration, and security control processes. J Adv Transp 2020:1–14. https://doi.org/10.1155/2020/6328016
12. Schultz M (2010) Entwicklung eines individuenbasierten Modells zur Abbildung des Bewegungsverhaltens von Passagieren im Flughafenterminal. Dissertation, Institut für Luftfahrt und Logistik, Fakultät Verkehrswissenschaften, Technische Universität Dresden, Dresden
13. Rossi R, Gastaldi M, Orsini F (2018) How to drive passenger airport experience: a decision support system based on user profile. In: Intelligent transport systems, special issue: selected paper from the Scientific Seminar of the Italian Association of Transport Academicians (SIDT) 2017, vol 12, no 4. pp 301–308
14. Kapski D, Samailovich T (2012) Defining the time of operations which occur when public traffic stops at a bus stop for alighting and boarding. In: Proceedings of the 12th international conference reliability and statistics in transportation and communication. Transport and Telecommunication Institute, Riga, pp 136–144
15. PTV AG (2020) VISSIM. Version 2020. PTV AG, Karlsruhe Available at https://www.ptvgroup.com/de/loesungen/produkte/ptv-vissim/
16. Helbing D, Molnár P (1995) Social force model for pedestrian dynamics. Phys Rev E 51(5):4282–4286
17. Sukennik P (2020) CoExist—Micro-simulation guide for automated vehicles—final. In: Charlotte Fléchon V. PTV Group (4.0). Last checked on 15 July 2021
18. Forschungsgesellschaft für Straßen- und Verkehrswesen (2006) Hinweise zur mikroskopischen Verkehrsflusssimulation. Grundlagen und Anwendungen. FGSV-Verlag, Köln

Building a Multi-sided Data-Driven Mobility Platform: Key Design Elements and Configurations

Andrea Carolina Soto, Marc Guerreiro Augusto, and Søren Salomo

Abstract Digital mobility platforms vary with respect to their design and configurations. Depending on the specific design choices, such platforms offer different value creation and value capture mechanisms. Building on research in the field of platform and existing data-driven mobility platforms, we develop an understanding of the elements and facets in which platforms are configured and construct a morphological box with these constitutional elements. To validate and exemplify our findings we draw on existing mobility data platforms and their configuration.

Keywords Multi-sided platforms · Platform design · Platform governance · Mobility data platform · Value creation · Platform economy

1 Introduction

Six in ten of the world's most valuable public companies are digital platforms with data-driven business models [37]. These platforms essentially leverage their network to enhance the matching between sides, for example, users derive more value from Amazon as more products are available to them. As Van Alstyne et al. [54] explain, greater scale generates more value, which attracts more participants, which in turn creates more value—thus creating a positive feedback loop or the so-called positive indirect network effects.

This new paradigm is transforming the global economy with both new entrants disrupting traditional industries such as Uber did with the taxi industry, and incumbents moving from pipeline business models into platforms [54], such as Apple before it turned the iPhone product into a platform with the introduction of various

A. C. Soto (✉) · S. Salomo
Institute of Technology and Innovation Management, Technische Universität Berlin, Berlin, Germany
e-mail: soto.ramirez@tu-berlin.de

M. Guerreiro Augusto
DAI-Labor, Technische Universität Berlin, Berlin, Germany

C. Antoniou et al. (eds.), *Proceedings of the 12th International Scientific Conference on Mobility and Transport*, Lecture Notes in Mobility,
https://doi.org/10.1007/978-981-19-8361-0_6

connected products and especially the AppStore for allowing new services to be added to the Apple platform ecosystem.

The automotive industry is also experiencing this paradigm shift. With the advent of cooperative connected and automated mobility (CCAM) and the increasing amount of data generated by both vehicles and the infrastructure, the car is no longer seen as a pipeline product but as a platform for digital products. The race to become the dominant player to monetize this data for future mobility solutions is tighter than ever. Nowadays, world-leading digital platform players are investing billions in the automotive industry [11] and established OEMs are making strategic alliances with tech companies to reconfigure their business around the car as a digital product. On the one hand we see internet giants such as Google with Waymo developing AI-driven/based solutions for autonomous driving, as well as Baidu creating an open-source autonomous driving platform called Apollo [20]. On the other hand, we see established OEMs like Volkswagen and Ford with Argo AI to quickly deploy self-driving technology.

Moreover, new dedicated data-driven mobility platforms are entering the scene and providing the digital infrastructure for the exchange and brokerage of mobility data. Platforms such as Caruso dataplace, a marketplace for multi-brand in-vehicle data of different vehicle manufacturers; HERE platform, for building, deploying and scaling location solutions; and mCloud, a German government's research platform for open data from the field of mobility and related topics are just three examples which leverage platform architecture to facilitate cross-side interactions [10] and orchestrate data matching amongst several players in the automotive industry.

While the three platforms mentioned above share core design principles, and so is the case for all digital platforms, we identified that they also vary with respect to their design and configuration. That is, the mechanisms used by platform owners for creating and capturing value. Scholars have provided the central concepts of platform strategy such as pricing mechanisms [25], regulation [8], governance [52], launch strategies [17], platform boundaries [21], network effects [44], competition [41], interfaces [23] and evolutionary dynamics [52].

When looking to understand the multiple design elements and facets platform owners are faced with, we found no single framework that would accommodate all choices into a single box. For instance, Choudary [10] proposed a "platform stack" which consists of a unified architectural framework to explain the different configurations of platforms by dividing their architecture into three different layers: network, infrastructure and data; albeit this framework is useful for understanding the difference across platforms and the similarities in their business model, it fails to pinpoint at individual design choices. Another example is [49] who developed a morphological analysis for platform design but whose scope only applied to marketplaces, deliberately leaving platforms with more than two sides out of the equation. Lastly, [53] propose a design framework involving four elements: platform architecture, value creation logic, governance, and platform competition, while this overarching framework for platform design facilitates a general understanding of design dimensions it lacks the available individual choices for platform owners.

The field of mobility was deliberately chosen given the combination of increasing mobility data being generated by vehicles and traffic infrastructure, and the importance of the collection of this data for optimizing the flow of traffic, increasing safety and protecting the environment. One example is the research project BeIntelli—AI in Mobility based on Platform Economy– emphasizing the conceptualization and development of a data platform and new ecosystems for autonomous mobility solutions in urban environments, which motivated this study ("BeIntelli" 2022). We identified several platform designs in this field with varying levels of data access, available interfaces, monetization strategies, amongst others, which support our conceptual model of the design space of platforms.

By developing a morphological box, we expect to show the relevance of platforms and exemplify that they vary in their configurations.

2 Problem Statement

As outlined above, our study addresses the need to understand the different platform design parameters by constructing a morphological box with these constitutional elements. To develop our model, we draw on conceptual underpinnings seeking case information in mobility data platforms that supports and extends our model.

The paper addresses two research questions addressing the key design elements of platforms:

RQ # 1: What are the available design choices and configurations when establishing/designing a platform?

RQ #2: What are the interdependencies in the platform design parameters? i.e., some fundamental design choices define the space, in which other parameters become viable?

Starting with the three dimensions of business model design defined by Teece [50], who describes a business model as "the design or architecture of the value creation, delivery, and capture mechanisms it employs", we review existing platform literature and extract individual design elements into one of the three categories (i) value creation, (ii) value delivery and (iii) value capture. This paper considers only those elements that appear to be of high relevance for digital multi-sided platforms.

The selection of elements and variables follows the design principles of morphological analysis. This approach has been identified to be suitable to accommodate multiple configurations of platform design elements given its flexibility to choose different combinations of attributes. The resulting morphological box can serve as an artifact to identify different platform business models.

This paper employs a systematic literature review as the methodology to extract the individual design elements of multi-sided platforms and coded the identified papers according to the three previously identified dimensions. The tool MAXQDA20 was employed during the coding process.

To validate and exemplify our findings, we draw on three mobility data platform use cases and match their business model to the proposed morphological box. The selected use cases have been identified via desk research using Google's search engine with the term "mobility data platforms". From the available results, we selected three use cases that complied with the following criteria: (1) data-driven platforms, (2) in the field of mobility, (3) with a growing user base and (4) comparably different from each other.

3 Theoretical Framework

3.1 Multi-sided Platform: Designs and Mechanisms

What is a platform?

The notion of platforms refers to different things depending on the theoretical stream it derives from. Scholars from various disciplines take different perspectives on how platforms create and capture value, these include management and economic scholars with a market-based perspective and information systems scholars with a technical perspective.

Management and economics literature characterizes the term "platform" as products, services, firms, or institutions that mediate transactions between two or more groups of agents [44, 16, 13, 22].

These can be further distinguished from traditional businesses, named "pipelines" or single-sided businesses [54] with a linear value chain from platforms. In pipeline-type businesses, there is a straight line between the company which makes a product or service and the customer who purchases or uses the product. In contrast, platform businesses enable complex interactions between several producers and consumers. More importantly, the value proposition greatly differs for both types. While pipeline businesses create value by orchestrating the value chain to transform an input into an output that's worth more, platform businesses create value by facilitating the interaction between users where each user's value increases due to adoption by further users [18] and so does the value of the platform.

This higher valuation on the user base over financial revenues has been reflected in the past decades on the market capitalization of platforms such as Google (Alphabet), Amazon, Facebook, Microsoft and Alibaba which are now considered within the most valuable public firms in the world [42] alongside platform "unicorns" (i.e., $1B +) such as Netflix and Shopify. These platforms essentially leverage their network to enhance the match quality between sides, for example, users derive more value from Amazon as more products are available to them. As Van Alstyne et al. [54] explain, greater scale generates more value, which attracts more participants, which creates more value. Nevertheless, the multi-sidedness of platforms has an inherent chicken-and-egg problem where no side will join without the other [26], making it one of the most difficult challenges for platforms to overcome.

Information systems literature defines "platform" as a set of stable components that provides core functionalities in a system by constraining the interfaces through which they operate [3, 6, 52]. For communication between components and interaction between users, application programming interfaces (APIs) are used. APIs also enable third parties to provide additional services [23] which create value for the platform.

Depending on the openness of interfaces, platform owners can restrict access to resources and information [53] which can influence the degree of innovativeness in the platform –the main value generating activity. In a study, [6] compared 21 handheld computing systems over a 14-year span and found that granting greater levels of access to complementors produced up to a five-fold acceleration in the rate of new handheld device development, but this effect was diminished as platform owners further opened their platform giving up ownership, thus jeopardizing innovative performance.

Researches in this field have specifically focused on digital platforms, seen as a major outcome of digitalization [56] and the rise of telecommunication technologies as these have driven innovation via lower costs, faster speed and larger scope of connections between platform sides [18].

There is a unified view in the research community that sees platforms as sociotechnical constructs that incorporate elements both from economics literature and information systems literature [39]. Platforms are market intermediaries between two or more sides and digital, modularly-architectured entities.

The platform design choices

Digital platforms vary with respect to their design and configuration. That is, the mechanisms used by platform owners for creating and capturing value in their specific industry. Although payment systems, ride sharing applications, data marketplaces and social networks are different on the surface, they are essentially all platforms which share crucial design principles to facilitate cross-side interactions [10].

Platform scholars have provided constitutional elements of platform strategy, clarifying platform success factors in terms of pricing [25], governance [8, 52], boundaries [21], network effects [44], and interfaces [23]. Yet, a holistic framework encompassing the multiple design elements of platform business models is lacking.

We see the opportunity of developing an understanding of the platform design elements by constructing a morphological box with these constitutional elements. To develop our model, we draw on conceptual underpinnings seeking case information in mobility data platforms which supports and extends our model.

Conceptual foundation for developing the design space

Building on the business model framework defined by [50], who describes a business model as "the design or architecture of the value creation, delivery, and capture mechanisms it employs", this paper aims at identifying the core elements of platform design, clustered under (i) value creation, (ii) value delivery and (iii) value capture. The purpose of using this framework is to be able to capture the complexity

Table 1 Platform design elements

Dimension	Design element	Sources
Value creation	Market structure	(Täuscher and Laudien 2018; Hagiu 2013; G. Parker and Van Alstyne 2014; Evans and Schmalensee 2016; K. J. Boudreau and Hagiu 2008; Evans 2012; Gawer 2021; Otto and Jarke 2019; Hein et al. 2020; Eisenmann, Parker, and Van Alstyne 2006; Choudary 2015; Strahilevitz 2005; Rochet and Tirole 2003; Cusumano, Gawer, and Yoffie 2019)
	Platform ownership	
	Platform type	
	Key activity	
	Price discovery	
	Access control	
	Trust system	
Value delivery	Key value proposition	(G. G. Parker, Van Alstyne, and Choudary 2016; Gawer 2021; Evans and Schmalensee 2016; Hagiu 2013; K. Boudreau 2017; Evans 2012; G. Parker and Van Alstyne 2014)
	Transaction content	
	Digital interface	
	Participants	
	Geographic scope	
Value capture	Key revenue stream	(D. McIntyre et al. 2021; Evans and Schmalensee 2016; Evans, Hagiu, and Schmalensee 2008; Gawer 2021; Osterwalder 2004; Hagiu 2009; Eisenmann, Parker, and Van Alstyne 2006; Rochet and Tirole 2003; Tiwana, Konsynski, and Bush 2010)
	Pricing Mechanism	
	Revenue source	
	Subsidy	
	Multihoming	
	Switching costs	

of designing business models in multi-sided markets, where the key activity of transacting across sides requires a more complicated and dynamic approach than that of a pipeline business model [54].

The selection of elements follows the design principles of morphological analysis. This approach has been identified to be suitable to accommodate multiple configurations of platform design elements given its flexibility to choose different combinations of attributes. Table 1 gives an overview of the selected business model elements that are derived from the literature review and the key authors for each dimension.

The first seven elements are part of the value creation dimension (value architecture and mechanisms), the subsequent five are part of the value delivery dimension (value proposition and boundaries), and the final four represent the value capture dimension (revenue and pricing model).

3.2 Value Creation

The value creation dimension refers to the value architecture and mechanisms that allow for creating the value proposition of the platform [49] and clearly affect the value of its ecosystem. Thus, a key part of the platform owners' strategy should be some regulation of third-party actions [26] and the orchestration of resources and processes.

For digital platforms, the value creation lies in facilitating exchange across sides [10, 26] and ensuring that there are enough agents of each type to make participation worthwhile for all [19]. This makes them subject to negative externalities arising from network effects, information asymmetry, uninsured risks, and congestion [8, 15].

The attributes of platform value creation identified are market structure, platform ownership, platform type, key activity, price discovery, exclusivity, access control and trust discovery.

Market structure (number of sides)

Defining the number of sides is an essential step in building a platform business [21]. There are two basic forms of digital platforms: two-sided platforms and multi-sided platforms [39]. Two-sided platforms mediate between two groups of users, e.g., buyers and sellers, and multi-sided platforms (MSP) bring together multiple sides or groups of users, e.g., developers, buyers, sellers and advertisers.

The decision of the number of sides to associate to the platform can have a great impact on the value creation of the platform's business model. Hagiu [26] argues that, while more sides lead to potentially larger cross-side network effects, larger scale and potentially diversified sources of revenues, there is a risk of too much complexity which can cause a 'lowest common denominator issue', i.e., pleasing many heterogeneous platform sides greatly constraints an MSP's ability to innovate.

Platform ownership

Whether the platform is owned by a single organization (also referred to as keystone firm), multiple organizations [39] or decentralized by peer-to-peer communities [27] it will affect the governance mechanisms, distribution of power and decision-making processes taking place among the different actors.

The implication of selecting a shared ownership versus. an individual ownership extends to having two different sets of governance frameworks: one for internal governance which is concerned with the group of collective platform owners and another one for external governance which coordinates the interaction between the platform owners, platform users and platform complementors [39].

Platform type

Broadly speaking, multisided platforms (MSPs) are technologies, products or services that create value primarily by enabling direct interactions between two or more customer or participant groups [13, 26]. This definition includes platforms such as Uber, eBay, Xbox, Facebook and Amazon which do not necessarily have much in common aside from their ability to create value by facilitating connections across multiple sides, subject to cross-side network effects [21]. In their assertion that not all platforms are the same, [22] distinguish between transaction platforms, innovation platforms, and hybrid platforms.

Transaction platforms are those which facilitate transactions of goods, services and data across sides. They are intermediaries, or online marketplaces or social networks [21]. As the more participants, functions, and digital content is available in transaction platforms, the more value is created. Transaction platform owners aim at increasing the volume of players on each side of the platform.

In contrast, innovation platforms act as the technological foundation upon which the members of one side can develop new complementary goods and services such as apps for the AppStore or video games for the Xbox. The complementary innovations developed add functionality or grant access to assets that make the platform increasingly useful [11, 52].

Hybrid platforms are the combination of both innovation and transaction platforms, as they not only facilitate transactions across multiple sides, but also enable the creation of complementary innovations. A good example of a hybrid platform is the HERE platform, where developers can create applications based on location data and monetize these innovations via its marketplace where buyers and sellers of location data and services transact [28].

Platform type is driven by the strategic decisions of business model, value proposition and number of sides [21].

Key activity

A platform's key activity can be distinguished between data services, network-community building and complementary innovations. Based on the platform stack proposed by [10], platforms can function across these three layers, but the degree to which each layer dominates can vary.

For instance, an innovation platform such as Android has a dominant key activity layer of complementary innovations, and a not so dominant data services layer which feeds the Google PlayStore recommendation algorithm to match consumer preferences with available apps. Another example is the transaction platform Airbnb with a dominant network-community layer enabling interactions between hosts and guests, and a not so dominant data layer which helps hosts analyze and visualize their transactions to better tailor their offering.

Price discovery

Building on the marketplace morphological box of [49], price discovery on digital platforms can either be set by the platform provider, the demand side or the supply side. When the price is not fixed by either of the platform participants it can arise from competitive pricing mechanisms such as auction, where the service or product is awarded to the highest bidder –mirroring the mechanism of financial markets, and negotiation, where both supply and demand agree on a price [2].

Access control—regulating negative network effects

For the platform, property rights are necessary for governance as they reduce the risk of negative network effects by limiting access to certain actors. Evans [15] describes the tools which multi-sided platforms use to optimize the value of their property to deal with platform access.

Platform owners may selectively exclude participants from their platform using the Bouncer's Right, and /or discourage unwanted participants from joining the platform by using social and psychological sanctions and club goods to sort undesirable entrants (also referred to as Exclusionary Vibe and Exclusionary Amenity). Selective exclusion and discouraging participants via social and psychological cues are often used conjunctively by a platform owner [47].

Trust system—regulating information asymmetry

Once the platform owner has decided the number of sides to include, the type of players to allow and the key activity to be performed, building up trust to ensure high quality within the platform is next. Platform users are encouraged to signal trust through a review system which can be enforced either between members or by the platform itself [44].

By providing a system of reviews, the platform owner limits the ability of bad users to take advantage of other users by exploiting asymmetric information [15]. For instance, Amazon and eBay provide ratings on sellers as buyers can rate their experience after they have made a purchase.

3.3 Value Delivery

The value delivery dimension refers to the elements which generate value for the platform sides. Platforms generate value by reducing economic friction [41] and by acting as "matchmakers" across sides supporting the transaction. This implies making strategic decisions over the platform's scope and digital interfaces to identify and exploit transaction opportunities and innovation complementarities [21].

The attributes of value capture identified are value proposition, transaction content, digital interface, participants and geographic scope.

Value proposition

Platforms generate value for all sides when they reduce transaction costs [41], search costs by acting as matchmakers [19], product development costs via programming interfaces and digital resources [26] or both. Platform owners may incorporate a great variety of features to reduce these costs.

More specifically, depending on the type of platform (i.e. innovation, transaction or hybrid) and key activity, platform owners need to make the decision whether to include many or just some of the features based on a cost–benefit analysis: balancing value created vs. cost of implementation [26].

Transaction content

A firm wishing to build a transaction platform will, therefore, aim to populate its sides with members that will exchange or transact goods, services, and/or data amongst themselves via the platform [21].

Digital interface

Technological or digital interfaces refer to how the 2-way exchange of data between the platform firm and each of its sides is specified. It both delimits the economic activities between sides [7] and specifies the characteristics of the connection.

Gawer [21] notes that scholars in strategy, innovation, and information systems research agree that decisions made on the type and openness (i.e., which data will be shared or withheld, and to and from whom) of technological interfaces play an important role in the degree of complementary innovation. It is not only about opening up the platform architecture, but considering the ease of development to encourage creativity and innovation while maintaining control over the evolution of the platform Vittorio Dal [55].

Typical digital interfaces are Application Programing Interfaces (APIs) which forge the relationship between application developers and users, and Software Development Kits (SDKs) that facilitate the efficient development of software that works on the platform [15].

Participants

The platform owner should specify the user segments that it primarily connects as participants, including Consumer-to-consumer (C2C); Business-to-Consumer (B2C) and Business-to-Business (B2B).

Given the multi–directionality of business relationships within a platform, the user segments are expected to be of mixed nature, meaning more than one segment is possible. For instance, Facebook connects its users amongst each other (C2C) but also connects advertisers with users (B2C).

Geographic scope

The platform's geographic scope could be broken down into local, regional and global coverage. Given the low geographic boundaries of digital goods, we expect most platforms to fall into the regional and global clusters. In his study on marketplace characterization, however, [49] note that as ventures grow, it is likely that they will expand their geographic scope.

The decision to limit the geographic scope at the beginning or launch of a platform has been further explored by Parker and Van Alstyne [41] with micro-market launch strategies. Limiting coverage to small, targeted communities can help achieve stronger network effects and, once adoption is reached, the platform can be expanded to include adjacent groups dragging new users into the platform. For instance, Facebook started connecting students within Harvard and eventually evolved to a global platform connecting more than 2.912 billion users by December 2021 [35].

3.4 Value Capture

The value capture dimension explains how the platform owner achieves the conversion of the benefits that users perceive in a platform into profitable revenue streams [33].

Pricing multi-sided platforms is more complex than traditional pipeline businesses [19], as the demand by one of the sides depends on the number of members on the other sides the platform serves. Platforms create and capture value once positive cross-side network effects are achieved.

By facilitating the interaction across sides, platform owners are able to regulate other aspects of the platform participation such as which side to charge pricing mechanisms, switching and multi-homing costs [33].

Key revenue stream

The key revenue stream specifies the platform's mode of generating revenues. Depending on the type of platform, the way to capture value may vary.

On one hand, transaction platforms primarily capture value by collecting transaction fees, charging for advertising or both [21], consequently, platform owners should aim at increasing members on each side who transact amongst themselves. Innovation platforms on the other hand generate revenues mostly by charging for access, usage or both and should carefully balance the optimal level of fees to overcome the chicken-egg problem of end-user and complementor skepticism at launch [16].

Pricing mechanism

Taking the pricing model proposed by Osterwalder [38], the mechanisms are characterized by fixed pricing, market pricing or differentiated pricing.

Firstly, a fixed pricing mechanism encompasses pricing strategies that do not vary depending on customer characteristics, are not value dependent and are not based on real-time conditions. This definition includes platforms which are sponsored by the platform provider and charge virtually zero for the services, data or products offered. Secondly, differentiated pricing mechanism depends on customer characteristics or is value dependent; for instance, a dataset which varies on the level of granularity. Lastly, a market pricing mechanism is dependent on real-time data of market conditions such as auctions or the stock market.

Revenue and subsidy sides

To maximize profits, platform owners need to make the decision on which side to subsidize and which to monetize [25]. Generally, multi-sided platforms have a 'subsidy side' which is a group of users who pay less than marginal cost and is highly valued by the 'money side' –who pays above the marginal cost [13]. This means, as more subsidy-side users join the platform, more money-side users join too, resulting in strong cross-side network effects.

The decision about the number and nature of the sides affects directly the pricing structure and the ability of the platform to capture value [21]. Some characteristics that can play a role in this decision are the ability to capture cross-side network effects, price-sensitivity (subsidize the most price-sensitive side), quality-sensitivity (subsidize the most quality-sensitive side), and output cost [13].

Multi-homing

When a user is affiliated with more than one platform, it is referred to as multi-homing [44]. When users multi-home, it diminishes the ability of platform owners to capture value. For example, when a driver is subscribed to Uber and other ride-hailing platforms simultaneously, Uber loses out on transaction fees when the driver drives customers on competing platforms instead.

Essentially, limiting a user's ability to multi-home is a widely used competitive strategy and can affect the platform's evolution. Tiwana et al. [52] define homing costs as the aggregate of adoption, operating, and opportunity costs incurred to maintain participation in a platform. When these are high, users are less incentivized to be affiliated with multiple platforms, and vice versa. Multi-homing is either allowed, partially allowed or forbidden.

Switching costs

To avoid platform users from leaving the platform, platform owners build up switching costs to make it costly for them to switch [33]. Take for example user reputation built from rating systems on one platform to signal trustworthiness [44]. If a seller on eBay, for example, would start selling on a competing platform she would need to start from zero as all her user ratings are left on her profile on eBay–thus signaling trustworthiness to buyers would be much harder than if she had stayed on eBay.

Building switching costs is important for the platform owner's ability to capture value.

3.5 Mobility Data Platforms

With the advancing digitization of means of transport, transport infrastructure and the continuous development of application-based mobility products and services, the fundamental and indispensable importance of the data platform component is becoming apparent Guerreiro [24]. Transport operators are challenged by today's services of integrated mobility, namely information provision, booking, ticketing and payment, all oftentimes converged from multiple discrete services in one single platform [36, 46].

Mobility as a Service (MaaS) can be seen as an on-demand model incorporating various discrete services in one place [36], thus multi-sided platforms can act as a model for MaaS [30]. The platform's attribute is the one as provider of basic functions and aggregator used for products and services. Basic functions are real time information, trip planning, booking or ticketing while not limited to those [30].

Bolt, FreeNow, and Uber, all operate a mobility platform addressing above introduced indirect network effects. The core challenge is to reach a critical mass and to attract users on both sides [27]. While organizations can act as owners and operators of the platform and as providers of the basic functions outlined above, they could also take on a different role by acting solely as data providers and processors themselves. This is to allow other sides to use these resources to test and validate existing solutions and develop new MaaS or autonomous mobility solutions Guerreiro [24].

Mobility data platforms have become central to enabling new mobility services and products–in particular, the development of entirely new mobility approaches such as cooperative, connected and automated mobility (CCAM) has the potential to reinterpret the mobility landscape. Guerreiro [24] then identified three core enablers for building autonomous mobility solutions: (1) digitalized and intelligent infrastructures, (2) scalable and adaptable software-stacks for mobility solutions, and digital platforms as the central mean for providing data a different granularity, AI models and services.

4 Methodology

Starting with the three dimensions of business model design defined by Teece [50], who describes a business model as "the design or architecture of the value creation, delivery, and capture mechanisms it employs", we review existing platform literature and extract individual design elements into one of the three categories (i) value creation, (ii) value delivery and (iii) value capture. This paper considers only those elements that appear to be of high relevance for digital multi-sided platforms.

The selection of elements and variables follows the design principles of morphological analysis. This approach has been identified to be suitable to accommodate multiple configurations of platform design elements given its flexibility to choose different combinations of attributes. The resulting morphological box can serve as an artifact to identify different platform business models.

This paper employs a systematic literature review as the methodology to extract the individual design elements of multi-sided platforms and coded the identified papers according to the three previously identified dimensions. The software MAXQDA20 was employed during the coding process, where the output is a database of 49 articles and 627 coded segments.

To validate and exemplify our findings, we draw on three mobility data platform use cases and match their business model to the proposed morphological box.

4.1 Morphological Box for Building Data-Driven Mobility Platforms

BM Dimensions	Design Elements	Specifications				
Value creation dimension	Market structure	Two-sided		Multi-sided		
	Platform ownership	Single firm	Alliance/consortium		Decentralized	
	Platform type	Transaction	Innovation		Hybrid	
	Key activity	Data services	Network-community building		Complementary innovations	
	Price discovery	Fixed priced	Set by sellers	Set by buyers	Auction	Negotiation
	Access control	Selective exclusion	Social behavior	Both	None	
	Trust system	User reviews	Review by platform		None	
Value delivery dimension	Key value proposition	Product development costs	Transaction costs	Search costs	Both	
	Transaction content	Data	Service		Goods	
	Digital interface	API	SDK	Both	None/other	
	Participants	C2C	B2C	B2B		None/other
	Geographic scope	Global	Regional		Local	
Value capture dimension	Key revenue stream	Membership/access fee	Usage fee	Transaction fee	Advertising	None
	Pricing mechanism	Fixed pricing	Market pricing		Differentiated pricing	
	Revenue side	Provider	Consumer	Third party	None/other	
	Subsidy side	Provider	Consumer	Third party	None/other	
	Multihoming costs	Allowed	Partially allowed		Forbidden	
	Switching costs	Low	High			

5 Testing the Morphological Box—A Case Study Analysis

5.1 Case 1: HERE Platform

HERE is a geodata service provider jointly owned by a *consortium of automotive OEMs*, their suppliers, and technology companies. It features a location data sharing platform, a marketplace and development platform where businesses can exchange and monetize their data, services and algorithms as well as build applications.

HERE platform creates value for its *multiple sides,* namely, application developers, data providers and data consumers by allowing location data to be transacted and innovated upon, resulting in a *hybrid platform.* Furthermore, the automotive sector represents the largest income stream for HERE platform through licensing map content and traffic services to car OEMs, thus most efforts are put into the *data services* layer. Yet, in the past years the company has been expanding its platform encouraging *complementary innovations* by investing in the development of the HERE platform, HERE workspace, HERE marketplace and HERE Studio. The price discovery is *set by sellers*, independently if it is a third-party data provider or HERE as a data provider –not acting as platform owner. The platform employs a *selective exclusion* approach to controlling access to the platform and encourages trust in the platform by only allowing data providers who have been previously approved as map data distributors. There is no peer review system with regards to data quality as it is solely the platform owner's job.

When it comes to value delivery, the platform's key value proposition is primarily reducing *search costs* via the HERE platform, and on a secondary basis, reducing *product development* and *transaction costs* via HERE studio and HERE marketplace. The key transaction content in the platform is location-related *data.* Developers can access location data and services, such as maps, routing, geocoding, search and create new services or applications via the software development tools (HERE *SDK*) and HERE REST *APIs.* The platform is targeted at *B2B* customers on a *global* level who are usually the owners of large, high-quality data and want to monetize it, such as AccuWeather, BMW, Continental and Mercedes Benz, and other businesses who would benefit from accessing it, such as Volta Trucks and what3words who partnered with HERE Technologies to deliver a tailored last-mile electric vehicle navigation system.

HERE platform captures value from *usage* fees. It offers two different plans depending on the number of transactions or the so-called 'pay-as-you-grow' mechanism. The price of the data is *fixed* as it is independent from market conditions or customer characteristics. The main revenue source for the platform lies on the *third parties*, in this case developers, who use the data provided to build new applications and pay for software development tools and data. The subsidized side are *data providers* who are critical to the platform's success in attracting developers who innovate with the data. Multi-homing is *allowed* and switching costs in the location-data platform ecosystem are high. When app developers look for location-related services such as maps, navigation, data, location intelligence, and other location services it is not easy to find all key resources in one platform and coupled with a vibrant developer ecosystem. While there are alternative players such as Google and TomTom, their offerings differ in terms of data types, e.g., Google Maps strength lies on POIs, location-based services integrated with its advertising platform for businesses, and available development tools.

5.2 Case 2: CARUSO Dataplace

CARUSO Dataplace is a neutral vehicle data platform born from an Independent Automotive Aftermarket (IAM) initiative. It provides open access to in-vehicle data across different brands.

CARUSO configures its value creation as a *multi-sided platform* where automotive brands, developers and data consumers come together. It is owned by a *consortia* of mobility players who have a vote on the platform's strategy and business model. CARUSO acts, essentially, as a reseller of automotive data and connects data providers with data consumers, which makes it a *transaction* platform. With regards to the key activity layer, *data services* are the key component of the platform, although *complementary innovations* are also encouraged by the platform to showcase use cases and signal usefulness. Prices for data points are *set by sellers*. The access control is based on the *selective exclusion* policy, while on the provider side only those who have previously signed a partnership with the platform can publish data, on the consumer side only those who have signed up to the platform and agreed to the terms and conditions are able to use the data via CARUSO's API. Any misuse of the data is handled by the platform owner. When it comes to trust between the parties, there is no review system to rate either side; platform users trust each other based on the previously signed agreement with the platform provider. For example, it is disclosed in advance that the Data Supplier does not guarantee a certain quality of requested In-Vehicle Data, the data is supplied on an "as-is" and "as available" basis and without any warranty or representation for quality, quantity, completeness, accuracy, availability, error free and fitness for any particular purpose [9].

On the value delivery dimension, CARUSO enables the reduction of *search costs* via its marketplace, *transaction costs* via the easy-to-use API, and *product development costs* by unlocking previously unavailable data to innovators on the consumer side. As previously mentioned, the platform features an API which allows data consumers to request in-vehicle data items, but only legal entities (*B2B*) can have access to the platform. CARUSO operates on a regional basis in several European countries.

CARUSO captures value by charging a *usage fee* depending on the level of in-vehicle data requested. This usage fee is further discriminated by a *differentiated pricing* mechanism where data fees vary depending on data volume and subscription package. As a data brokering platform, CARUSO monetizes its consumer side and subsidizes the provider side. Multihoming is *allowed* and switching costs are *high* as the data types available at CARUSO are not widely available in competing platforms.

5.3 *Case 3: mCLOUD*

The German Federal Ministry for Digital Affairs and Transport (BMVI) owns, governs and provides the "mCLOUD": platform. It offers open-source data in the domain of mobility and related fields, which is provided by institutions from the government, industry, research and others.

In terms of the value creation dimension, it can thus be interpreted as a *multi-sided* platform with *single-organization* ownership. Also, mCLOUD is a great example of a *hybrid platform type.* One of the mCLOUD's principles is "easy accessibility to data", such as the abandonment of mandatory user accounts. This policy facilitates data *transactions* across all user groups. Furthermore, a related support program "mFUND" uses mCLOUD in order to provide the data base for *innovations* to a large variety of stakeholders coming from different domains. When it comes to the key activity layer, clearly *data services* are the main driver of a dataset-providing platform, though certainly also value for the other two dimensions *community* and *innovation* is created indirectly. Price discovery does not happen in this use case, due to the platform being a free-to-use public service funded by tax money, the platform provider sets a fixed price of 0€. In the access control layer, the platform runs on a *selectively exclusive* policy. Data providers only get data publishing access rights after a successful review of their requested contribution by the platform administrators. Concerning this matter, the fit of the proposed dataset to BMVI-related topics is evaluated. In addition, the administrators urge the publishers to follow their guidelines in terms of metadata and data quality management. No additional trust system in terms of reviews is used. The publications are hand-picked by the platform but cannot be publicly commented on.

Related to value delivery, the key value proposition of mCLOUD is both *search and transaction cost reduction.* The platform acts as a matchmaker between instances that need and provide certain mobility-specific *data.* Furthermore, it enables their easy integration into automated systems by providing information about the involved *APIs* for getting and posting data [34].

Furthermore, as already mentioned, the user base of such a platform is manyfold and covers both B2B and B2C applications, while the latter might mainly use it in case of a general informative interest. The geographic scope of mCLOUD is *local*, as most data providers offer data from inside Germany and the whole webpage is only available in German language. However, in the future data from other countries could join the portfolio, especially since the BMVI very much focuses on topics that affect the whole European Union [5]. Hence, in the long run the focus could shift towards providing a *global* platform.

Finally, value capturing barely happens in this example. Due to its non-profit orientation, a key revenue stream, revenue source and pricing are simply not present.

The subsidy side is characterized by a complete funding by the platform provider, thus indirectly by all taxpayers. Multihoming is *allowed* as the interfaces are easy to access and to integrate and users can access other mobility platforms without having made a large investment on mCloud. Switching costs are presumably *low*, because it is not costly for a user to switch to a different platform with a similar offering in terms of data types.

6 Conclusions and Limitations

6.1 Scientific Contributions

This paper addresses data-driven MSPs (multi-sided platforms) and their key design elements and configurations. Literature has investigated the central concepts of platform strategy and individual elements, but a lack of a holistic approach for platform design is evident. Furthermore, an empirical test with a focus on data-driven mobility platforms has been neglected. Therefore, our research attempts to bridge this gap by providing three main contributions.

First, based on the literature analysis this paper provides a comprehensive picture of data-driven mobility platform design and categorizes design choices in a morphological box. Second, by using [50] business model design framework comprising value creation, value delivery and value capture, we were able to allocate individual design elements to these three dimensions and find interdependencies between those elements. Lastly, by testing the morphological box with the data-driven mobility platforms CARUSO, Here and mCloud, this paper contributes to illustrating a newly arising type of platform and its key design choices.

Regarding RQ #1 (i.e., available choices for platform design), this study identified eighteen elements which are key to platforms (see Table 1) and further clustered them into three dimensions (i) value creation, (ii) value delivery and (iii) value capture. The first seven elements are part of the value creation dimension and describe the value architecture and mechanisms to orchestrate resources and processes. The subsequent five elements are part of the value delivery dimension and define the components which create value for the platform sides as well as the interfaces to deliver it. The remaining six elements represent the value capture dimension and explain the conversion from value proposition into revenues for the platform.

As for RQ #2 (i.e., interdependencies between platform design elements), we found the ‘key activity’ layer in the value creation dimension to be fundamental in the design of a platform. Whether a platform’s key activity is data services, complementary innovations, community building or a combination of the three, it dictates the transaction content and digital interfaces. For example, if the platform specializes in *data services,* then *data* is the content of the transaction for which there must be a digital interface available (*API / SDK*). Moreover, when faced with the choice on how to monetize i.e., ‘key revenue stream’, platform providers need to critically assess which side they are charging and which side they are subsidizing as it will impact their ability to capture value. We see that, for data-driven mobility platforms, the most common monetization is 'usage fee'.

Furthermore, we identified data-driven mobility platforms to be *multi-sided*. New data-driven business models in the mobility space require resources from many different players, such as car manufacturers, infrastructure providers, part suppliers, etc., which come together in a single platform and unlock the value of previously siloed data. Also, while the underlying activity of data-driven mobility platforms is facilitating data transactions, we see that when an innovation layer is added to the platform it promotes complementary innovations and acts as a differentiator factor from other similar platforms.

In addition, in these platforms the access and usage rules are partly limited by the platform owner, unlike in two-sided markets where anyone who signs up to the platform can provide and consume data, goods or services.

6.2 Managerial Implications

This research can aid practitioners in designing platform businesses. The proposed morphological box formalizes the key design elements of digital multi-sided platforms which is useful to both companies looking to change their pipeline business model into a platform, and for new companies creating a platform business model from scratch. Moreover, the morphological box may be used by platform providers as an evaluation tool to visualize different configurations of competitor platforms and adapt their business model to remain competitive.

As platform practitioners use the proposed morphological box to design their platforms' value creation, delivery and capture mechanisms, we expect new business models to arise and platform owners to adjust their business strategy according to the design choices.

6.3 Future Research and Limitations

The research agenda on digital platforms is at a decisive point. The digital economy is increasingly being driven by platforms and further research on this phenomenon is relevant both to platform practitioners, policy makers and society at large.

This study has focused on specific design elements of digital platforms configured under a business model perspective taking value creation, delivery and capture as clustering categories. We believe, however, that there are other ways to structure platforms when using a different theoretical lens which may add valuable insights to platform practitioners.

We empirically tested our box with selected cases in Germany, but further research is needed into other geographies which may result in different configurations. Furthermore, data-driven mobility platforms are still a new phenomenon which makes gathering a large sample more complex, to which we suggest taking a wider approach and empirically test the framework using data-sharing platforms at large.

This paper does not intend to evaluate which platform configuration is superior but instead show the available choices that platform owners have when designing a platform. With the case studies, we illustrate how these design choices result in varied business models without favoring one or the other.

Data sharing platforms are still in their early stages of their development, but we expect more players to enter the scene in the coming years as they realize the monetization and innovation opportunities that platforms offer. For the specific case of mobility, as the industry digitizes and more cars are equipped with telematic units and V2X communication, more opportunities will arise.

Acknowledgements We would like to thank our colleague Konrad Katzschke for his contribution and useful comments on our research. Additionally, special thanks to Ketan Sunil Motlag who supported us throughout the project.

Funding This research has received funding from the German Federal Ministry for Digital Affairs and Transport (BMVI) as part of the BeIntelli project—AI in Mobility based on Platform Economy. The funding reference number is 01MM200004.

References

1. Afuah A, Tucci CL (2012) Crowdsourcing as a solution to distant search. Acad Manag Rev 37(3):355–375. https://doi.org/10.5465/amr.2010.0146
2. Bakos Y 1998 The emerging role of electronic marketplaces on the internet. Commun ACM 41(8):35–42
3. Baldwin CY, Jason Woodard C (2008) The architecture of platforms: A unified view. SSRN Electron J. https://doi.org/10.2139/ssrn.1265155
4. BeIntelli (2022). BeIntelli. https://be-intelli.com/
5. BMDV (2021) Strategie für nachhaltige und intelligente mobilität der europäischen commission. https://bmdv.bund.de/DE/Themen/EU-Politik/Uebergreifende-EU-Themen/Verkehrsstrategie-der-Europaeischen-Kommission/verkehrsstrategie-europaeische-kommission.html
6. Boudreau K (2010) Open platform strategies and innovation: granting access vs. devolving control. Manage Sci 56(10):1849–1872. https://doi.org/10.1287/mnsc.1100.1215
7. Boudreau K (2017) Platform-based organization and boundary choices: 'opening-up' while still coordinating and orchestrating. SSRN Electron J. https://doi.org/10.2139/ssrn.2929725
8. Boudreau KJ, Hagiu A (2008) Platform rules: multi-sided platforms as regulators. SSRN Electron J. https://doi.org/10.2139/ssrn.1269966
9. Caruso (2022) Terms & conditions. Corporate website. CARUSO Dataplace. https://www.caruso-dataplace.com/dataplace_terms_and_conditions/
10. Choudary SP (2015) Platform scale: How an emerging business model helps startups build large empires with minimum investment. Platform thinking labs

11. Cusumano MA (2020) Self-driving vehicle technology: progress and promises. Commun ACM 63(10):20–22. https://doi.org/10.1145/3417074
12. Cusumano MA, Gawer A, Yoffie DB (2019) The business of platforms: strategy in the age of digital competition, innovation and power. Harper business
13. Eisenmann T, Parker G, Van Alstyne MW (2006) Strategies for two-sided markets. Harv Bus Rev. https://hbr.org/2006/10/strategies-for-two-sided-markets
14. Evans DS (2003) Some empirical aspects of multi-sided platform industries. SSRN Electron J. https://doi.org/10.2139/ssrn.447981
15. Evans DS (2012) Governing bad behavior by users of multi-sided platforms. SSRN Electron J. https://doi.org/10.2139/ssrn.1950474
16. Evans DS, Hagiu A, Schmalensee R (2008) Invisible engines: How software platforms drive innovation and transform industries. The MIT Press
17. Evans DS, Schmalensee R (2010) Failure to launch: critical mass in platform businesses. SSRN Electron J. https://doi.org/10.2139/ssrn.1353502
18. Evans DS, Schmalensee R (2016) The new economics of multisided platforms. Harvard Business Review Press
19. Evans DS, Schmalensee R (2017) Multi-sided platforms. In: Palgrave Macmillan (ed) The new Palgrave dictionary of economics. Palgrave Macmillan UK, London, pp 1–9.https://doi.org/10.1057/978-1-349-95121-5_3069-1
20. FirstMile VC (2019) Decoding the autonomous driving landscape: software will indeed eat the (Automotive) world. Medium (blog). https://medium.com/@firstmilevc/avlandscape-8a21491f1f54
21. Gawer A (2021) Digital platforms' boundaries: The interplay of firm scope, platform sides, and digital interfaces. Long Range Plan 54(5):102045. https://doi.org/10.1016/j.lrp.2020.102045
22. Gawer A, Cusumano MA (2014) Industry platforms and ecosystem innovation: platforms and innovation. J Prod Innov Manag 31(3):417–433. https://doi.org/10.1111/jpim.12105
23. Ghazawneh A, Henfridsson O (2013) Balancing platform control and external contribution in third-party development: The boundary resources model: control and contribution in third-party development. Inf Syst J 23(2):173–192. https://doi.org/10.1111/j.1365-2575.2012.00406.x
24. Guerreiro A, Keiser JM, Albayrak S, Masuch N, Hessler A (2021) Towards intelligent infrastructures and AI-driven platform ecosystems for connected and automated mobility solutions. In: Infrastructure supporting automated driving. Book of Abstracts: 2634. Hamburg. https://itsworldcongress.com/the-book-of-abstracts/
25. Hagiu A (2009) Two-sided platforms: product variety and pricing structures. J Econ & Manag Strat 18(4):1011–1043. https://doi.org/10.1111/j.1530-9134.2009.00236.x
26. Hagiu A (2013) Strategic decisions for multisided platforms. MIT Sloan Manag Rev. https://sloanreview.mit.edu/article/strategic-decisions-for-multisided-platforms/
27. Hein A, Schreieck M, Riasanow T, Setzke DS, Wiesche M, Böhm M, Krcmar H (2020) Digital platform ecosystems. Electron Mark 30(1):87–98. https://doi.org/10.1007/s12525-019-00377-4
28. HERE Technologies (2022) HERE Platform. Corporate website. HERE Technologies. https://www.here.com/platform
29. High Mobility (2020) How vehicle manufacturers can successfully engage in the data business. Medium (blog). https://medium.com/high-mobility/how-vehicle-manufacturers-can-successfully-engage-in-data-business-3bd66c67d8ae
30. Jittrapirom P, Caiati V, Feneri AM, Ebrahimigharehbaghi S, Alonso González MJ, Narayan J (2017) Mobility as a service: A critical review of definitions, assessments of schemes, and key challenges. Urban Plan 2 (2): 13–25. https://doi.org/10.17645/up.v2i2.931
31. Lindner M, Straub S, Kühne B (2021) How to share data? data sharing platforms for organizations. Study commissioned by the federal ministry for economic affairs and energy. German Federal Ministry for Economic Affairs and Energy
32. McIntyre DP, Srinivasan A (2017) Networks, platforms, and strategy: emerging views and next steps: Networks, platforms, and strategy. Strateg Manag J 38(1):141–160. https://doi.org/10.1002/smj.2596

33. McIntyre D, Srinivasan A, Afuah A, Gawer A, Kretschmer T (2021) Multisided platforms as new organizational forms. Acad Manag Perspect 35(4):566–583. https://doi.org/10.5465/amp.2018.0018
34. mCloud (2022) Beschreibung der importschnittstelle. https://www.mcloud.de/documents/20147/174370/harvester.pdf/c250c69e-6ce4-55be-4223-8031a9fed807?t=1619793837482
35. Meta Platforms Inc (2021) Meta reports fourth quarter and full year 2021 results. Financial statement. Meta. https://s21.q4cdn.com/399680738/files/doc_financials/2021/q4/FB-12.31.2021-Exhibit-99.1-Final.pdf
36. Mulley C, Nelson JD, Wright S (2018) Community transport meets mobility as a service: On the road to a new a flexible future. Res Transp Econ 69(September):583–591. https://doi.org/10.1016/j.retrec.2018.02.004
37. Orlik T, Jimenez J, Sam C (2021) World-dominating superstar firms get bigger, techier, and more chinese." Bloomberg 2021. https://www.bloomberg.com/graphics/2021-biggest-global-companies-growth-trends/
38. Osterwalder A (2004) The business model ontology a proposition in a design science approach. Doctoral Dissertation, Lausanne: Université de Lausanne, Faculté des hautes études commercialess
39. Otto B, Jarke M (2019) Designing a multi-sided data platform: findings from the international data spaces case. Electron Mark 29(4):561–580. https://doi.org/10.1007/s12525-019-00362-x
40. Parker GG, Van Alstyne MW, Choudary SP (2016) Platform revolution: How networked markets are transforming the economy and how to make them work for you. WW Norton and Company
41. Parker G, Van Alstyne MW (2014) Platform strategy. SSRN Electron J. https://doi.org/10.2139/ssrn.2439323
42. PwC (2021) Global top 100 companies by market capitalisation. PwC
43. de Reuver M, Sørensen C, Basole RC (2018) The digital platform: A research agenda. J Inf Technol 33(2):124–135. https://doi.org/10.1057/s41265-016-0033-3
44. Rochet J-C, Tirole J (2003) Platform competition in two-sided markets. J Eur Econ Assoc 1(4):990–1029. https://doi.org/10.1162/154247603322493212
45. Schreieck, M, Wiesche M, Krcmar H (2016) Design and governance of platform ecosystems–key concepts and issues for future research. In: ECIS 2016. https://www.platform-economy.de/wp-content/uploads/2020/11/Design-and-governance-of-platform-ecosystems%E2%80%93key-concepts-and-issues-for-future-research.pdf.
46. Stehlin J, Hodson M, McMeekin A (2020) Platform mobilities and the production of urban space: Toward a typology of platformization trajectories. Environ Plan A: Econ Space 52(7):1250–1268. https://doi.org/10.1177/0308518X19896801
47. Strahilevitz LJ (2005) Information asymmetries and the rights to exclude. Mich Law Rev
48. Strasser M, Weiner N, Albayrak S (2015) The potential of interconnected service marketplaces for future mobility. Comput Electr Eng 45(July):169–181. https://doi.org/10.1016/j.compeleceng.2015.06.008
49. Täuscher K, Laudien SM (2018) Understanding platform business models: A mixed methods study of marketplaces. Eur Manag J 36(3):319–329. https://doi.org/10.1016/j.emj.2017.06.005
50. Teece DJ (2010) Business models, business strategy and innovation. Long Range Plan 43(2–3):172–194. https://doi.org/10.1016/j.lrp.2009.07.003
51. Schiller T, Kummer P, Berdichevskiy A, Weidenbach M, Sadoun J (2020) Future of automotive sales and aftersales—impact of current industry trends on OEM revenues and profits until 2035. Industry report. Deloitte. https://www2.deloitte.com/content/dam/Deloitte/global/Documents/Consumer-Business/gx-deloitte-future-of-automotive-sales-aftersales.pdf
52. Tiwana A, Konsynski B, Bush AA (2010) **Research commentary**—platform evolution: coevolution of platform architecture, governance, and environmental dynamics. Inf Syst Res 21(4):675–687. https://doi.org/10.1287/isre.1100.0323
53. Tura N, Kutvonen A, Ritala P (2018) Platform design framework: conceptualisation and application. Technol Anal & Strat Manag 30(8):881–894. https://doi.org/10.1080/09537325.2017.1390220

54. Van Alstyne MW, Parker GG, Choudary SP (2016) Pipelines, platforms, and the new rules of strategy. Harv Bus Rev
55. Dal Bianco V, Myllärniemi V, Komssi M, Raatikainen M (2014) The role of platform boundary resources in software ecosystems: A case study. In: IEEE. https://d1wqtxts1xzle7.cloudfront.net/51729706/PlatformBoundaryResources-with-cover-page-v2.pdf?Expires=1652780349&Signature=Lsvb2S4cBwMDHLt1Zyo-7RV4~IwD5FOwUAgO-h-gvqKZkkeNE~DbdLMn7v4gLLUsNxlXpwXTxsA8C5rYxlu7~HrSmKIc0QDfGglJZ1oIl-0a2IATVCAIBojSd-npPhfoSwvTb5dQi6ldkNgpPT6TTR04F8zs3MDetZuAwb6urJ~a57tvH~MpXtm7Tt6oTXwXms9YF2WMfMx7MyuvzRlektt9vGKpo6RN78wndAMrhUYaSm5cfy5THRThxX70bRC1hfjYwi-ImcvoPk7eaYQvLYN19b1kVxx69aKuRbVa7GyIl9zePZLjptQCsATXZM954t1BqDQT40bJ83QybOgrNg__&Key-Pair-Id=APKAJLOHF5GGSLRBV4ZA
56. Yoo Y, Boland RJ, Lyytinen K, Majchrzak A (2012) Organizing for innovation in the digitized world. Organ Sci 23(5):1398–1408. https://doi.org/10.1287/orsc.1120.0771

Generating Standardized Agent-Based Transport Models in Germany

Check for updates

Torben Lelke, Lasse Bienzeisler, and Bernhard Friedrich

Abstract The paper presents a standardized method to generate agent-based transport models. In this context, the paper depicts an approach to processing multiple public and nationally available datasets and illustrates how they can be fused to create a synthetic population of agents. We then present a methodology to supply the generated agents with individual attributes and assign a schedule of activities as well as individual mobility behavior. Subsequently, we apply the methodology to the city of Hanover, and the resulting model is compared to and evaluated with real-world data. In the course of this evaluation, it becomes apparent that the pipeline generates a sufficiently accurate model that can be used to study the traffic in the city.

Keywords Scenario creation · National dataset · Data fusion · Modeling pipeline · MATSim

1 Introduction

In recent years, agent-based transport models have gained popularity because these models allow microscopic modeling and analysis of traffic that is continuous in space and time. Additionally, agent-based simulation facilitates modeling individual decisions of travelers. The agent-based simulation method is therefore beneficial for modeling new mobility solutions such as Mobility-as-a-Service or micro-mobility. In these approaches, personal decisions on the utility of different opportunities are essential to the participants' choices, which traditional macroscopic models cannot replicate in enough detail. However, agent-based models are not yet widely used in planning practice. This is, at least partly, due to the complex and extensive process that is necessary to create these models [11].

Agent-based transport models rely on a synthetic population of the specified planning area. A synthetic population is a simplified microscopic representation

T. Lelke (✉) · L. Bienzeisler · B. Friedrich
Institute of Transportation and Urban Engineering, TU Braunschweig, Hermann-Blenk-Straße 42, 38108 Braunschweig, Germany
e-mail: t.lelke@tu-braunschweig.de

C. Antoniou et al. (eds.), *Proceedings of the 12th International Scientific Conference on Mobility and Transport*, Lecture Notes in Mobility,
https://doi.org/10.1007/978-981-19-8361-0_7

of the actual population in which only attributes of interest are reproduced. Each person living in that area has to be represented in the model, including the location of this person's residence and characteristics such as sex, age, or other information about the person's standard of living. In most synthetic populations, these attributes match aggregated statistical measures of the actual population rather than precisely reproducing each individual [2]. In addition to these attributes, each agent is assigned a travel schedule holding information about the facilities the person visits during the modeled day and which activities are performed there. Additionally, the modes the agents use to travel to the specified locations are usually included. Due to the richness of information in such agent-based transport models, the necessary amount of data and the effort to process this data are comparably high. Especially data about attributes of interest is challenging to collect and will often represent only a subset of the actual information. Therefore, modelers need to find ways to expand on the available data. Another problem is a lack of consistency among data formats, resulting in a necessity to harmonize data from different sources [2].

2 Literature Review and Research Objective

There have been several case studies dealing with the creation of agent-based transport models with different approaches to generating agents and assigning travel behavior in the form of daily schedules using openly available data [1, 6, 12, 16]. However, most of these existing approaches are often focused on just one specific planning area and use data that is, at least in part, unique to this area. Therefore, these methods are transferable to other planning areas in terms of their methodologies but not directly applicable elsewhere.

To further facilitate agent-based modeling, Hörl and Balac recently presented a consistent modeling pipeline for agent-based transport models [9] and applied it to the Île-de-France region around Paris [10]. Their *equasim* pipeline is structured into sequential stages and delivers a complete methodology from processing input data to a simulation in the open-source simulation framework MATSim [8]. The core of this approach is the *Population Pipeline*, which creates the synthetic population and travel demand. To generate the synthetic population, Hörl and Balac apply direct sampling from a local census to distribute the agents within a defined spatial zoning system. The agents in the resulting synthetic population contain attributes such as age, sex, and information about their employment status. Following these attributes, each agent is assigned a fitting activity schedule from an available household travel survey for the Île-de-France region. The authors then estimate the locations for these activities with dedicated algorithms for primary activities such as home, work, and education as well as secondary activities such as leisure or shopping. The methodology presented by Hörl and Balac was already successfully applied to California, Sao Paulo, and Zurich [9].

The objective of the research presented in this paper is to generate a lightweight approach capable of generating a synthetic population for agent-based transport models for the entirety of Germany in a standardized way. Our approach is similar to the one taken by Hörl and Balac. However, we focus not only on a standardized methodology but also on standardized data input for all models created with this methodology. This means that the underlying data should be the same for every conceivable planning area. Naturally, the models resulting from such a methodology would be less precise due to a lack of detailed information about inner-city particularities. However, we argue that it is not always necessary to aim at the greatest possible detail for all aspects of the model and invest substantial work into model creation. It is essential to find a trade-off between the required model detail and the amount of work necessary to achieve this detail. For example, for preliminary studies or small projects, it could be sufficient to have a slightly less detailed population model capable of approximately reproducing traffic conditions. Additionally, we wanted to generate a methodology that is easy in application to allow less experienced users of agent-based simulations to build the necessary models and facilitate agent-based modeling in planning practice.

After presenting the underlying data sources and the resulting methodology in the first part of this paper, we evaluate the approach. To do this, we apply the methods to a planning area and investigate the quality of the synthetic population and the resulting modeled traffic conditions. In addition, we want to explore whether our model can sufficiently approximate the traffic conditions inside the chosen planning area.

3 Data Sources

The proposed methodology for generating standardized agent-based transport models is based on a selection of public datasets that are available on a national scope. These datasets must include information for every possible planning area inside Germany. In the following section, we describe these datasets briefly.

The main data source we use to generate the synthetic population is the german national census [15]. In the census, which was last performed in 2011, the number of inhabitants for each home address in Germany was counted. These persons were then aggregated into households. Each home address was assigned to a 100 × 100 m grid cell to comply with data privacy regulations. Additionally, data privacy regulations mandated the issuing of cells with just one inhabitant as empty cells. The grid cells are all identified with a unique identification number consistent with the *European INSPIRE* directive [5]. Through this dataset, precise statements about the number of inhabitants and households in a resolution of one hectare can be given for the entirety of Germany. In total, the dataset contains over 36 million grid cells, of which approximately 3.2 million are inhabited.

The assignment of travel behavior to the agents is based on the German national mobility survey *Mobilität in Deutschland* (MiD) [14]. This survey, which was last conducted in 2017, is a representative data source on Germany's transport demand and everyday mobility. The study consists of a nationwide household travel survey

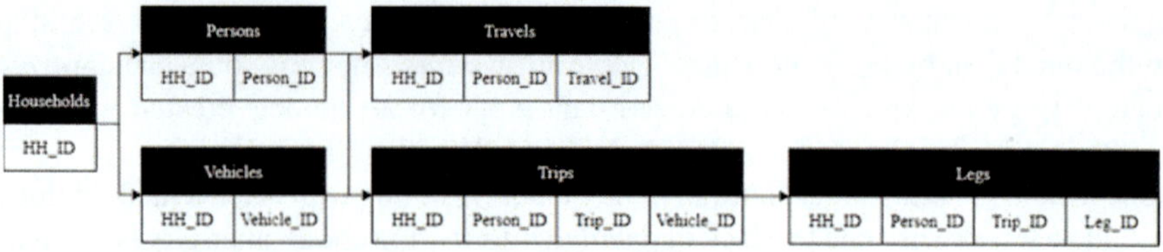

Fig. 1 Datasets and linking variables included in MiD 2017

and a total of 60 regional top-up samples and thus comes to a total number of more than 300,000 respondents, who documented more than one million trips. The data consists of several subsets. These subsets are provided with key variables that enable the individual data records to be linked. Thus, with the help of a household's identification number, it is possible to infer its occupants, which in turn are provided with an identification number appearing in the dataset of the recorded trips. These linkage possibilities allow the reconstruction of the mobility behavior of an entire household. Only the datasets on households, persons, and trips are used in this work. The various datasets of the MiD 2017 and their linking options are shown in Fig. 1. In the datasets for households and persons, each entry is weighted by how many entities it represents in Germany. Therefore, an exemplary household with a weight of 150 represents 150 households with the same structure in Germany.

To differentiate regions inside of Germany in terms of mobility behavior, we use the German *Regional Statistical Spatial Typology for Mobility and Transport Research* (RegioStaR) [7]. This spatial typology assigns each German municipality with different regional types that state how urban or rural the region is and how it is placed in relation to other central places around it. In the first step, urban regions are defined as all municipalities located within the surrounding area of large cities with more than 100,000 inhabitants or of larger metropolitan regions such as the Ruhr region. All municipalities outside these areas are considered rural. The further differentiation of the urban and rural regions is based on the *central place theory* (e.g., [3]). By considering the various municipal functions, this spatial planning concept defines so-called central places, which take on hierarchically increasing supply and development functions. Thus, metropolises within urban regions or central cities of rural regions are captured. For statistical evaluations, as they are also carried out in this work, the developers of RegioStaR recommend using the summarized regional statistical spatial type (RegioStaR7).

We use spatial shapes tagged with information about the land use in the corresponding area to locate the activities. In the *CORINE Land Cover* (CLC) dataset [4] for Germany, these areas are defined through automated classification based on aerial images. The CLC data distinguishes land use into 44 classes, which can be amalgamated into five overarching superclasses. Since only the CLC classes in settlement areas are relevant for the location of activities in urban transport models, we consider solely these areas in our methodology. We use the newest CLC version of 2018 with a resolution of five hectares. Additionally, points of interest inside the planning area are identified through standardized *OpenStreetMap* (OSM) queries.

The presented data sources are all centrally stored in a SQL database that allows for dedicated queries of individual parts of each dataset. Therefore, all data processing steps only need to be performed once.

4 Methodological Approach

Our methodology consists of three consecutive steps: population generation, plan assignment, and activity localization. The only external input needed for our pipeline is a geometry defining the planning area. The following section will explain each step and the data we used to perform the necessary computations.

The first step in the standardized modeling approach is the generation of a basis for the synthetic population by creating the households and persons in the defined planning area. To do this, we created a new national dataset by fusing the aforementioned data. The spatial reference structures in this new dataset are the grid cells taken from the census dataset. This spatial reference is then used to assign a RegioStaR regional type and a CLC land use to each cell. Finally, we use an OSM query to identify whether an education facility (school or university) is located in each cell. The result is a national enhanced census dataset of grid cells containing the total number of inhabitants and the number of households along with information about the land use, the region type, and geometry.

We generate the base population from this dataset by extracting all grid cells that lay inside of the geometry of the defined planning area. For each of these grid cells, we then generate the corresponding number of households. Each household is assigned the regional type of the cell and a random coordinate within the cell as the concrete location of residence of the household members. Afterward, we sequentially assign the specified number of persons living in the cell to one of the households until all inhabitants are distributed. When distributing the agents to the households, we set the maximum number of inhabitants in a household to seven. We use this limit to avoid statistical outliers in the distribution of household sizes and the subsequent compilation of individuals based on MiD households. A particular case in the distribution of residents among households are cells in which the census data does not contain information on the number of households. In such cells, a separate household is created for each agent. The agents created in this step are, however, not yet the complete representation of a person as they have not been assigned any attributes or an activity schedule to define their travel behavior.

The second step of our pipeline is the plan assignment. To generate and assign the activity and travel patterns for the agents in the synthetic population, we use a direct sampling method regarding the weights associated with each household and person in the corresponding MiD datasets. In a preliminary step, we use the same data to assign individual attributes to the agents in the synthetic population. To do this, we take the number of agents assigned to a synthetic household and this household's corresponding regional type and select an equivalent household from the MiD-dataset matching these values. Afterward, we extract all persons linked to his household in

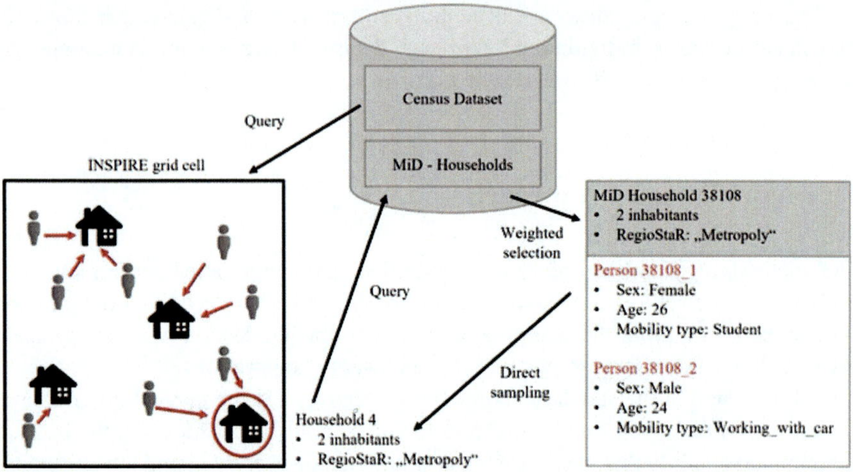

Fig. 2 Direct sampling method used for the creation of the synthetic population and the assignment of attributes

the dataset and directly copy their attributes and the reported activity schedule to the agents in the synthetic household. This process is presented in Fig. 2.

The extracted schedules consist of all trips the person has performed during the reporting day and information about the activities at the start and end of these trips. Additionally, the data from MiD includes a distance range for each trip and information about the used mode of transport. The reported activities are aggregated into five different activities we consider for our methodology. These activity types include the primary activity types home, work, and education, as well as the secondary activity types shopping and leisure. For some households in the MiD dataset, not all inhabitants participated in the travel survey. Therefore, it is impossible to sample their activity schedule directly. In such cases, we still copy all attributes of the original person from the MiD survey to the agent. We then use these attributes to find a person with similar characteristics and the same household type within the MiD data. To choose the most fitting one, we again use a weighted selection and assign the corresponding activity schedule among the identified possible replacement persons. At the end of the plan assignment step, the model contains all agents in the specified planning area with individual attributes and a matching schedule of activities and trips. In addition to the generated agents, we also create a specific number of vehicles for each household that matches the reported number of available vehicles from the MiD survey.

The trips assigned to the agents are not yet localized to the planning area. Since the schedules do not necessarily all originally stem from the planning area, we can not directly adapt travel distances from the survey. To realistically model the traffic demand with the standardized agent-based model, we need to assign a location to each activity in an agent's plan. We use the reported trip lengths from the MiD survey to generate a maximum movement radius in which the activities can occur. This radius is defined by all trips between two activities at the agent's location of residence. We sum all the reported travel distances of these trips and divide the resulting value by two. Through this process, we consider the hypothetical case that the person drove in a straight line to one activity, and all other activities were situated directly along this line. We search for cells whose associated land use and points of interest allow the performance of the targeted activity. To obtain such a location, we first search for a suitable cell within the distance range specified in the MiD. Missing values are imputed so that no values are missing and a corresponding distance can be assigned to each trip. If no suitable cell is available in the defined distance range, a cell within the movement radius of the trip closest to the reported trip length is selected instead. If no cell suitable is found within the movement radius either, we then search for a cell in the entire planning area and select the one closest to the movement radius. A random coordinate is then chosen within the identified cell and used to create the activity. We also assign the reported mode and the start and end times for the activities from the MiD directly to each of the agent's trips.

5 Methodology Application and Evaluation

To analyze the validity of models created with our standardized methodology, we generate an agent-based transport model for the city of Hanover and evaluate it against locally available data. We split this analysis into two steps. In the first step, we analyze the capability of our methodology to generate a synthetic population of the city that reflects the population as a whole as well as spatial inhomogeneities of population groups inside the city. Selected values from the generated synthetic population are compared to the matching statistical data from Hanover [13]. In the second step, we analyze the validity of the transportation model itself. This analysis includes an evaluation of aggregated values such as the modal split and a more detailed look at the simulated traffic flows on individual streets.

Figure 3 depicts a visual comparison of population densities in the model and Hanover, where darker shades of blue indicate high densities. As can be seen, the city of Hanover is very diverse regarding the population distribution inside the city boundaries. The city core, which is mainly a mixed-use of commercial and residential, is surrounded by primarily residential areas with a high population density. Additionally, the city of Hanover includes small agglomerations with a low population density, similar to rural towns. As shown, our methodology can recreate this diversity within the city in terms of population density.

Not only the placement of the agents' home locations is essential for an agent-based transport model, but also the distribution of the agents' other attributes. To analyze how well our methodology performs in this regard, we focus on the distribution of the agents' ages. First, we collect all age values from the generated synthetic population and then compare the age distribution with the statistical reports from Hanover [13]. This comparison can be seen in Table 1. It shows that the distribution from the standardized methodology overall matches the one from Hanover. However,

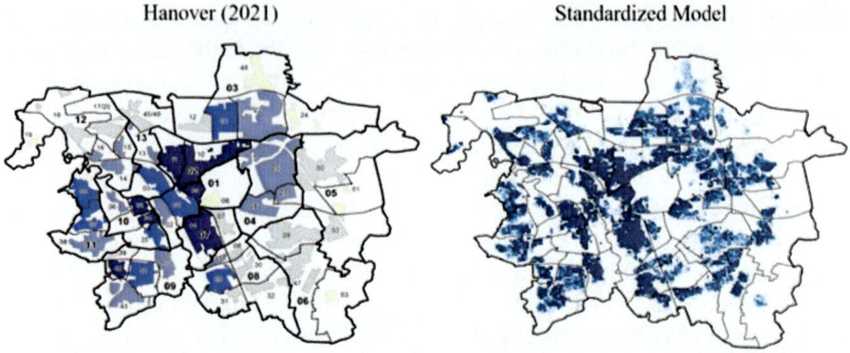

Fig. 3 Comparison of population densities in Hanover and the standardized model

it is also evident that the survey-based attribute assignment is not able to model the age distribution of Hanover with complete accuracy. The number of agents aged between 18 and 30 is too low, and in contrast, the number of agents aged older than 45 is too high. These differences become even more evident when analyzing the average age per district. The aforementioned spatial diversity within Hanover also leads to differences in the average age of particular districts. Some districts, especially the ones near the university, have a population that is on average younger, while other districts have an older population. Comparing the average age of the districts with the values from the standardized methodology, we find that these inhomogeneities are not included in the generated synthetic population (see Fig. 4).

In the second evaluation step, we aim to analyze the capabilities of our methodology to recreate the traffic conditions in Hanover correctly and thus generate a valid transportation model. To evaluate how well the model represents the mobility behavior of people and the resulting traffic, we perform a simulation with MATSim. The necessary input network is automatically generated with data from *OpenStreetMap*. For this, we export all links with the keys 'highway' for the roads and

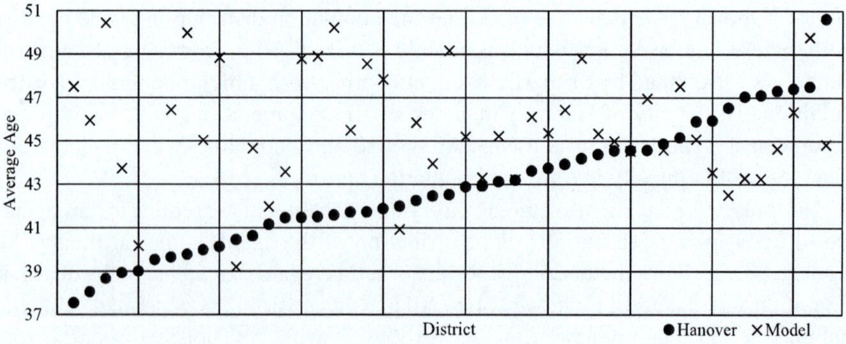

Fig. 4 Average age per district in Hanover and the generated model

Table 1 Age distributions for Hanover and the generated model

Age group	Share Hanover [%]	Share model [%]	Deviation [%]
0–3	2,90	3,39	0,49
4–6	2,80	2,45	−0,35
7–10	3,40	2,92	−0,48
11–18	6,60	5,06	−1,54
19–30	17,80	12,05	−5,75
31–45	21,60	21,57	−0,03
46–60	20,30	21,94	1,64
61–65	5,90	6,67	0,77
66–75	8,80	13,14	4,34
Older	10,00	10,81	0,81

'railway' for the railway tracks of public transport. We then assign fixed values for capacity and free speed to all road links in the network, which depend on the specified OSM road type. The public transport in Hanover and long-distance train connections passing through Hanover central station are also implemented using GTFS through standardized methods. As with the road links, we assign a free speed to the public transport links in the network to realistically reproduce the hierarchy of long-distance and regional travel routes through Germany. A capacity value is not assigned to the public transport links, since the capacities here are defined through the vehicle types, which travel in a fixed timetable. Since this evaluation aims to investigate the quality of the model, we do not allow the agents to change their.

Modes of transport or the start and end times of their activities. The simulation is run over 250 iterations to allow the agents to determine their optimal route in the network. After completing this simulation, the output files generated by MATSim are evaluated concerning specific parameters, which characterize the city's transportation system.

Figure 5 depicts the modal split, which results from the agents' travel schedules, in comparison with the modal split of Hanover. The presented data for Hanover is taken from the MiD top-up sample for Hanover and thus presents the status quo in 2017. It can be seen that the standardized methodology can reproduce the modal split in the planning area, although an exact match with the modal split in Hanover is not achieved.

Additionally, we evaluated how well the simulation matches real-world traffic volumes. Therefore, we compared the simulated traffic volumes with traffic counts from 59 roads spread across Hanover. Figure 6 shows this comparison in a scatter plot. The results show that while the counts are generally met well in the simulation, the real-world values are still underestimated by 16% on average. This fact can be explained by the focus of the methodology on individual traffic only. The missing

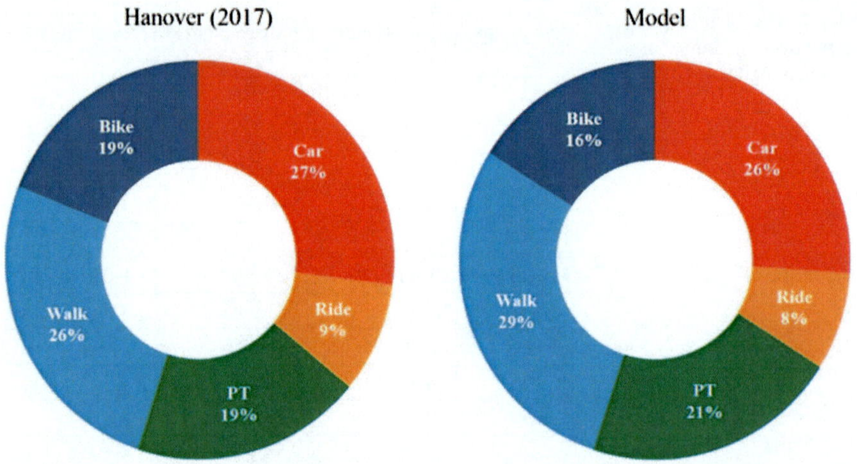

Fig. 5 Modal split in Hanover and the standardized model

16% of cars on the specified roads can be assumed to be commercial and freight vehicles, which are not considered in the model.

The application of the methodology to Hanover and subsequent evaluation has shown that the model is generally valid but still presents some deviations in comparison with real-world data. These recognized deviations can be explained by the survey-based direct sampling method we used within our modeling pipeline due to the lack of more detailed datasets available on a national scope. However, further analysis of the simulated traffic showed satisfactory results. Even though the synthetic

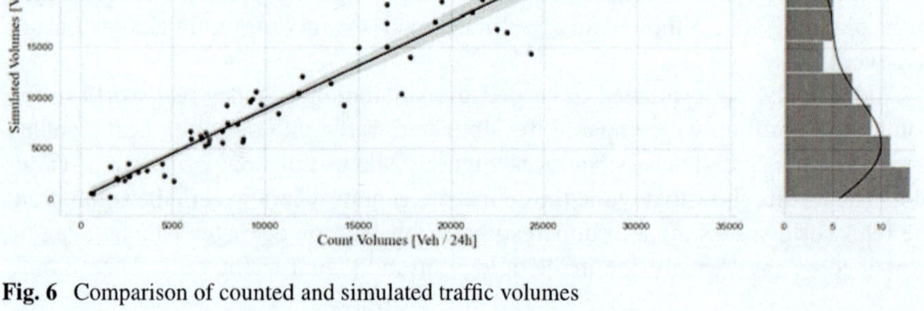

Fig. 6 Comparison of counted and simulated traffic volumes

population is somewhat lacking in detail, the traffic in Hanover is still simulated realistically.

6 Conclusion and Outlook

In a general conclusion, it can be stated that our standardized methodology is capable of generating realistic agent-based models for any planning area within Germany. Therefore, using the presented methodology, it is possible to create an agent-based model for any conceivable area inside of Germany without the need for additional input data. The application of the method to the city of Hanover and the subsequent evaluations showed that the modeling of the population does not reach the accuracy a model with specific data for the respective planning area would achieve. However, the analysis of traffic parameters from the generated transport model has shown that despite the deviations in the population model, the reconstruction of traffic volumes and other traffic parameters in the city is nevertheless sufficient.

The presented methodology contributes to the existing works in that it offers a lightweight solution to the generation of generally valid agent-based transport models that is easily applicable to different planning areas. Based on the conducted evaluation in this paper, we consider the pipeline suited for basic studies focusing on the transportation system as a whole and especially for comparative studies between several cities. We can not conclusively state the applicability of our presented methodology to more rural regions within the scope of this work. However, a preliminary application and evaluation of the methodology in the smaller city of Gifhorn have presented similar results to the ones shown in this paper. Additionally, the methodology allows inexperienced users an easier introduction to the topic of agent-based transport modeling. Experience from working with students has shown us, that a straightforward and easy-to-understand model creation process allows users to focus on the agent-based simulation process at first. When these users then need a more detailed model for in-depth analysis, the other established methods offer a range of methods to work with city-specific data.

In the future, we also want to address some of the presented methodology's shortcomings to further improve the accuracy of the population model and the resulting transport model. To make the synthetic population more detailed and implement some of the spatial diversities regarding personal attributes inside of cities, we aim to integrate locally available data sources into the modeling process. Therefore, we need to study the availability of such data in different municipalities to stay in line with the direct applicability to other planning areas the methodology currently offers. While performing such an evaluation of municipal data availability, we also want to research data sources that can be used to assign additional attributes to the agents in the population model that may improve the assignment of fitting activity schedules. The implementation of commercial and freight traffic is also be targeted as a future development for the methodology to model all types of traffic. However, the issue of data availability is even more prevalent in this regard because of lacking central data collection efforts for commercial data.

Acknowledgements The scientific research published in this article is granted by the Federal Ministry of Education and Research Germany for projects USEfUL and USEfUL XT (grant ID 03SF0547 & 03SF0609). The authors cordially thank the partners and funding agency.

References

1. Bienzeisler L, Lelke T, Wage O, Thiel F, Friedrich B (2020) Development of an agent-based transport model for the city of hanover using empirical mobility data and data fusion. Transp Res Procedia 47:99–106. https://doi.org/10.1016/j.trpro.2020.03.073
2. Chapuis K, Taillandier P (2019) A brief review of synthetic population generation practices in agent-based social simulation. In:Proceedings submitted to SSC 2019, social simulation conference
3. Christaller W (1933) Die zentralen orte in süddeutschland: Eine ökonomisch-geographische untersuchung über die gesetzmäßigkeit der verbreitung und entwicklung der siedlungen mit städtischen funktionen. Gustav Fischer, Jena
4. Copernicus Program (2018) CORINE land cover. https://land.copernicus.eu/pan-european/corine-land-cover
5. Directive 2007/2/EC Establishing an infrastructure for spatial information in the European community (INSPIRE). European parliament and council. http://data.europa.eu/eli/dir/1994/62/oj
6. Erath A, Fourie PJ, Van Eggermond MAB, Ordonez Medina SA, Chakirov A, Axhausen KW (2012) Large-scale agent-based transport demand model for Singapore. p 40. https://doi.org/10.3929/ETHZ-B-000306926
7. Federal Ministry of Transport and Digital Infrastructure (2020) Regional statistical spatial typology for mobility and transport research (RegioStaR). https://www.bmvi.de/SharedDocs/DE/Artikel/G/regionalstatistische-raumtypologie.html
8. Horni A, Nagel K, Axhausen KW (2016) The multi-agent transport simulation MATsim. Ubiquity Press. https://doi.org/10.5334/baw
9. Hörl S, Balac M (2021) Introducing the eqasim pipeline: From raw data to agent-based transport simulation. Procedia Comput Sci 184:712–719. https://doi.org/10.1016/j.procs.2021.03.089
10. Hörl S, Balac M (2021) Synthetic population and travel demand for Paris and Île-de-France based on open and publicly available data. Transp Res Part C: Emerg Technol 130:103291. https://doi.org/10.1016/j.trc.2021.103291
11. Kagho GO, Balac M, Axhausen KW (2020) Agent-based models in transport planning: current state, issues, and expectations. Procedia Comput Sci 170:726–732. https://doi.org/10.1016/j.procs.2020.03.164
12. Kickhöfer B, Hosse D, Turner K, Tirachini A (2016) Creating an open MATSim scenario from open data: The case of Santiago de Chile. https://doi.org/10.13140/RG.2.2.25394.40649
13. Landeshauptstadt Hannover (2021) Strukturdaten der Stadtteile und Stadtbezirke 2021. https://www.hannover.de/content/download/865574/file/Strukturdaten%202021.pdf
14. Nobis C, Kuhnimhof T (2018) Mobilität in Deutschland—MiD ergebnisbericht. Studie von infas, DLR, IVT und infas 360 im Auftrag des Bundesministeriums für Verkehr und digitale Infrastruktur (FE-Nr. 70.904/15). https://www.mobilitaet-in-deutschland.de/pdf/MiD2017_Ergebnisbericht.pdf

15. Statistische Ämter des Bundes und der Länder (2015) Zensus 2011. https://www.zensus2011.de/DE/Home/Aktuelles/DemografischeGrunddaten.html?nn=3065474
16. Ziemke D, Kaddoura I, Nagel K (2019) The MATSim open berlin scenario: A multimodal agent-based transport simulation scenario based on synthetic demand modeling and open data. Procedia Comput Sci 151: 870–877. https://doi.org/10.1016/j.procs.2019.04.120

Simulation of Car-Sharing Pricing and Its Impacts on Public Transport: Kyoto Case Study

Yihe Zhou, Riccardo Iacobucci, Jan-Dirk Schmöcker, and Tadashi Yamada

Abstract The wide-ranging implementation of car-sharing has led to many positive impacts but also has been a controversial issue as it can cause congestion if it replaces public transport trips. Therefore, we are aiming to investigate the competition and cooperation between car-sharing and public transport with a focus on the car-sharing pricing policy. We consider shared autonomous vehicles (SAVs) for which a range of pricing scenarios are possible and simulate different scenarios with a given public transport networks. An important aspect of SAVs is the relocation strategy to cater future demand optimally and this is considered in our simulation. We consider distance specific pricing, as well as origin–destination specific pricing for car-sharing. We create a model of Kyoto city with the real-world subway lines and trip data and simulate choices between public transport and SAVs. The performance of the whole network is evaluated by the cost for passengers, profit for the car-sharing operator as well as the spatial distribution of car-sharing prices and its modal share. Results show that if a car-sharing service is introduced in the network, cost for passengers can be decreased. The benefit is spatially heterogenous and depends on the public transport network. A main result is that the benefit is significantly reduced if the SAV operator can pick prices for all ODs freely instead of being bound to a distance-based fare. Hence, it is necessary for city planners to introduce some pricing regulation, in order to balance and maximize the benefits for both passengers and car-sharing operators.

Keywords Shared autonomous vehicles · Mode choice · Transit network · Pricing

1 Introduction

Car-sharing services have expanded rapidly during recent years worldwide. According to an estimation of the global car-sharing market trends, the global fleet size of car-sharing has increased from 11,501 to 198,418 vehicles, that is increased

Y. Zhou (✉) · R. Iacobucci · J.-D. Schmöcker · T. Yamada
Department of Urban Management, Kyoto University, Kyoto, Japan
e-mail: zhou.yihe.55r@st.kyoto-u.ac.jp

C. Antoniou et al. (eds.), *Proceedings of the 12th International Scientific Conference on Mobility and Transport*, Lecture Notes in Mobility,
https://doi.org/10.1007/978-981-19-8361-0_8

17 times, during 2006–2018 [1]. In Japan, the fleet size of car-sharing was estimated at about 33,000 vehicles in 2018 [2] and these are mainly sharing services where the vehicles have to be dropped off at the same location where they have been picked up. After the emergence of car-sharing services, the impact on public transport has been a critical issue which has been studied by many researchers. Some studies concluded positive effects while others take the opposite point of view (e.g. [3, 4, 5, 6]). With regard to positive effects, a car-sharing service can complement public transit as it can solve last mile problems and take over some non-profitable trips from public transport. Car-sharing can also increase ridership of public transport by attracting users from private cars to mixed mode trips such as combining car-sharing and metro lines. On the other hand, competition between public transport and car-sharing is regarded to be a negative effect since car-sharing attracts regular users of public transport leading to the reduction of public transport ridership and the increasing number of vehicles can exacerbate congestion and cause longer travel times for public transport [7]. The impact of car-sharing and other on-demand public transport services varies with different city types [8, 9]. Moreover, some researchers stated that increasing car-sharing usage causes congestion. It has been found that ride sharing services such as Uber and Lyft add mileage to city streets, which in some cases can be signficant [7].

Therefore, we are aiming to investigate the competition and cooperation between car-sharing and public transport, especially to reduce the travel cost of passengers. Further, the effect of different pricing strategies of car-sharing is investigated by considering both fixed distance-based pricing and flexible origin–destination (OD) based pricing. In our work, car-sharing vehicles are considered as relocated shared autonomous cars which means vehicles can pick up users anywhere in the service area and users do not need to return the vehicles to the parking station. Hence, this service is similar to taxi but the cost will be lower [10]. Our study will show that the pricing flexibility of the shared service can have a profound impact on public transport usage as well as user benefits.

2 Literature Review

There is a large body of literature studying factors determining car-sharing demand (e.g. [11, 12, 13]). A comprehensive literature review of methods used to study car-sharing systems is presented by Jorge and Correia [14]. Among others it is found that population size and the share of car-sharing members have positive feedback impacts on car-sharing usage [11]. Also relevant for our work is the finding that passengers with discretionary trips are more willing to use free-floating car-sharing. Compared to station-based car-sharing, free-floating car-sharing can better bridge gaps in the public transport network. Ogata et al [13] conducted a survey on mode choices between car-sharing, public transport and private cars with the option to also combine the modes. Then they develop a choice model combining revealed preference and stated preference information. The results show that greater transportation inconvenience and less cost of car-sharing are important characteristics. Moreover, user experience also has a positive impact on mode choices. Zhang et al. [15] investigate the dynamics of car-sharing service adoption during the initial years with a focus on new adopters. The potential demand of stations is also forecasted considering the synergistic effect between stations and the importance of strategic locations in the city. A spatial difference in the adoption process is found depending on the predominant land-use in the area.

Also, the impact of car-sharing on private car usage has been broadly investigated. The early-stage impact of free-floating car-sharing (FFCS) on private car ownership is investigated through a survey in the FFCS service area in London [16]. The impacts include deterrence of buying a car and disposal or sale of the owned car. Customers who use the FFCS service as a substitute for other transportation modes are less likely to own a car, whereas users whose trip purpose is to attend business meetings are more likely to remain car owners. Furthermore, high bus service frequency and FFCS usage has a positive relation with car ownership impact, while the opposite result is found for national rail trains and private cars. The survey-based study by Haustein [17] also concluded that there is a positive impact of FFCS membership on car ownership decline.

There are less studies discussing public transport network issues and car-sharing. The relation between public transport and car-sharing is investigated by Jin and Schmöcker [8] considering shuttle and feeder modes of car-sharing in a relatively simple virtual network with a north–south and east–west public traffic line. Their study showed how shared modes contribute to lower social cost. Further, as for the difference between two shared modes, they discussed that a feeder mode can be beneficial in most newly developed commuter towns, including dense towns, while shuttle services are more appropriate in larger towns. A limitation of their study is the assumption of a uniform demand distribution. There are also several studies of the impact of car-sharing on public transport based on usage statistics. According to the studies of US cities, the introduction of Transportation Network Companies such as Uber and Lyft induce a decline of rail and bus ridership [3, 4]. Further, the effect of shared autonomous vehicles (SAV) on private car and public transport is investigated by Iacobucci et al. [18]. The results show the decline of both car ownership and public transport ridership if the SAV service is unregulated.

The performance of a car-sharing system highly depends on the imbalance of vehicles, thus a proper relocation strategy is crucial. An integrated framework is proposed for investigating the effect of temporal and spatial flexibility on the system performance which suggests that flexibility has positive impact on the profitability of the system [19]. There are studies focusing on the optimization framework of one-way car-sharing considering relocation operations [20, 21]. A user-based relocation by incentives is developed by Stokkink and Geroliminis [22].

Further, a proper pricing strategy is crucial for attracting more users. It is found that the price structure has an impact on the car-sharing demand as the minimal usage occurs in the scenario of normal price; i.e., current price in the carsharing service area, while half-price leads to maximal usage [12]. Particular pricing strategies of car-sharing services also have been explored in recent contributions. There are several studies focusing on the pricing schemes with the aim of improving the spatial vehicle distribution [23, 24]. The price of ride sourcing services is optimized by considering its negative impact caused by congestion in multimodal transportation networks [25].

The potential of using autonomous vehicles (AVS) in sharing mode is highlighted by Mounce and Nelson [26]. The development of AVS has been happening since the 1980s as many companies such as Google and Volvo have started testing AV systems [27]. The advantages of AVS are that they can ameliorate waiting time and parking, as well as lead to a more regulated and predictable urban traffic network. De Almeida Correia and van Arem [28] suggest that AVs will lead to reduced generalized transport costs, more trip satisfaction with increased traffic congestion. Moreover, SAVs can be used as the link to public transport particularly for rural area. The cost advantages of car-sharing over taxi are investigated by Dong et al. [10] since these two modes are highly similar and compete for the same market. The study finds that for medium and long trips, the travel-cost advantages of car-sharing over taxi is larger than for short trips and that car-sharing is most attractive for neither very long nor very short trips. Moreover, the travel-cost advantages of car-sharing are primarily observed before rush hours on weekdays evenings and on weekends except after 21:00.

In the subsequent study we aim to pick up on some of the gaps that we perceive exist in the literature. For one, the impact of free-floating car-sharing with relocated autonomous vehicles needs to be evaluated from the viewpoint of passengers, car-sharing and public transport operators by using a mode choice model including the public transport network and the car-sharing service. Moreover, there is a lack of simulations with real trip data on the above-mentioned aspects. In addition, our main contribution is that we compare two pricing strategies for the car-sharing service. Origin–destination specific pricing is the unregulated pricing option whereas in the distance-specific pricing option, the SAV pricing follows taxi regulations.

3 Methodology

3.1 Notation

The notation used in this study is shown in Table 1.

3.2 Base Model

3.2.1 Model

The simulation is conducted based on the model used in Iacobucci et al.'s research and in this section we summarize the main features of this base model [29]. There

Table 1 Notation

Notation	Unit	Meanings
$p_C^{ij}(t)$	%	Probability of choosing car-sharing from node *i* to *j* in the coming time horizon *t*
C_P^{ij}	$	Cost of public transport of one trip from node *i* to *j*
$C_C^{ij}(t)$	$	Cost of car-sharing of one trip from node *i* to *j* in time period *t*
$n_{ij}(t)$	trip	Predicted demand for car-sharing from node *i* to *j* in time period *t*
$\hat{n}_{ij}(t)$	trip	Latent demand from node *i* to *j* in time period *t*
b	$/min	Static base tariff of car-sharing
r	$/min	Operational cost of a car-sharing vehicle
$b_{ij}(t)$	$/min	Dynamic price of car-sharing for a trip from node *i* to *j* in time period *t*
b_0	$	Minimal fare of car-sharing
d_{ij}	min	Distance between nodes *i* and *j*
$q_{ij}(t)$	trip	Predicted relocation flow of cars from node *i* to *j* in time period *t*
VOT	$/h	Value of time
F_s^{ij}	$	Subway fare from node *i* to *j*
F_b	$	Bus fare (assumed to be a flat fare)
T_s^{ij}	hour	Travel time by subway from node *i* to *j*
T_b^{ij}	hour	Bus travel time from node *i* to *j*
T_w^{ij}	hour	Walking time from node *i* to *j*
TC_p	$	Total cost of passengers who take public transport
TC_c	$	Total cost of passengers who take car-sharing
TC	$	Total travel cost of all passengers
$\overline{TC}$	$	Average total travel cost
R	$	Revenue of the car-sharing operator
L	$	Total relocation cost of the car-sharing operator
P	$	Total profit of the car-sharing operator
N_C	trip	Total number of trips choosing car-sharing
N	trip	Total number of trips
M_C	%	Modal share of car-sharing
N_C^i	trip	The number of trips at node *i* choosing car-sharing
N^i	trip	The number of trips at node *i*
M_C^i	%	Modal share of car-sharing at node *i*

are mainly three modules, pricing optimization, relocation optimization and trip assignment. In the model, trip demand is integrated into nodes, and the model uses OD trip demand data as inputs including OD node pairs and starting time of trips. The distance between each node is calculated in advance. The model uses a relocation strategy based on [18] and the relocation optimization is performed every 10 min. The model uses predicted demand and vehicle locations in a given time slot as input data and outputs relocation actions for future time slots. The strategy decomposes the problem into three simpler stages. The first stage is the prediction of inventory imbalance which means the difference between the needed numbers of vehicles and the available number of vehicles, aiming to transfer vehicles from nodes with positive imbalance (feeder nodes) to nodes with negative imbalance (receiver nodes) to maintain balance between nodes. The second stage is using an optimization model to design the matching pairs of feeders and receivers as well as to decide the number of vehicles needed to be relocated with the aim of reducing the imbalance of inventory and minimizing relocation distance. In the final stage, the simulator conducts the relocation tasks assuming that the vehicles can be transferred as soon as they are available at feeder nodes.

Dynamic prices of car-sharing for the next time interval are optimized based on expected demand as relocation costs need to be considered. The detailed pricing strategy will be discussed in Sect. 3.2.2. The simulator will provide prices of public transport and car-sharing as well as estimated waiting times of the passenger and if the request is accepted by passengers, the trip assignment module will allocate vehicles to each trip.

Mode choice of each trip is based on the probability which is calculated using a binary logit model according to the cost of each mode. The probability of choosing car-sharing for one trip is shown as Eq. (1).

$$p_C^{ij}(t) = \frac{\exp\left(-C_C^{ij}(t)\right)}{\exp\left(-C_C^{ij}(t)\right) + \exp\left(-C_P^{ij}(t)\right)} \tag{1}$$

Here, C_C is the cost if a passenger chooses car-sharing while C_p is the cost of public transport. We note that due to the logit structure the model becomes stochastic even though a unique optimal re-location strategy is found. From the choice probability, expected demand of car-sharing, n_{ij}, can be calculated by multiplying the probability with latent demand $\hat{n}_{ij}$, that is, the total trips in the coming time horizon t are:

$$n_{ij}(t) = p_C^{ij}(t)\hat{n}_{ij}(t) \tag{2}$$

3.2.2 Pricing Strategy of Car-Sharing

In general, a range of pricing structures are possible. The prices can differ depending on time of day, traffic conditions or be use specific and depending among others on when and how a trip was booked. We consider only temporal as well as spatial aspects, with a focus on the latter. Within spatial pricing there are mainly two pricing strategies of interest for car-sharing. The first one is distance specific pricing as common for taxi services. Here the price of car-sharing is calculated by multiplying the distance between OD nodes with a static base tariff as shown by Eq. (3).

$$c_C^{ij} = bd^{ij} \tag{3}$$

Carsharing operators might have, however, additional flexibility. This leads to a second type of strategy which we call "dynamic pricing". Here the price per kilometer for each OD pair depends on the time of day and the OD node as in (4) where t is the time and i, j are origin and destination node respectively.

$$b_{ij}(t) = f(t, i, j) \tag{4}$$

Dynamic price is optimized to maximize the net profit, P, between total revenue, R, and total relocation cost, L, in the coming time horizon with the constraints of vehicle distribution based on the imbalance of each node. The optimization problem is shown as Eq. (5). The first two terms describe the revenue by subtracting the cost of moving vehicles from the total revenue. The third term describes the relocation costs.

$$\max_{B(t)} P = R - L = \sum_{i,j} b_{ij}(t) n_{ij}(t) d_{ij} - \sum_{i,j} r n_{ij}(t) d_{ij} - \sum_{i,j} r q_{ij}(t) d_{ij} \forall t \tag{5}$$

In Eq. (5) $B(t) = \left[b_{ij}(t)\right]$ is denoting the matrix of optimal prices to be found for each zone pair ij. If the entries of $\boldsymbol{B}$ (t) are not restricted this describes the (spatially) dynamic pricing and the case where $\boldsymbol{B}(t)$ is restricted to a multiplier of the distance between zones this describes the fixed "distance-based" pricing. $n_{ij}(t)$, and $q_{ij}(t)$ denote the time dependent predicted demand and relocation flows respectively. Relocation cost per min, r, as well as travel distance d_{ij} are considered to be time independent. We also set a fixed base fare b_0 so that all entries in B are larger than this minimal fare.

Further, it is assumed that demand depends on the price as higher pricing leads to lower demand and lower price increases demand. Hence, besides OD pair and time, the optimized price also depends on the price of the alternative mode. This means if the alternative mode price is higher, the model can adjust the car-sharing price to be higher in order to increase revenues. For the price optimisation model, in order to reduce the difficulty and to obtain a scalable method, the estimated demand for each OD pair is approximated as a linear function of the price. In our study, price optimization is performed every 10 min.

3.3 Extension to Non-uniform Public Transport Network and Improved Choice Model

In the study with the base model [29], besides car-sharing service, alternative modes are presumed to be spatially uniformly distributed and the cost is calculated by multiplying distance with a unit price. In this study the public transport network and choice model are improved by explicitly describing the public transport.

We consider a nested choice structure where passengers first inform themselves about the best public transport option for their OD relation. Then a binary logit model is applied for the choice between carsharing and the best public transport option considering on-board costs, walking time, waiting time and fare. As to the mode choice, we consider subway, bus and walking besides car-sharing. In this study our main purpose is to investigate the spatial difference of car-sharing pricing impacts on the traffic network considering the inequalities of public transport access. For instance, travel may be inconvenient in peripheral area of a city and on contrary travelling within the city center are is much more convenient. Therefore, adding other mode choices such as taxi would weaken the inequalities we show in this paper.

The distance between each node and between nodes and subway stops are calculated and the closest subway stop for both origin node and destination node of one trip is found. Passengers can either take bus or walk from the origin/destination node to the subway stop. Then, based on these settings, we can obtain the generalized cost of two public transport choices with Eq. (6). We select the one with minimal cost for passengers as the best public transport option. Then the cost of this public transport mode is substituted into Eq. (1) and compared with the car-sharing cost. The procedure of mode choice is shown in Fig. 1.

$$C_P = \begin{cases} F_s + F_b + (T_s + T_b) * \text{VOT} & \text{subway} + \text{bus} \\ F_s + (T_s + T_w) * \text{VOT} & \text{subway} + \text{walking} \end{cases} \tag{6}$$

Here, F_s and F_b mean subway fare and bus fare respectively. Total time needed for traveling by subway and bus are represented by T_s and T_b respectively. T_w indicates the walking time from an OD node to a subway stop if passengers choose to use subway with walking.

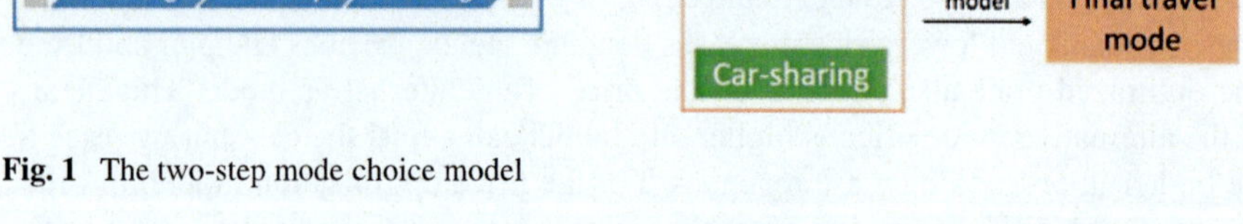

Fig. 1 The two-step mode choice model

3.4 Evaluation

Firstly, we evaluate the generalized cost of passengers. We consider the total cost for travellers based on the mode they chose. To do so, we obtain the public transport cost TC_p as the total cost for those who chose public transport as their traveling mode, as well as the total cost of car-sharing TC_C for those who chose car-sharing. The overall cost for all passengers is then obtained as the sum as in Eq. (7). Based on mode choice, we can estimate the average travel cost $\overline{TC}$ as in Eq. (8). Moreover, we consider the outcomes from the car-sharing operator perspective by evaluating revenues, relocation cost and profit for two types of pricing strategy. Profit is calculated by subtracting relocation cost from revenue as in Eq. (9). In addition, we also estimated the modal share of car-sharing M_c as Eq. (10). Furthermore, we consider the spatial difference of car-sharing price C_C^{ij} and modal share M_C^i based on average value of trips originating from each node as shown by Eqs. (11) and (12).

$$TC = TC_P + TC_C \tag{7}$$

$$\overline{TC} = \frac{TC}{N} \tag{8}$$

$$P = R - L \tag{9}$$

$$M_C = \frac{N_C}{N} \tag{10}$$

$$C_C^{ij} = \frac{\sum_j C_C^{ij}}{N^i} \tag{11}$$

$$M_C^i = \frac{N_C^i}{N^i} \tag{12}$$

To sum up, the framework of our methodology is shown in Fig. 2. To note is that we consider the public transport costs as fixed. The inputs of the model are further trip data including OD pair and starting time. This information is used for estimating latent demand of the coming time horizon and based on the latent demand, the optimized dynamic price of car-sharing can be obtained with Eq. (5). Also, iteratively price affects demand as shown by Eqs. (1) and (2). Then the costs of public transport and car-sharing are substituted into the mode choice module and accordingly trip assignment is conducted. Based on the needed vehicles of each node, relocation is employed which is also the constraint for the price optimization. After the assignment is completed, the performance is evaluated by the measures derived with Eqs. (6)–(12).

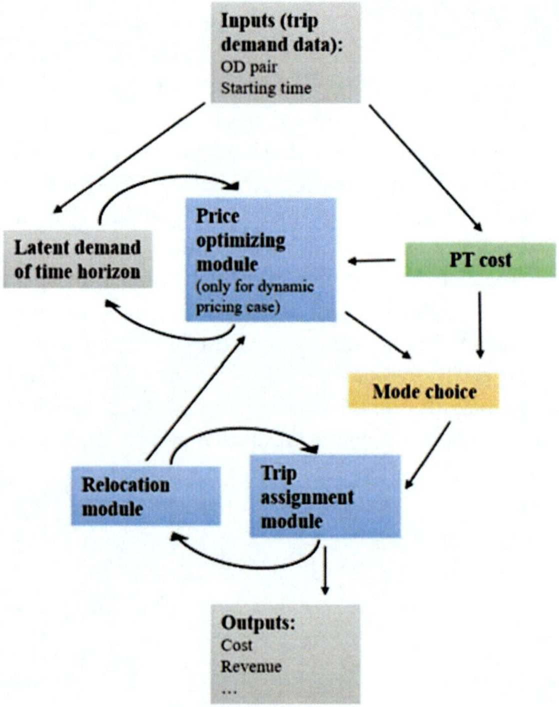

Fig. 2 Simulation framework

4 Kyoto City Case Study

4.1 Introduction

The model is applied to Kyoto city. A map of Kyoto city with its main landmarks indicated as blue diamonds is shown in Fig. 3. The city is surrounded by mountains in the north as well as East and West meaning that the city boundaries are fairly clearly defined. As to the public transport system, there are two subway lines, Karasuma line in north–south direction and Tozai line in east–west direction. The bus system is the main public transport mode and spreads over the whole Kyoto area. Moreover, we note that currently there is no car-sharing service operated.

Fig. 3 Kyoto city with some of its main landmarks (diamonds) and subway stops (triangles)

4.2 Data

4.2.1 Mobile Spatial Statistics

"Docomo" is a major mobile phone service provider in Japan and we obtained aggregated spatial statistics regarding flows between different mesh areas. The data information includes date and time of departure and arrival, as well as the OD node. The data consists of the number of trips moving between areas for each hour, however, if there are only a few trips from one mesh, the data are not provided for privacy reasons. The data available to us includes trips moving between grid meshes of size 500 m^2. We also note that only those trips are recorded where the stay time at the destination is more than one hour so that short time, and often short distance, trips will be underestimated.

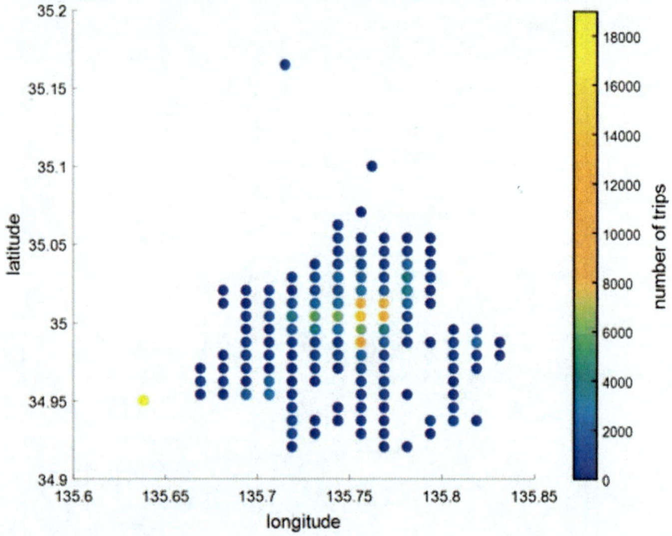

Fig. 4 Number of trips originating from each node

4.2.2 Preparation of Demand Data

The 1057 meshes that constitute Kyoto city are aggregated into 196 nodes for which we have data (large parts of the area are mountainous with few trips). We take the trip data of a "normal" Wednesday in 2016 (19.10.2016) and consider the trips between these nodes. This means that through traffic and trips to other nodes is omitted. Furthermore, we also remove the trips which start and arrive at the same node. As a result, there remain in total 199,840 trips. The data record the amount of movements every hour, however in this model the time interval is in minute, thus it is needed to convert the starting time to minute by dividing trips within each hour into minutes uniformly. It is noteworthy that since we have removed some trips there are 57 nodes from which there is no trip originating. Therefore, in the following discussion, we only consider 139 nodes. The number of trips originating from each node is shown in Fig. 4. As can be expected we can find that there are more trips in the central part of Kyoto.

4.3 Network

We now describe the remaining simulation settings. The base tariff of the car-sharing service is set as \$0.3 per min which is lower than the taxi tariff of Japan, and relocation cost is set as \$0.1 per min. Minimal fare b_0 is assumed as \$5. In the network, we consider two public transport lines which are the subway system of Kyoto city, Karasuma line in north–south direction and Tozai line in east–west direction. In total, there are 31 stops shown as red triangles in Fig. 5 and passengers can transfer between two lines. Given the latitude and longitude of nodes and stops, distance between each node as well as between nodes and stops are calculated by the Haversine formula. The fare is referring to the actual value between stops according to the fare table of Kyoto city. Note that we only consider single ticket fares for public transport. Travel time is also obtained from published information and includes waiting time considering an average service of headway of 7 min and random passenger arrivals. With regard to the bus system, there is a flat fare system in the central part of Kyoto with a charge of Yen 230 (around US$2.3). Travel time is determined by distance given an estimate of 8 km/h taking boarding/alighting times into consideration. We presume the bus system equally spreads over the whole area thus, passengers do not need to wait or go to bus stops which is reasonable due to the extensive bus network in Kyoto. Hence, the bus system in our study is more similar to walking with a higher speed and fare. Therefore, this assumption is favorable to public transport. Walking speed is set as 4 km/h. All parameters for public transport mode are summarized in Table 2. With regard to value of time, we set it as 15$/h which is reasonable considering the income and salaries in Kyoto. Figure 6 shows the average cost of public transport for all trips originating from each node. It can be found that along two public transport lines, the price is relatively lower. And the highest cost happens at left bottom corner and top side which are far from stops. In the right bottom, there is a green dot whose price is higher than surrounding nodes. In the trip data, trips originating from this node only arrive at one node which locates at the left bottom; hence, the price is higher.

4.4 Results

We firstly consider the general evaluation from the viewpoint of passengers as shown in Table 3. As described in Sect. 3.2.1, the model is stochastic due to the logit model, thus we show the results with a range in Tables 3 and 4 by running the program for six times. From the result we can conclude that both pricing scenarios with the car-sharing service can decrease the total cost of passengers compared to the base scenario which means passengers can save cost if the car-sharing service is operated in the network. Especially, distance-specific pricing of the car-sharing service is more beneficial for passengers despite lower modal share which enables the cost to reduce significantly due to its lower price. Dynamic pricing is far less beneficial to passengers compared to distance specific pricing. The reason is that the objective function is to maximize the profit of the car-sharing operator, so that some non-profitable trips, such as trips in the periphery which are likely inducing relocation movements, will not be served at a beneficial price for travelers. The general evaluation for the car-sharing service is shown in Table 4. The table confirms that the profit of dynamic pricing is higher than that of distance-specific pricing since this strategy prices each OD just below the general cost of a public transport trip.

The spatial distribution of costs encountered by travelers for car-sharing are illustrated in Figs. 7 and 8. We remind that for distance pricing the price/km is the same

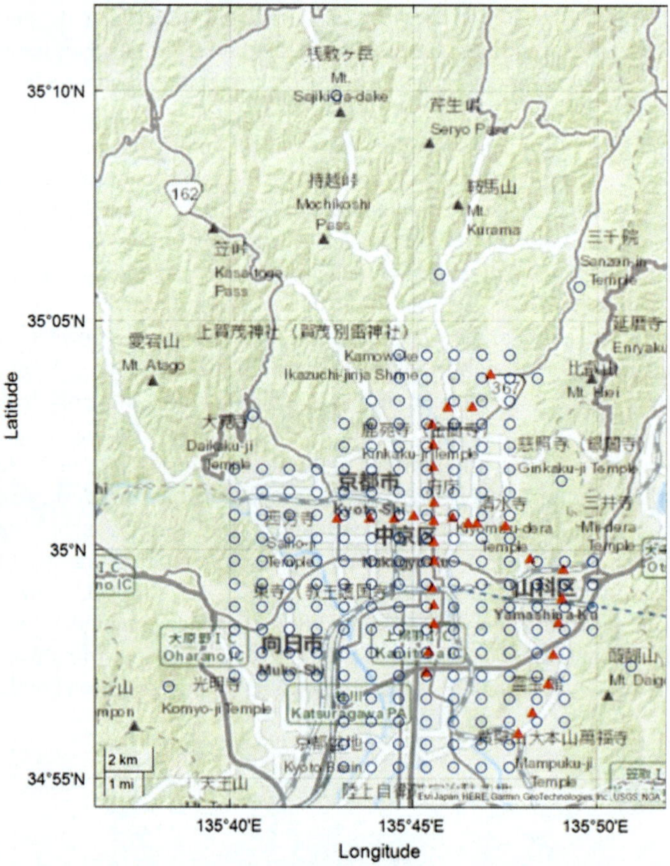

Fig. 5 The 196 nodes and 31 subway stops considered in the study

Table 2 Parameters of public transport modes

	Fare (F)	Total time (T)	Access
Subway (s)	Real fare between stations	Real travel time + Assumed waiting time	Find the closest station
Bus (b)	\$2.24 (JPY230)	8 km/h	Equally spread over the whole area
Walking (w)	0	4 km/h	

independent of location so that the effects shown in Fig. 7 reflect the distanced travelled by carsharing. Costs along the subway lines are in generally lower, reflecting that passengers near to the subway do not benefit as much from carsharing but generally the differences in carsharing usage and cost are fairly homogeneous except for some location at the peripheries where travel distances to most destinations are long.

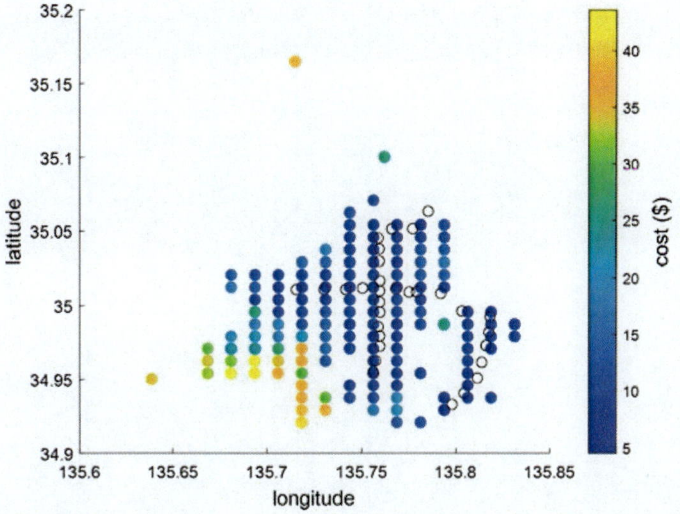

Fig. 6 Cost of public transport

The situation is different in Fig. 8. With dynamic pricing the south-west corner of Kyoto experiences high cost for carsharing. The area is not well served by public transport and the carsharing operator is taking advantage of this by charging more. Further, along the subway lines the price is particularly low, so that the car operator does not lose all demand for trips starting from these nodes.

To illustrate these points further, Figs. 9 and 10 show the modal share in terms car-sharing usage. In the distance-based pricing car-sharing usage is very low along the subway lines. In the dynamic pricing case the market share is slightly higher, as the operator wants to secure some market share for this attractive market since these are often trips within the city center that do not require relocating the vehicles. Further, in the south-west corner the carsharing usage is particular high as it is an attractive alternative to the bus. In many nodes the car-sharing percentage even exceeds 90%. In the dynamic pricing the residents in this part of Kyoto do not benefit as much from the introduction of car-sharing. To note is also that in both scenarios nodes that are too far, such as the outlier in the north of Kyoto, do not use carsharing much confirming some of the discussing in the literature review regarding appropriate distances for carsharing being those of "mid-size distances".

4.5 *Sensitivity: Demand Distribution, Base Tariff and Fleet Size*

To investigate the sensitivity with respect to the demand distribution, we modify the real trip data (Scenario R). We compare this case to the case of uniform demand

Table 3 Total and average cost for passengers by mode

	TC_P($)	TC_c ($)	TC ($)	$\overline{TC}$ ($)
Without car-sharing	2,822,950		2,822,950	14.126
Distance-specific pricing	1,617,829−1,631,711	728,520−741,516	2,355,081−2,370,023	11.785−11.860
Dynamic pricing	2,039,005−2,059,421	750,715−768,173	2,806,741−2,810,137	14.045−14.062

Table 4 Revenues, relocation costs, profits and model share of the car-sharing service

	R($)	L($)	P($)	M_C
Distance-specific	254,686–259,771	25,096–25,330	229,557–234,545	28.6–29.18%
Dynamic	591,770 – 606,609	17,231–17,441	574,505–589,185	34.41– 35.69%

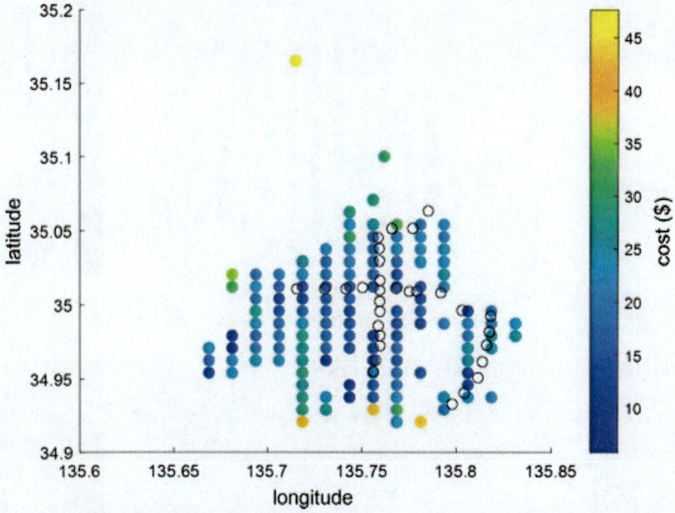

Fig. 7 Distance specific price of car-sharing

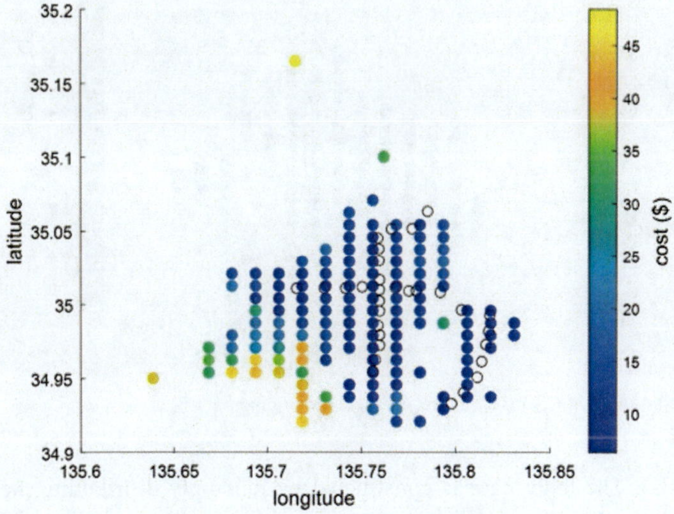

Fig. 8 Dynamic, origin–destination specific price of car-sharing

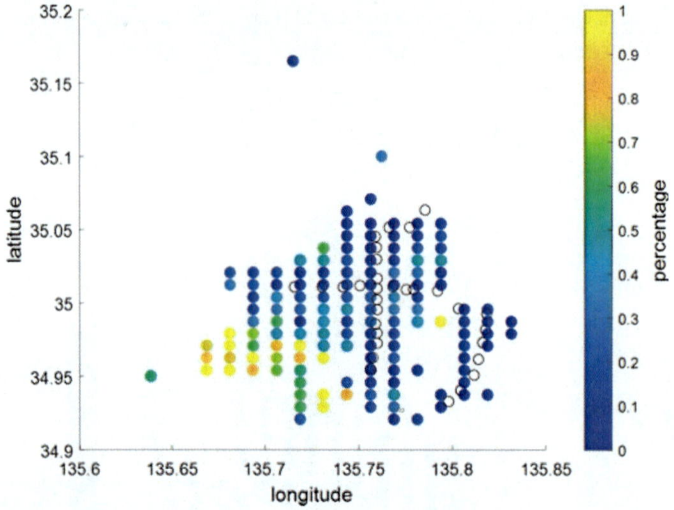

Fig. 9 Modal share of CS for distance specific pricing

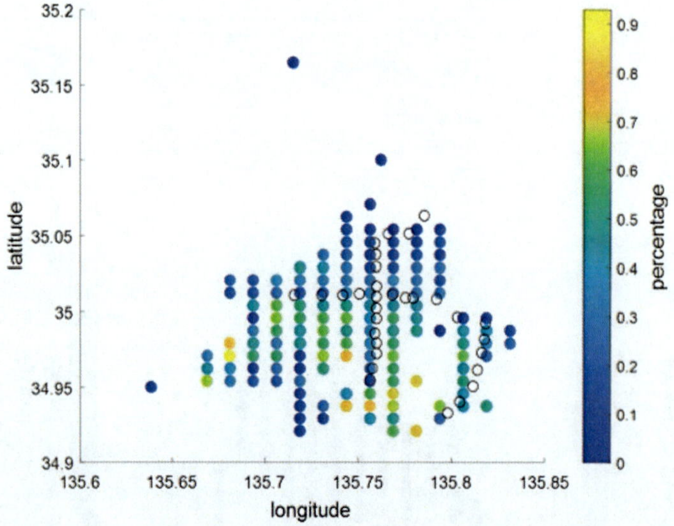

Fig. 10 Modal share of CS for dynamic, OD specific pricing

(Scenario U). The latter case is constructed by randomly distributing the origins and destinations of all trips. We further create an "intermediate case" where we use half the randomly created trips and half the observed trips (Scenario (R + U)/2). Table 5 shows the average trip length which increases as demand becomes uniform distribution. This indicates that uniform distribution tends to have longer average trip length, and as a result average cost of passengers will also increase.

Table 5 Average trip length depending on demand distribution

Distance specific pricing			Dynamic pricing		
R	(R + U)/2	U	R	(R + U)/2	U
3.1994	4.3533	5.5763	1.4463	1.8946	3.4500

We test these trip data with different fleet sizes and base tariffs of car-sharing. However, with the strategy of dynamic pricing, the price of car-sharing depends on the alternative modes and the time, therefore with regard to the sensitivity of the base tariff case we only investigate its impact for the distance specific pricing scenario. We set the three levels for fleet size as 500, 1000, 2000 and the four levels for the base tariff as 0.3, 0.5, 1, 2 $/min. The resulting average travel cost of passengers for the three types of demand distribution are shown in Table 6 and for different base fares in Table 7.

If we compare the average travel cost for different fleet sizes, we can find that a larger fleet size tends to reduce the travel cost. A reason is that larger fleets mean shorter waiting time which means time cost can be reduced. Table 7 shows that a higher base tariff in general increases average cost but interestingly we observe a price increase when the base traffic reduces from 0.5 to 0.3. On closer inspection we find that this is due to the fact that longer trips are less sensitive to the base tariff. Hence, when the base tariff is low there will be more long trips choosing car-sharing which take more time and leads to longer waiting time. Overall, we conclude hence that a too low base fare can lead to an inefficient use of the car-sharing fleet from a social welfare perspective.

Table 6 Average generalized cost depending on demand distribution and fleet size (b = 0.3$/min)

m	Distance specific pricing			Dynamic pricing		
	R	(R + U)/2	U	R	(R + U)/2	U
500	13.349	15.574	18.173	14.215	16.594	19.026
1000	11.802	14.764	17.534	14.062	16.508	19.010
2000	9.807	13.363	16.397	13.836	16.388	19.005
Without CS	14.126	16.550	18.974			

Table 7 Average generalized cost depending on demand distribution and base tariff (m = 1000)

b	Distance specific pricing		
	R	(R + U)/2	U
0.3	11.802	14.764	17.534
0.5	11.666	14.637	17.467
1	11.772	14.511	17.518
2	12.482	15.083	18.369

5 Discussion and Conclusion

We investigated the competition and cooperation between a car-sharing service and public transport by testing a range of scenarios. We implemented our model to Kyoto city with the real-world subway lines and trip data. Our focus was the effect of the two kinds of car-sharing pricing strategy, distance-specific pricing and dynamic pricing. We firstly evaluated the generalized performance of the whole network from the perspective of passengers and the car-sharing operator. Then we estimated how the price and modal share of car-sharing varies in different areas of the city.

Our main findings are as follows: Firstly, we show how much the car-sharing service can reduce the generalized travel cost for passengers. Especially, distance specific pricing can largely reduce the total travel cost for travelers. Instead, from the perspective of a car-sharing service provider, who would like to maximize the profit, the additional flexibility of dynamic pricing can lead to significantly higher profits. The operator can use dynamic pricing to charge passengers according to their public transport (in-)accessibility. Therefore, in order to balance and maximize the benefits for both passengers and car-sharing providers, a proper pricing regulation is necessary.

Furthermore, the case study illustrated how car-sharing can be used as a substitute for public transport in areas with less accessibility, even though the effect is minor for places too far from the center. Generally, car-sharing does not attract many passengers along nodes with good public transport which is encouraging from a sustainability perspective. In conclusion, also from a sustainability perspective, in order to obtain collaboration instead of competition between car-sharing and public transport, a regulated car-sharing and a good public transport network is required. We finally suggest that our "paradoxical" finding that a lower car-sharing fare can lead to higher generalized cost is worth considering in policy. If an operator is forced to provide services at a too low fare, it can have the effect that trips that are also relatively easy to make by public transport are replaced by car-sharing and take capacity away to serve trips for which the usage of car-sharing is more beneficial.

Clearly, in future we hope to validate our simulation by using cities where car-sharing has already received a significant market share. Another limitation of our study is that trip chaining as well as sociodemographics and other factors influencing willingness to take car-sharing are not considered. Such data could help to add more realism to the mode choice model used in our study. Our work could also be advanced by considering the role of private cars.

References

1. Shaheen S, Cohen A (2020) Innovative mobility: carsharing outlook; carsharing market overview, analysis, and trends, s.l. Transportation sustainability research center, UC Berkeley
2. Statista Research (2022) Corporate type car sharing fleet in Japan FY 2018-2024. Available from: https://www.statista.com/statistics/1030770/japan-corporate-car-sharing-market-size/. [Accessed January 2023]

3. Erhardt GD, Mucci RA, Cooper D, Sana B, Chen M, Castiglione J (2022) Do transportation network companies increase or decrease transit ridership? Empirical evidence from San Francisco. Transp 49:313–342
4. Graehler M, Mucci A, Erhardt GD 2019 Understanding the recent transit ridership decline in major US cities: Service cuts or emerging modes, s.l. In: 98th annual meeting of the transportation research board. Washington DC
5. Henao A, Marshall WE (2019) The impact of ride-hailing on vehicle miles traveled. Transportation 46:2173–2194
6. Shaheen SA, Cohen AP, Chung MS (2009) North American carsharing: 10-Year retrospective. Transp Res Rec 2110(1):35–44
7. Schaller Consulting 2017 Empty seats, full streets fixing Manhattan's traffic problem, Available from: http://schallerconsult.com/rideservices/emptyseats.htm
8. Jin Z, Schmöcker JD (2019) Shared and mass transit operational strategies in radial cities with many-to-one-demand. In:Presented at 3rd international symposium on multimodal transportation. Singapore
9. Luo S, Nie YM (2019) Impact of ride-pooling on the nature of transit network design. Transp Res Part B: Policy Pract 129:175–192
10. Dong X, Cai Y, Cheng J, Hu B, Sun H (2021) Understanding the Competitive Advantages of Car Sharing from the Travel-Cost Perspective. Int J Environ Res Public Health 17(13):4666
11. Becker H, Ciari F, Axhausen KW (2017) Modeling free-floating car-sharing use in Switzerland: A spatial regression and conditional logit approach. Transp Res Part C: Methodol 81:286–299
12. Ciari F, Balac M, Balmer M (2015) Modelling the effect of different pricing schemes on free-floating carsharing travel demand: a test case for Zurich, Switzerland. Transp 42:413–433
13. Ogata R, Schmöcker J-D, Nakamura T and Kuwahara M (2022). On the Potential of Carsharing to Attract Regular Trips of Private Car and Public Transport Users in Metropolitan Areas. Transportation Research Part A 163:386-404
14. Jorge D, Correia G 2013 Carsharing systems demand estimation and defined operations: a literature review. Eur J Transp Infrastruct Res 13 (3)
15. Zhang C, Schmöcker JD, Kuwahara M, Nakamura T 2020 A diffusion model for estimating adoption patterns of a one-way carsharing system in its initial years. s.l.:Transp Res Part A: Policy Pract 136:135–150
16. Le Vine S, Polak J 2019 The impact of free-floating carsharing on car ownership: Early-stage findings from London, s.l. Transp Policy 75:119–127
17. Haustein S (2021) What role does free-floating car sharing play for changes in car ownership? Evidence from longitudinal survey data and population segments in Copenhagen. Travel Behav Soc 24:181–194
18. Iacobucci R, Bruno R, Boldrini C 2022 A multi-stage optimisation approach to design relocation strategies in one-way car-sharing systems with stackable cars. IEEE Trans Intell Transp Systems
19. Boyacı B, Zografos KG (2019) Investigating the effect of temporal and spatial flexibility on the performance of one-way electric carsharing systems. Transp Res Part B: Methodol 129:244–272
20. Boyacı B, Zografos KG, Geroliminis N (2015) An optimization framework for the development of efficient one-way car-sharing systems. Eur J Oper Res 240(3):718–733
21. Boyacı B, Zografos KG, Geroliminis N (2017) An integrated optimization-simulation framework for vehicle and personnel relocations of electric carsharing systems with reservations. Transp Res Part B: Methodol 95:214–237
22. Stokkink P, Geroliminis N (2021) Predictive user-based relocation through incentives in one-way car-sharing systems. Transp Res Part B: Methodol 149:230–249
23. Brendel AB, Brennecke JT, Zapadka P, Kolbe LM 2017 A decision support system for computation of carsharing pricing areas and its influence on vehicle distribution. In: Seoul, Thirty eighth international conference on information systems
24. Jorge D, Molnar G, de Almeida Correia GH (2015) Trip pricing of one-way station-based carsharing networks with zone and time of day price variations. Transp Res Part B: Policy Pract 81:461–482

25. Gómez-Lobo A, Tirachini A, Gutierrez I 2022 Optimal prices for ridesourcing in the presence of taxi, public transport and car competition. Transp Res Part C: Emerg Technol 137
26. Mounce R, Nelson JD 2019 On the potential for one-way electric vehicle car-sharing in future mobility systems. Transp Res Part A: Policy Pract 120:17–30
27. Urmson C, Whittaker W 2008 Self-driving cars and the urban challenge. IEEE Intell Syst 23(2)
28. de Almeida Correia GH, van Arem B (2016) Solving the user optimum privately owned automated vehicles assignment problem (UO-POAVAP): a model to explore the impacts of self-driving vehicles on urban mobility. Transp Res Part B: Methodol 87:64–88
29. Iacobucci R, Bruno R, Schmöcker J-D (2021) An integrated optimisation-simulation framework for scalable smart charging and relocation of shared autonomous electric vehicles. Energies 14(12):1–22
30. Iacobucci R, Pruckner M, Schmöcker JD (2022) A large scale simulation of the electrification effects of SAVs.In: 6th Conference on Sustainable Urban Mobility—CSUM2022, 31 August–2 September. Skiathos, Greece

Analysing Long-Term Effects of the Covid-19 Pandemic on Last-Mile Delivery Traffic Using an Agent-Based Travel Demand Model

Anna Reiffer, Jelle Kübler, Lars Briem, Martin Kagerbauer, and Peter Vortisch

Abstract E-commerce demand has increased steadily over the last decades and this trend has accelerated even more since the start of the Covid-19 pandemic. This entailed that user groups such as older people who previously only shopped in-store were incited to shop online to reduce risk of infection leading some to switch to online shopping as the main shopping channel. This study analyses the long-term effects of increased online shopping and subsequent delivery demand due to the Covid-19 pandemic using an agent-based travel demand model. We analyse the simulation of two scenarios for the model area Karlsruhe, Germany: one scenario simulates the parcel delivery demand before the pandemic and the other scenario simulates the demand during the pandemic of the synthetic population. Our results show that there have been shifts in both socio-demographic characteristics of online shoppers and spatial distribution of parcel delivery demand induced by the Covid-19 pandemic. The scenario simulation based on the pandemic related data shows that not only the influence of income has shifted but also the effects of age on e-commerce activity has changed due to the pandemic. The findings are of interest to transport planners and delivery service providers as they highlight the importance of recognising that the Covid-19 pandemic not only induced a shift in socio-demographic profiles of online shoppers but that this shift also entails a change in the spatial distribution of parcel deliveries.

A. Reiffer (✉) · J. Kübler · L. Briem · M. Kagerbauer · P. Vortisch
Institute for Transport Studies, Karlsruhe Institute of Technology, Kaiserstr. 12, 76131 Karlsruhe, Germany
e-mail: anna.reiffer@kit.edu

C. Antoniou et al. (eds.), *Proceedings of the 12th International Scientific Conference on Mobility and Transport*, Lecture Notes in Mobility,
https://doi.org/10.1007/978-981-19-8361-0_9

1 Introduction

E-commerce is one of the fastest growing market segments and online shopping has increased steadily over the last years [1]. This trend has accelerated even more since the start of the Covid-19 pandemic, as policy measures included temporary closings of retail stores and general stay-at-home recommendations to decrease the spread of the virus. While many of the behavioural changes induced by the pandemic are expected to return to normal, the sudden disruption of daily routines may have lead people to change their habits which are usually hard to break out of [2]. In the case of e-commerce, this entailed that user groups such as older people who previously only shopped in-store were incited to shop online to reduce risk of infection leading some to switch to online shopping as their main shopping channel [3, 4]. While the Covid-19 pandemic has highlighted generational differences in consumer shopping behaviour [5], it forced consumers to adopt new technologies in order to reduce the risk of infection [6]. Although many studies highlight managerial implications of the changes in consumption patterns, no study has analysed the effect of shifts in e-commerce activity induced by the pandemic on parcel delivery demand.

This study analyses the long-term effects of increased online shopping and subsequent delivery demand due to the Covid-19 pandemic using the agent-based travel demand model *mobiTopp* [7, 8] and the urban logistics model extension *logiTopp* [9, 10]. To analyse the long-term effects of the Covid-19 pandemic on e-commerce demand and subsequent delivery traffic, we simulated two scenarios for the model area Karlsruhe, Germany with a synthetic population of 303,809 agents: one scenario simulates the parcel delivery demand before the pandemic and the other scenario simulates the demand during the pandemic. The travel demand for both scenarios is based on pre-pandemic data.

The rest of this paper is structured as follows: We first provide an overview over the data on which our analyses are based. Subsequently, we present the modelling framework mobiTopp and it's last mile logistics extension logiTopp. We go on to present and discuss the results of the scenario simulations and conclude the paper with overarching implications.

2 Materials and Methods

In this section, we provide an overview over the data used to model the behavioural differences in online shopping behaviour before and during the Covid-19 pandemic. Subsequently, we describe the modelling framework we used for our analyses.

2.1 Data

The data for the parcel order and destination choice models was collected using a survey among participants in Germany. Participants were recruited through an online access panel and the sample was stratified by age, gender, and population of the place of residence. The survey was completed by a net sample of 1,002 participants. The survey included questions regarding

- socio-demographic information,
- mode choice behaviour,
- weekly activity patterns, and
- online shopping.

For all questions except those regarding socio-demographics, we asked participants to reflect on their behaviour during the Covid-19 pandemic and retrospectively provide information of the time before the pandemic. For the purpose of this study, we focused on the socio-demographic variables and changes in online shopping behaviour.

Figure 1 shows the changes in online shopping frequency from the time before the pandemic to the time during the pandemic. From the graph we can see, that respondents considerably increased the frequency with which they shopped online. Overall, only a minority (8%) of respondents decreased the frequency of online shopping, whereas a third increased the frequency. While the share of respondents shopping online on a daily basis only increased by 1%, a considerable change can be detected for those who shopped online on a weekly basis: Before the Covid-19

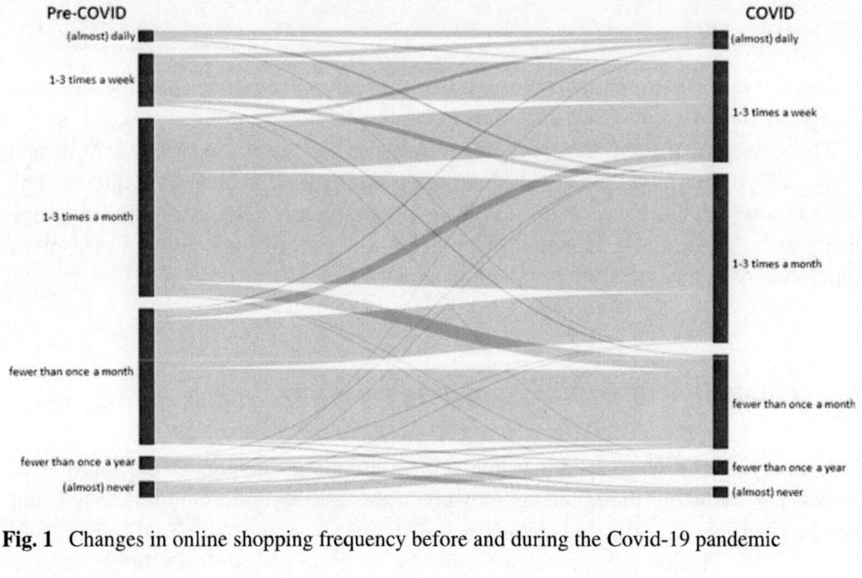

Fig. 1 Changes in online shopping frequency before and during the Covid-19 pandemic

Table 1 Socio-demographic information of online shoppers (OS) versus non-OS pre-covid and during covid

	Survey (%)	Pre-covid		During covid	
		OS (%)	Non-OS (%)	OS (%)	Non-OS (%)
Gender					
Female	50	38	62	49	51
Male	50	38	62	51	49
Age					
Under 25 years old	10	53	47	62	38
25—under 45 years old	31	47	53	61	39
45—under 65 years old	41	33	67	44	56
Over 65 years old	18	25	75	38	62
Income					
Below 2,500 €	45	33	67	44	56
2,500—below 4,000 €	26	38	62	52	48
4,000 and more	28	45	55	57	43
Work status					
Not working	41	31	69	44	56
Working	59	58	42	54	46

pandemic, 13% of respondents attested that they shopped online 1–3 times a week. This share almost doubled to 25% during the pandemic.

The summary of socio-demographic information of the entire survey sample and grouped by online shoppers versus non online shoppers is presented in Table 1. The results show that there was a shift in all socio-demographic characteristics of online shoppers, with the most prominent changes in the ratio of those aged between 25 and 44 years and over 65 years.

2.2 Estimation of Online Shopping Demand

The survey data is used to create the online shopping demand as the number of parcels per week and modelled agent using a poisson regression model. To account for the overdispersion in the data, we opted for a negative binomial generalised linear model. For the estimation, we used the function *glm.nb* from the R package *MASS* [11].

To simulate and analyse these shifts between the two scenarios, we estimated two sets of parameters based on the responses regarding online shopping frequency before and during the Covid-19 pandemic. While the modelling framework is designed such that activity patterns can be regarded in the online order model, the Covid-19 pandemic induced changes in activity patterns of participants to a degree where it is insensible to account for them in a model aimed at simulating the post-pandemic state. We therefore only included socio-demographic characteristics as independent variables to determine online shopping frequency as the dependent variable. As this study focuses on the application of the model and the effects that can be discerned, presentation of the parameters is outside the scope of this paper, however, we intend to present the underlying behavioural models of logiTopp in a later publication. Currently, the parameters are available upon request to the corresponding author. The integration of the model of online shopping demand into the modelling framework is described in the following section.

2.3 Modelling Framework

To analyse the changes in online shopping behaviour induced by the Covid-19 pandemic on last-mile deliveries, we integrated the behavioural changes into the modelling framework *mobiTopp* and it's last-mile logistics extension *logiTopp* [9, 10]. This section provides a concise overview of the modelling framework.

mobiTopp

The travel demand modelling framework *mobiTopp* [7, 8] consists of two modules: a long- and a short-term module. In the long-term module, a synthetic population of households and their individual agents are generated. Each agent is assigned socio-demographic attributes including: age, gender, work status, highest degree of education, income, place of work/education, drivers license, commuter ticket and membership to mobility service providers like bike-sharing or car-sharing. Households are assigned household-attributes accordingly including: a number of household members, a number of cars, a home location and a net income. To ensure intrapersonal and intra-household consistency, households and persons are drawn from a population pool provided by the German Mobility Panel [12] which is a national household travel survey. Additionally, activity schedules are generated for each agent including work, business, education, shopping, leisure, service and home activities. These activity schedules contain activities for the entire simulation period of one week [13].

All trips towards these activities are simulated in the short-term module. For each new activity, a destination and a travel mode are chosen. The simulated travel times may differ from the estimated travel times used in the long-term module when planning an activity schedule. Therefore, the activity schedule can be updated before each trip to consider the actual travel times. These steps are repeated for each activity and are simulated for all agents simultaneously.

logiTopp

We integrate last-mile deliveries into the *mobiTopp* framework in the form of the logistics extension called *logiTopp* [9, 10]. *logiTopp* takes advantage of *mobiTopp*'s agent based approach and the simultaneous simulation and interaction of these agents to model parcel orders of individual agents and simulate their last-mile delivery. *logiTopp* is implemented in Java and available as an open-source extension of *mobiTopp* on GitHub [14]. *logiTopp* mostly extends the short-term module of *mobiTopp*. Before the simulation of the short-term module starts, delivery agents are selected from the population and assigned to one of the modelled distribution centres (DCs). In this way, both the delivery agent's private and commercial trips are simulated.

To determine the parcel demand and the specific attributes of these parcels, a parcel demand model is applied to each potential recipient. *logiTopp* supports both private persons and businesses as recipients, however, in this paper we will focus on the demand of private parcels (b2c). As a first step, the parcel demand model determines the number of parcels expected by an agent for the simulated time period of one week. The specific attributes of these parcels are determined by the following steps: Selecting a delivery location (home, work or parcel locker); selecting a "Courier, Express and Parcel" service provider (CEPSP); selecting a DC from where it will be delivered and selecting an arrival date at the DC.

The model steps used in this study are shown in Fig. 2 on the left. The step 'choose number of parcels' is implemented by the means of the model described above (Sect. 2.2). The delivery location is selected based on a discrete choice model similar to the one described in [9, 10] The DC and the corresponding CEPSP are selected based on their relative share in the survey area. The arrival date is selected equally distributed throughout the week, except for Sundays, on which no parcels are distributed in Germany.

The parcel demand and the delivery agents are generated before the actual simulation of trips and activities. This simulation is carried out by the *mobiTopp* framework. The *logiTopp* extension comes into play, once a delivery agent arrives at their workplace (a DC). The model components used during the simulation are shown on the right in Fig. 2. Upon arrival at the DC, the agent's delivery tour is planned. We use a route first, cluster second heuristic [15] to approximate optimal delivery tours by using a 2-approximation TSP (travelling salesperson problem) algorithm provided by the JGraphT library [16]. The delivery agent's work activity is split into multiple delivery activities in the planned order. They are enclosed by an initial 'load' activity and a final 'unload' activity. These activities and the intermediate trips are simulated simultaneously to those of private agents.

Each DC specifies two rules to define under which conditions parcels can be delivered and what happens if a parcel cannot be served. In our model, home deliveries can be received by the recipient themselves, an other household member or a neighbour. Deliveries to parcel lockers are always successful, while work deliveries can only be received personally. Checking the presence and absence of the recipient as well as the other household members or neighbors is possible due to the agent based nature of *mobiTopp* and the simultaneous simulation of all agents and their activities. In case of an unsuccessful delivery, the DC can decide to update the parcel's destination to

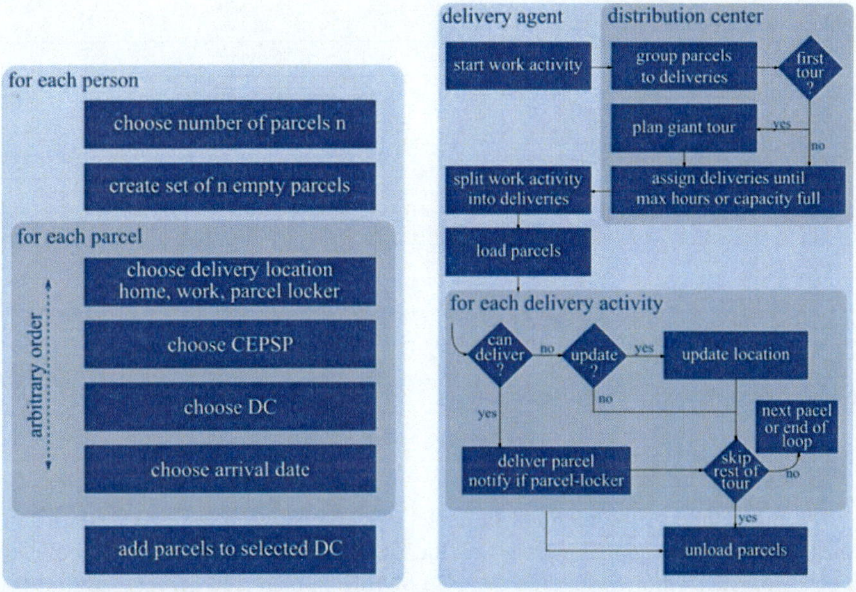

Fig. 2 Steps of the parcel demand model on the left; Steps of the delivery process on the right

a parcel locker. A common policy in our survey area are three delivery attempts, however, two CEPSP only perform one delivery attempt, which is reflected by the individual policies per DC in our model.

3 Results and Discussion

In this section, we describe the model results of the two simulation scenarios. We applied the previously described modelling framework to Karlsruhe, a city in the Southwest of Germany. The synthetic population includes 303,808 agents in 170,013 households. The results of the two simulated scenarios are summarised in Table 2.

From the table we can see that in the scenario based on the Covid-19 data, both the number of online shopping participants and the number of parcels increased. The number of simulated parcels increased by 9.94%, while the number of agents participating in online shopping increased slightly less by 8.36%, indicating that the online shopping frequency per online shopper changed. These changes are discernible in the distribution of parcel order frequency: Regarding all online shopping participants, the number of those who only ordered one parcel during the simulated week decreased by 2% in the Covid-19 based scenario, while those ordering two or three parcels increased slightly. The number of vehicles needed to deliver the simulated parcels increased by 2%.

Table 2 Simulation results by scenario

	Pre-covid	Covid
Number of parcels	211,284	232,291
Online shopping participants	121,436	131,593
Distribution of parcel order frequency		
1 parcel	59%	57%
2 parcels	24%	26%
3 parcels	9%	10%
4 parcels	4%	4%
5 parcels	2%	2%
6 or more parcels	1%	1%
Number of delivery vehicles	196	200

The magnitude of the increase in parcel demand between the two scenarios is consistent with the report presented by the German federal association parcel and express logistics who reported an increase of 10% in delivered parcels induced by the Covid-19 pandemic [17]. The relatively small increase in delivery vehicles is somewhat surprising, but can be explained by the fact that the capacity of the vehicles is–as of yet–not based on shipment sizes and thus does not necessarily reflect the actual number of vehicles but rather a magnitude of vehicles needed.

Because we account for socio-demographic variables in the parcel order model, we are able to analyse these characteristics for agents who order parcels in the simulation. Figure 3 shows the socio-demographic variables gender, age group, income group and work status for the pre-Covid simulation, the scenario based on Covid-19 data as well as the entire synthetic population (online shoppers and non-online shoppers). Overall, the shifts between socio-demographic variables of online shoppers in the simulation are similar to those from the survey data (see Table 1) confirming the general validity of the choice models.

Considering the gender of online shoppers, there is no discernible difference between the two simulations as in both cases males are slightly more likely to order and receive a parcel in the simulation week compared to the entire population. The barplot considering the distribution among the different age groups shows that in the Covid-19 based scenario, especially those over the age of 65 years are more likely to shop online compared to the pre-pandemic scenario. By contrast, the opposite holds true for the age group of those under 25 years old. Considering middle two age groups, no considerable changes in their percentage shares can be identified. This indicates, that new online shoppers are older and previously did not feel the need to shop online instead of in-store. Considering the income group of online shoppers, the graph shows that the Covid-19 pandemic shifted towards those with a lower income. For both scenarios but for the pre-pandemic scenario especially, those shopping online are present with a disproportional high income compared to the rest of the population. The graph considering the work status of agents who ordered and

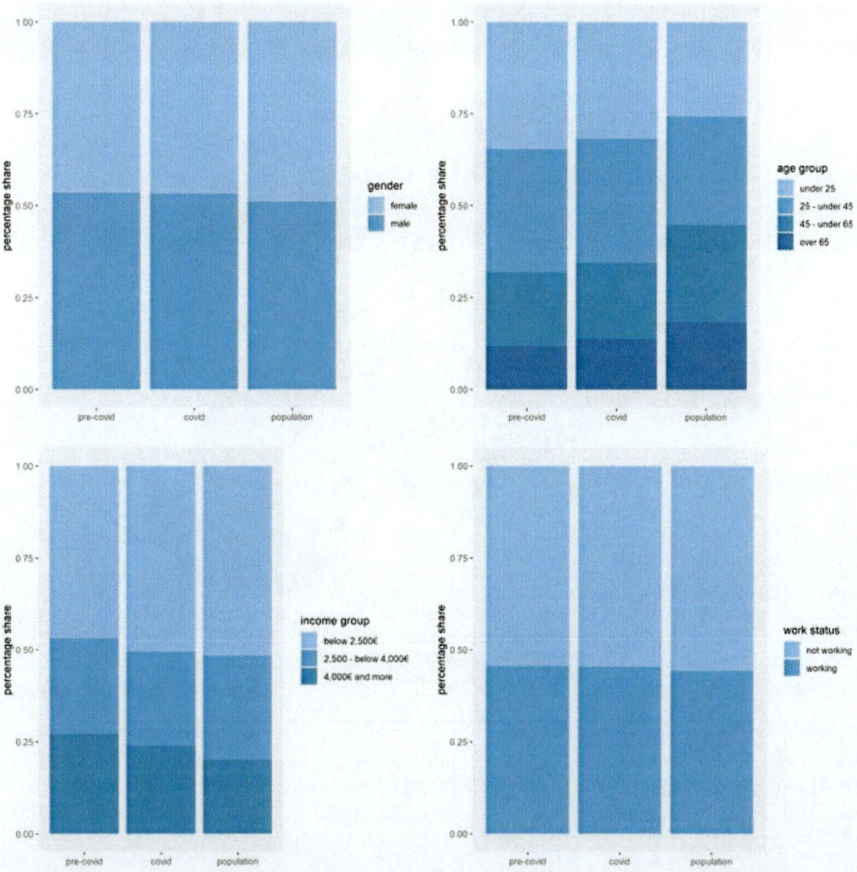

Fig. 3 Socio-demographics of online shopping agents by scenario: gender (top left), age group (top right), income group (bottom left), work status (bottom right)

received a parcel in the simulated week shows that there is neither a difference in distributions between the scenarios nor the overall population.

To assess if the changes in socio-demographics of online shoppers also result in different delivery areas, we also analysed changes in the spatial distribution of deliveries. Figure 4 shows the absolute difference of parcel orders and subsequent deliveries in the travel analysis zones (TAZs) between the two scenarios. A positive value (bright green to yellow) means that more parcels were ordered in the Covid-19 scenario. From the graph we can see that especially in the North, the Northeast and the South of the city there are several TAZs in which deliveries increased in the Covid-19 scenario compared to the pre-pandemic scenario. Considering the age distribution of the respective city districts, the delivery increase in the Northern and Southern parts of the city can be traced back to the fact that in these city districts, there is a high prevalence of older inhabitants. Analysis of the Northeastern part of

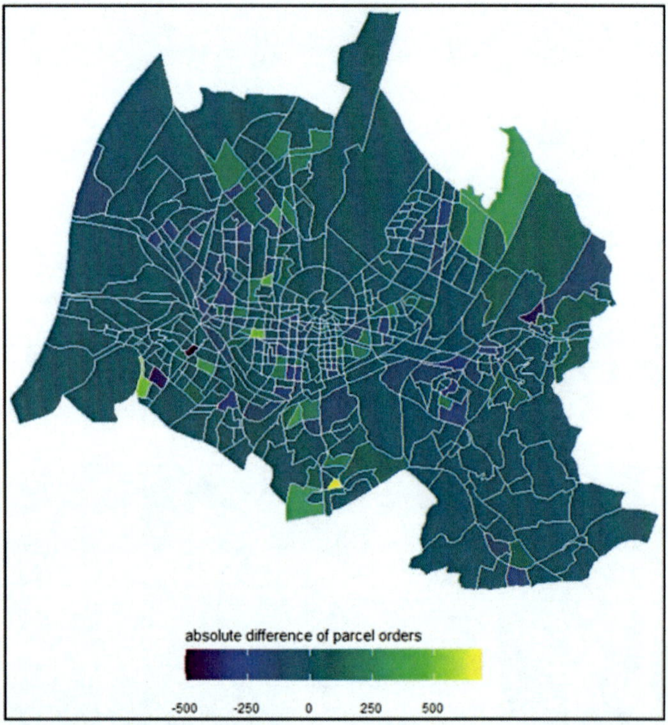

Fig. 4 Absolute difference of parcel orders and deliveries between the two scenarios. Positive values represent an increase of deliveries in the Covid-19 scenario

the city revealed that the increase in online shopping activity is most likely explained by the distribution of the net household income in the city district.

Our results show that there have been shifts in both socio-demographic characteristics of online shoppers and spatial distribution of parcel delivery demand induced by the Covid-19 pandemic. Results of the pre-pandemic scenario are consistent with studies on socio-demographic influence on consumers' e-commerce activity. Literature shows that prior to the pandemic, income has had a positive effect on online shopping behaviour [18, 19]. The results of our scenario analysis based on data collected during the pandemic show that the pandemic has somewhat shifted this interaction towards lower income groups. Considering the influence of e-commerce on consumption equality, previous research has shown that e-commerce can reduce consumption inequality [20]. Our results indicate that this effect has been positively influenced by the pandemic as people with a lower income have started or increased e-commerce activity. The scenario simulation based on the pandemic related data shows that not only the influence of income has shifted but also the effects of age on e-commerce activity has changed due to the pandemic. While the pre-pandemic scenario is consistent with previous literature suggesting that age has a negative influence on e-commerce [19], our results indicate that with the pandemic, also people aged 65 and older started or increased their use of e-commerce. The presented shift in socio-demographic characteristics is especially interesting considering that previous findings have highlighted that the influence of these factors on e-commerce activity cannot be changed through policy measures [19]. These findings were contradicted in times of the Covid-19 pandemic and we could observe that severe policy measures can have an effect on socio-demographic profiles of online shoppers. However, these measures were unprecedented and will not be imposed under regular circumstances. Nevertheless, the temporary closings of retails shops and the increased risk of infection during shopping trips lead a lot of people to switch shopping channels even if they had not (frequently) shopped online before.

While the previously discussed results could simply be studied through analysis of the survey data alone, the application of our model allows us to assess the results on a spatial level as well. The need to do so has been highlighted in previous studies, most recently Cheng et al. [21] analysed variables influencing the spatial distribution of parcel delivery demand. Among other variables, household characteristics play an important role for the correct prediction of spatial distribution of delivery trips. Taken together with the results of the scenario simulations of our study, our findings suggest that with the shift in socio-demographic profiles also came a change in spatial distribution of parcel delivery demand.

Many studies have highlighted the impact of accessibility measures on e-commerce activity and parcel demand [21–26]. Due to the fact that the accessibility to in-store shopping was reduced rather by policy measures to decrease the spread of the virus than geographical or transport related factors, we did not account for accessibility in our study. However, as data becomes available from times with fewer or any pandemic related policy measures, we will update the model accordingly in future work. Another limitation of our study is the restriction of online shopping of durable goods. Online grocery shopping has accelerated during as a result of the pandemic [4]. Because of the different mechanisms of delivering durable versus non-durable goods, simulating online grocery shopping behaviour and deliveries not only requires additional data but also an extension of the modelling framework. This extension will be part of future versions of *logiTopp*.

4 Conclusion

This study examines the effect of behavioural changes in online shopping induced and parcel delivery demand induced by the Covid-19 pandemic.

The scenario-based analysis shows an increase in parcel delivery demand changes consistent with surveys. The results of this study shows that the increase in demand can be traced back to people shopping online at higher frequencies and more people shopping online. There are two overarching implications of our findings: First, the extension of user groups of e-commerce by older people and those with a smaller income could lead to an acceleration of increase in e-commerce demand and that projected numbers will be reached faster than anticipated. And second, considering that the model area is a relatively small city, these results are disconcerting for larger cities and especially so for mega-cities. In population-dense areas, public space is scarce and the involvement of multiple providers delivering parcels uncoordinated in regards to their competition can lead to redundant delivery trips. Future research should thus focus on possible solutions to mitigate these problems, e.g. white-label deliveries in certain areas. The presented model framework and simulated scenarios help to assess future e-commerce demand and possible solutions to mitigate the problems caused by delivery traffic.

These findings are of interest to transport planners and delivery service providers as they highlight the importance of recognising that the Covid-19 pandemic not only induced a shift in socio-demographic profiles of online shoppers but that this shift also entails a change in the spatial distribution of parcel deliveries.

References

1. Statista. e-Commerce Report 2019. Statista Digital Market Outlook-Market Report (ed), Hamburg
2. Schlich R, Axhausen KW (2003) Habitual travel behaviour: evidence from a six-week travel diary. Transportation 30(1):13–36
3. Hao N, Akoorie M (2021) The impact of online shopping on consumers' habits in the supermarket industry in new zealand: Pre-and post-covid-19. In: ICL, pp 5–26
4. Shamshiripour A, Rahimi E, Shabanpour R, Mohammadian AK (2020) How is covid-19 reshaping activity-travel behavior? evidence from a comprehensive survey in chicago. Transp Res Interdiscip Perspect 7:100216
5. Eger L, Komárková L, Egerová D, Mičík M (2021) The effect of covid-19 on consumer shopping behaviour: Generational cohort perspective. J Retail Consum Serv 61:102542
6. Sheth J (2020) Impact of covid-19 on consumer behavior: will the old habits return or die? J Bus Res 117:280–283
7. Mallig N, Kagerbauer M, Vortisch P (2013) mobitopp - a modular agent-based travel demand modelling framework. Proc Comput Sci 19:854–859
8. Mallig N, Vortisch P (2017) Modeling travel demand over a period of one week: the mobitopp model (2017). arXiv preprint arXiv:1707.05050
9. Reiffer A, Kübler J, Briem L, Kagerbauer M, Vortisch P (2021) Integrating urban last-mile package deliveries into an agent-based travel demand model. Proc Comput Sci 184:178–185
10. Reiffer A, Kübler J, Briem L, Kagerbauer M, Vortisch P (2021) An integrated agent-based model of travel demand and package deliveries. Int J Traffic Transp Manag 3(2):17–24
11. Venables WN, Ripley BD (2002) Modern applied statistics with S, 4th edn. Springer, New York. 0-387-95457-0
12. Ecke L, Chlond B, Magdolen M, Vortisch P (2020) Deutsches mobilitätspanel (mop) – wissenschaftliche begleitung und auswertungen bericht 2019/2020: Alltagsmobilität und fahrleistung
13. Hilgert T, Heilig M, Kagerbauer M, Vortisch P (2017) Modeling week activity schedules for travel demand models. Transp Res Rec 2666(1):69–77
14. Kübler J, Reiffer A (2022) Logitopp, 3
15. Beasley JE (1983) Route first-cluster second methods for vehicle routing. Omega 11(4):403–408
16. Michail D, Kinable J, Naveh B, Sichi JV (2020) Jgrapht-a java library for graph data structures and algorithms. ACM Trans Math Softw (TOMS) 46(2):1–29

17. Esser K, Kurte J (2020) kep-studie 2021 - analyse des marktes in Deutschland
18. Akman I, Rehan M (2010) The predictive impact of socio-demographic and behavioural factors on professionals' e-commerce attitudes. Sci Res Essays 5(14):1890–1898
19. Pérez-Amaral T, Valarezo A, López R, Garín-Muñoz T, Herguera I (2020) E-commerce by individuals in spain using panel data 2008–2016. Telecommun Policy 44(4):101888
20. Luo X, Wang Y, Zhang X (2019) E-commerce development and household consumption growth in China. World Bank policy research working paper, (8810)
21. Cheng C, Sakai T, Alho A, Cheah L, Ben-Akiva M (2021) Exploring the relationship between locational and household characteristics and e-commerce home delivery demand. Logistics 5(2):29
22. Forman C, Goldfarb A, Greenstein S (2005) Geographic location and the diffusion of internet technology. Electron Commer Res Appl 4(1):1–13
23. Krizek KJ, Li Y (1926) Handy SL (2005) Spatial attributes and patterns of use in household-related information and communications technology activity. Transp Res Record: J Transp Res Board 1:252–259
24. Farag S, Weltevreden J, van Rietbergen T, Dijst M, van Oort F (2006) E-shopping in the netherlands: does geography matter? Environ Plann B Plann Des 33(1):59–74
25. Ren F, Kwan M-P (2009) The impact of geographic context on e-shopping behavior. Environ Plann B Plann Des 36(2):262–278
26. Weltevreden JWJ, van Rietbergen TON (2007) E-shopping versus city centre shopping: the role of perceived city centre attractiveness. Tijdschrift voor Economische en Sociale Geografie 98(1):68–85

How Far Are We From Transportation Equity? Measuring the Effect of Wheelchair Use on Daily Activity Patterns

Gregory S. Macfarlane and Nate Lant

Abstract The mobility needs of individuals with travel-limiting disabilities has been a transportation policy priority in the United States for more than thirty years, but efforts to model the behavioral implications of disability on travel have been limited. In this research, we present a daily activity pattern choice model for multiple person type segments including an individual's wheelchair use as an explanatory variable. The model results show a strong negative impact of wheelchair use on out-of-home travel, exceeding the impact of other variables commonly considered in such models. We then apply the estimated model within an activity-based model for the Wasatch Front region in Utah; the results suggest a shift in tour making of sufficient scale—among both wheelchair users and those in their households—to warrant further scrutiny and analysis.

Keywords Transportation equity · Travel behavior · Transportation equity · Travel behavior

1 Introduction

In 1990, the United States Congress passed the Americans with Disabilities Act (ADA), seeking to protect individuals with qualifying disabilities from discrimination while using public services including transportation systems (Title II), in public accommodations (Title III), and various other specific situations. For transportation service providers, ensuring equal access for wheelchair users is a critical design constraint for vehicles as well as stations and the surrounding areas [14]. Buses and

G. S. Macfarlane (✉) · N. Lant
Civil and Construction Engineering Department, Brigham Young University, 430 Engineering Building Provo, Utah 84602, USA
e-mail: gregmacfarlane@byu.edu

C. Antoniou et al. (eds.), *Proceedings of the 12th International Scientific Conference on Mobility and Transport*, Lecture Notes in Mobility,
https://doi.org/10.1007/978-981-19-8361-0_10

trains had to be redesigned with low floors and access ramps; elevators and ramps needed to be installed in stations alongside escalators and stairs. Even today, most traditional automobiles remain inaccessible to wheelchair users—at least without substantial modification. This last challenge is a particular concern for emerging mobility providers—who often use private vehicles owned by individual operators—and for public transit agencies who are beginning to cooperate with such providers to operate first/last mile services [20, 27].

Though the ADA only requires agencies to provide reasonable accommodation on public conveyances and does not try to establish equity in outcomes, the passage of 30 years provides a convenient time to consider what gaps and challenges persist for wheelchair users in accessing and using the transportation system. Specifically, what gap exists in the observed travel behavior outcomes of wheelchair users *vis a vis* the non wheelchair using population, all else equal? And more importantly, how should this gap be applied within travel forecasting models and related planning activities?

In this paper, we investigate the degree to which daily activity patterns are influenced by an individual's use of a wheelchair. This involves two separate analyses: first, we model daily activity pattern choice using responses to the 2017 National Household Travel Survey [30], incorporating the individual's wheelchair use as an explanatory variable. Second, we apply the behavioral estimates obtained from the choice analysis in a modified activity-based model for the Wasatch Front metropolitan region in Utah to estimate the population-level effects of introducing wheelchair status in a regional travel demand model.

The paper proceeds in a typical fashion. A literature review discusses prior attempts to evaluate and quantify the travel behavior of users with disabilities. A section describing the methodology of the choice analysis and model application is followed by a discussion of the results from both analyses. The paper concludes with a discussion of limitations in this analysis and associated avenues for future research and policy intervention.

2 Literature

Recent analysis [7] of the 2017 NHTS [30] suggests there are 13.4 million individuals in the US with travel-limiting disabilities—as defined by a specific question in the NHTS and encompassing sight, ambulatory, hearing, and other disabilities. Of these individuals, 20% (or 2.7 million) self report as using wheelchairs. With the relative aging of the U.S. population, the share of Americans with all disabilities as well as those using wheelchairs is likely to rise substantially [17, 28]. The question of how the travel behavior of individuals with travel-limiting disabilities varies from the travel behavior of individuals without these disabilities has been addressed previously in a number of studies.

With regards to travel patterns, surveys of the general population and surveys specifically targeted at individuals with disabilities both reveal significant and meaningful differences compared to individuals without disabilities. Specific findings include that individuals with disabilities leave their homes on fewer days if they leave at all [28], make fewer daily trips [7, 26] make fewer work trips and more healthcare maintenance trips [13], rely more on others for their travel [28] and have considerably restricted mode choices [24, 25]. These differences in mobility and activity patterns have important and observed negative implications for the individual's access to opportunity for employment [18, 24] and social interaction [3, 32]. Some recent studies have also looked at the meaningful role of developmental [34] and intellectual [15] disabilities on travel patterns.

The underlying reasons why wheelchair users and others with disabilities exhibit different travel patterns than other individuals are varied, but are likely to include both technical and attitudinal barriers. Technical barriers include poor access to private vehicles [31], poorly maintained sidewalk and pedestrian infrastructure [16], lack of physical access to TNC vehicles [25], bus ramp complication and malfunction [32], and numerous other problems across many modes. Attitudinal barriers include feeling embarrassment, for instance when a safety tone alerts all passengers that the wheelchair ramp is being deployed drawing attention not only to the user and his or her disability, but also to the fact that the transit vehicle is being delayed [32]. Additionally, some wheelchair users and others with disabilities have faced outright discrimination from private transportation service operators [3].

In spite of this relatively mature literature and the variety of findings on the topic, the research to date might be described as fragmented in both scope and application. That is, the literature reviewed here typically considers a single specific disability within the wide variety of relevant disabilities manifested in the NHTS results. But more importantly in the context of this research, the literature consists largely of ad-hoc studies conducted on specially collected datasets rather than holistically considering how disability manifests within an established transportation decision-making process. Specifically, there has not to our knowledge been a rigorous evaluation of the travel behavior of individuals with disabilities within the framework of a travel activity model.

As a result, recent attempts to simulate or model services aimed at this population have needed to make simplifying assumptions. In an attempt to model demand for a modern mobility system targeted at wheelchair users in Berlin, [5] simply assumed demand for this service would be similar to current demand for the regional paratransit system, augmented by a mode shift from taxi. Though paratransit patrons in the United States, Germany, and other similar contexts may use wheelchairs, the use of a wheelchair alone does not usually qualify a user for paratransit service. Further, in the initial modeling by [5] there was no link between the trips and the daily activities of the wheelchair users; there was not even a good understanding of likely trip origin, destination, or length distribution.

Including wheelchair users or wheelchair use status in a regularized travel model framework would help to fill two important gaps in the current literature. First, the comparison would help to illuminate the travel behavior characteristics of this important population within a framework that is readily understandable vis a vis other population segments. Second, researchers engaged in policy and planning work for this population could replace simplifying assumptions with plausible daily activity patterns rooted in observed behavior.

3 Methods

To begin the process of evaluating the effect of wheelchair use on travel behavior, we have developed a two-stage methodology. First, we estimate a daily activity pattern choice model using data from the 2017 NHTS. We then apply the coefficients obtained from that model in an activity-based model representing the Wasatch Front Region in Utah.

Activity-based models are a relatively mature construct in travel behavior research and in practical demand forecasting [23]. Activity-based models attempt to recreate the long- and short-term decision patterns of synthetic individuals using a chain of econometric and statistical choice models. The specific sub-models included in this chain can vary between specific implementations, but a recent open-source project—ActivitySim [1]—implements a popular set of models developed by [9]. Specifically, the ActivitySim demonstration model is a implementation of the "Travel Model One" model for the Metropolitan Transportation Commission (MTC, San Francisco Bay) [12]. For simplicity and potential future comparison with other models, we apply the ActivitySim model in this research.

3.1 Choice Model

The first step in the ActivitySim model chain is a *daily activity pattern* model of the type described by [6]. This model allows individuals to choose one of three daily activity patterns:

- Mandatory (M) daily patterns revolve around school and work activities that are typically considered non-discretionary. These activities and the travel to them anchor an individual's daily schedule, though other tours are possible.
- Non-Mandatory (NM) daily patterns involve only discretionary activities: shopping, maintenance, etc.
- At-Home (H) daily patterns describe the schedule and activities of individuals who never leave the home during the travel day.

The choice between the daily patterns is described with a multinomial logit model [10], where the utility functions for each option are determined by an individual's socioeconomic characteristics and person type segment. The specific innovation of the [6] model is that the daily activity patterns are coordinated, or that the choice of one individual in a household influences the choice probability of other household members.

Data for this study comes from the 2017 NHTS [30], which includes responses from across the United States involving rural, urban, and suburban areas. We restrict the data to households residing in an metropolitan statistical area (MSA) between one and three million in population. There are 76,367 individuals in 36,497 households that responded to the NHTS from these areas, though not all of these records are useful due to missing or incomplete data in key variables.

The NHTS releases public data in separate tables for persons, households, trips and vehicles; to determine the daily activity pattern for a given individual it was necessary to transform the trips table into a table of activities. We did this by reconstructing a schedule for each person from the reported trip origin and destination activity codes. We then determined whether each reported tour (a chain of activities away from the individual's home) contained a mandatory school or work activity. If any tour contained a mandatory activity, the person's entire daily activity pattern was classified as "mandatory"; if not, the daily activity pattern was "non-mandatory." By identifying respondents in the persons table without records in the trips table, we can determine individuals with a "home" daily activity pattern.

The NHTS has a number of questions where respondents can indicate a disability for themselves or other household members. Each respondent is asked "Do you have a condition or handicap that makes it difficult to travel outside of the home?" If the answer is yes, several follow-up questions are asked, including "Do you use any of the following medical devices? Select all that apply." The list of medical devices respondents can indicate includes canes, walkers, seeing-eye dogs, crutches, motorized scooters, manual wheelchairs, motorized wheelchairs, or something else (other). For this study, we identify wheelchair users as respondents who report using a manual wheelchair, mechanical wheelchair, or motorized scooter.

The specific variables included in the daily activity pattern choice models are based initially on the variables used in MTC Travel Model One [12]. The variables available in the NHTS include the age of the person and the household income treated as categorical ranges; gender, work, and college degree status are treated as binary values. Automobile availability is included via a binary "sufficiency" variable where a household with at least as many vehicles as adults is considered "auto sufficient." Descriptive statistics of the model variables are given in Table 1.

We adopt the person type segmentation strategy employed by ActivitySim; segmentation allows for heterogeneity in available alternatives and utility coefficients between individuals with highly divergent expected behaviors. For example, full time workers and pre-driving age school children will have strongly different responses to income, automobile availability, and other variables in determining their most likely daily pattern. ActivitySim classifies persons into seven person segments, though we only consider four types in this study, defined as follows:

- Full-time workers (FW)—reported working "full-time" at their primary job.
- Part-time worker (PW)—reported working "part-time" at their primary job, as well as any person who reported being a "non-worker" or "retired" who nevertheless reported a work or school activity.

Table 1 Model estimation data: descriptive statistics

		Full-time worker (N = 16188)		Non-worker (N = 3723)		Part-time worker (N = 4028)		Retired (N = 10060)	
		Mean	Std. Dev.	Mean	Std. Dev.	Mean	Std. Dev.	Mean	Std. Dev.
Bachelors or more		0.6	0.5	0.4	0.5	0.5	0.5	0.4	0.5
		N	Pct.	N	Pct.	N	Pct.	N	Pct.
Age	05–39	5762	35.6	1450	38.9	1356	33.7	8	0.1
	40–64	9570	59.1	2239	60.1	1705	42.3	1833	18.2
	65–79	836	5.2	33	0.9	903	22.4	6321	62.8
	80+	20	0.1	1	0.0	64	1.6	1898	18.9
Wheelchair	FALSE	16161	99.8	3609	96.9	4010	99.6	9619	95.6
	TRUE	27	0.2	114	3.1	18	0.4	441	4.4
Income	< $25,000	872	5.4	961	25.8	663	16.5	1825	18.1
	$25,000–$50,000	2235	13.8	632	17.0	713	17.7	2437	24.2
	$50,000–$100,000	5312	32.8	953	25.6	1203	29.9	3245	32.3
	> $100,000	7476	46.2	1102	29.6	1333	33.1	1975	19.6
Sex	Male	8820	54.5	1192	32.0	1487	36.9	4488	44.6
	Female	7368	45.5	2531	68.0	2541	63.1	5572	55.4
	I prefer not to answer	0	0.0	0	0.0	0	0.0	0	0.0
	I don't know	0	0.0	0	0.0	0	0.0	0	0.0
Works from Home	–1	0	0.0	3721	99.9	267	6.6	10060	100.0
	–7	2	0.0	0	0.0	1	0.0	0	0.0
	–8	1	0.0	0	0.0	1	0.0	0	0.0
	–9	728	4.5	0	0.0	0	0.0	0	0.0
	01	1718	10.6	0	0.0	940	23.3	0	0.0
	02	13739	84.9	2	0.1	2819	70.0	0	0.0

- Non-working adults (NW)—reported "unemployed" as their primary activity of the previous week, as well as individuals over 18 who were not classified elsewhere.
- Retired (RT)—reported "retired" as their primary activity of the previous week, or who are over the age of 65 and reported that they were not workers.

The other three person types are university students, schoolchildren under driving age, and driving-age schoolchildren. A limited number of individuals who could plausibly be considered university students responded to the NHTS, so we cannot estimate reliable choice models. Among schoolchildren of any age, too few report using wheelchairs to justify including these segments in this study.

3.2 ActivitySim Implementation

After obtaining an estimate for the relationship between wheelchair use and daily activity patterns, we wish to understand the impact of this variable on overall transportation demand forecasting. To do this, we can place the enhanced daily activity pattern model including this relationship within an Activitysim implementation. In this case, we use an implementation of ActivitySim in the Wasatch Front region of Utah; this implementation includes the Salt Lake City, Provo, and Ogden metropolitan areas.[1]

The synthetic population for this implementation is generated with PopulationSim [21], which uses American Community Survey Public Use Microdata Sample (ACS PUMS) [29] as seed table. Any variables from ACS PUMS can in principle be included in synthetic population and therefore in the model. The ACS PUMS includes a "disability" variable but not specific information on wheelchair use. The NHTS, however, does explicitly contain a wheelchair use variable as described earlier.

Using the NHTS data, we estimated a binary logit regression model where the probability for wheelchair use was determined to be a function of age,

$$P_{\text{wheelchair}|\text{disability}} = -2.59 + 0.014 * \text{age} \quad (1)$$

Other specifications of this regression equation did not result in substantively different model fit. For each person in the synthetic population with a disability as sampled from the ACS PUMS seed table, we determined the probability they would use a wheelchair based on the model in Eq. (1). A random draw allocated these individuals into using or not using a wheelchair. Of the total synthetic population, those using a wheelchair consisted of 0.8% of all individuals.

With the synthetic population described above, we modified the ActivitySim daily activity pattern model to consider wheelchair use. We then compared two complete runs of ActivitySim on the same synthetic population of 2.47 million individuals, with and without the wheelchair use variable activated. As described earlier, the

[1] This implementation is not the official regional travel demand model; the calibration and development of this model is described in [19].

daily activity pattern model in ActivitySim is coordinated, meaning that when an individual has an increased likelihood of choosing an at-home tour, other members of the household will have an increased likelihood as well [6]. Thus it is important to evaluate not only the daily activity patterns of individuals who use wheelchairs, but also their household members.

4 Results

4.1 Choice Analysis

We estimated the daily activity pattern choice models using mlogit for R [8, 22]. As described above, the alternatives for daily activity pattern choice are a Mandatory pattern where the individual's day involves a work or school tour, a Non-Mandatory pattern where only discretionary trips are taken, and a Home pattern where the individual does not leave home during the day. In the models estimated for this study, the Home pattern serves as the reference alternative with a utility of zero. Retired and otherwise non-working individuals choose only between Non-Mandatory and Home daily activity patterns.

The model estimates are presented in Table 2. The estimated coefficients are of the expected sign, though not all are significant. Some predictors that proved to be insignificant, such as automobile availability for full-time workers, were excluded from the estimated models. The overall model fit—as indicated by the McFadden ρ^2 with respect to a market shares (constants only) model—is not strikingly high. Were the purpose of this research to identify the best fit model of activity pattern choice for each person segment we would undertake an exercise to include, exclude, and identify potential transformations for different sets of variables. In this case, however, the goal of these models is simply to provide a plausible comparison point for the behavior of individuals using wheelchairs against the behavior of individuals in other person type segments.

The model coefficients indicate expected behavior for most included variables. All else equal, the model intercepts imply that full- and part-time workers are more likely to choose a mandatory daily pattern than either a non-mandatory home pattern. Men of all person types are less likely to choose non-mandatory patterns, and college graduates are more likely to choose any kind of out-of-home pattern. The effect of age group and income are less impactful for most trip purposes, with the exception of non-mandatory patterns being more likely to be chosen by individuals between the ages of 65 and 79. The strong coefficient seen on the choices of full-time workers over 79 years old is not significant, and is rather a relic of very few full-time workers of that age in the data set.

The use of a wheelchair is strongly significant for all person segments, and indicates wheelchair users are substantially less likely to make mandatory tours even if employed. For non-workers and retirees who use wheelchairs, their propensity to

Table 2 Daily activity pattern model estimates

		Full-time worker	Non-worker	Part-time worker	Retired
(Intercept)	Mandatory	2.083 (15.207)**		1.545 (10.300)**	
	Non-mandatory	1.137 (7.854)**	0.591 (6.359)**	0.338 (2.088)	−1.169 (−1.506)
Uses wheelchair	Mandatory	−1.851 (−3.328)**		−3.315 (−3.906)**	
	Non-mandatory	−0.625 (−1.315)	−0.721 (−3.647)**	−1.866 (−3.560)**	−1.258 (−11.924)**
Male	Mandatory	0.008 (0.140)		−0.040 (−0.347)	
	Non-mandatory	−0.148 (−2.378)	−0.271 (−3.477)**	−0.219 (−1.828)	0.235 (4.798)**
College graduate	Mandatory	0.353 (5.716)**		0.360 (3.033)**	
	Non-mandatory	0.648 (9.834)**	0.501 (6.028)**	0.584 (4.815)**	0.349 (6.521)**
Income $25,000–$50,000	Mandatory	−0.055 (−0.373)		0.169 (0.876)	
	Non-mandatory	−0.371 (−2.344)	−0.167 (−1.506)	0.444 (2.196)	−0.095 (−1.358)
Income $50,000–$100,000	Mandatory	−0.175 (−1.272)		−0.160 (−0.971)	
	Non-mandatory	−0.312 (−2.146)	−0.052 (−0.506)	0.190 (1.096)	0.115 (1.643)
Income > $100,000	Mandatory	−0.206 (−1.495)		−0.326 (−1.994)	
	Non-mandatory	−0.233 (−1.610)	−0.088 (−0.844)	0.127 (0.737)	0.036 (0.444)
Age 40–64	Mandatory	−0.006 (−0.104)		0.669 (5.094)**	
	Non-mandatory	0.026 (0.386)	0.433 (5.788)**	1.055 (7.805)**	2.241 (2.886)**
Age 65–79	Mandatory	0.179 (1.147)		0.395 (2.614)**	
	Non-mandatory	0.801 (5.081)**	1.690 (2.998)**	0.721 (4.647)**	2.132 (2.752)**
Age 80+	Mandatory	17.592 (0.004)		2.067 (2.293)	
	Non-mandatory	17.247 (0.004)	14.648 (0.008)	2.154 (2.391)	1.536 (1.980)
Works from home	Mandatory	−1.542 (−18.502)**		−1.340 (−9.830)**	
	Non-mandatory	−0.044 (−0.558)		0.112 (0.869)	
	N	15 895	3648	3912	9482
	AIC	26 550.18	4470.948	6965.196	10 689.72
	Log likelihood	−13 253.09	−2225.474	−3460.598	−5334.858
	ρ^2	0.031	0.019	0.055	0.026

Coefficients represent utility change relative to stay at home pattern
t-statistics in parentheses, * $p < 0.5$, ** $p < 0.01$

make non-mandatory tours is also substantially diminished. As a note, we considered making wheelchair use an independent person type segment; the results suggest that full-time workers who use wheelchairs are more similar to other full-time workers than they are to non-workers who also use wheelchairs. This justifies including wheelchair use as a variable within each person type segment. More notable than the significance of this variable alone, however, is the fact that its magnitude is so strong. Indeed, wheelchair use appears to be more influential on the choice of daily pattern than any other variable included in these models.

4.2 Activity-Based Model

At the individual level, wheelchair use appears highly predictive. It still remains to be seen, however, what the effect of including this variable in a forecasting model will do to aggregate trip making. To examine this question, we placed the wheelchair use variable coefficients shown in Table 2 into the ActivitySim daily activity pattern model. We allowed the other coefficients to retain their values as originally specified, assuming that wheelchair use is orthogonal to the other variables.

Overall, 8,886 individuals of the 2,487,002 in the region changed their daily activity patterns changed from the base scenario after including wheelchair use as a utility variable. Though this is only 0.357% of individuals, it is worth considering the distribution of this change in more detail. Table 3 presents an aggregate summary of individuals who kept and changed their plans. Individuals with no wheelchair users in their household are not affected, other than through some random simulation error. For wheelchair users themselves however, the changes are relatively large. For example, 43% of the wheelchair users who were modeled as making mandatory tour patterns in the base scenario are now expected to make non-mandatory or stay-at-home patterns. Similarly, wheelchair users who previously chose non-mandatory patttens are considerably more likely to stay at home.

The coordinated nature of the ActivitySim daily activity pattern model also allows household members of wheelchair users to be affected. For these people, the changes are smaller in proportion but potential meaningful in aggregate: 30.7% of the household members who stay at home in the new scenario incorporating wheelchair use previously made out-of-home tours in the base scenario.

In addition to these aggregate numbers, it is also worth considering where the affected households are concentrated. Wheelchair use and disability in general is not distributed evenly through the region. Figure 1 shows the concentration of households where at least one member changed its chosen daily activity pattern between the

Table 3 Daily activity pattern change

Group	Base	Considering wheelchair use		
		Home	Mandatory	Non-mandatory
Wheelchair users	H	3369	20	459
	M	932	1642	308
	N	3584	23	10261
Household members	H	4511	213	631
	M	759	15409	301
	N	1235	415	13119
Not affected	H	309965	2	
	M	2	1460582	
	N		2	659258

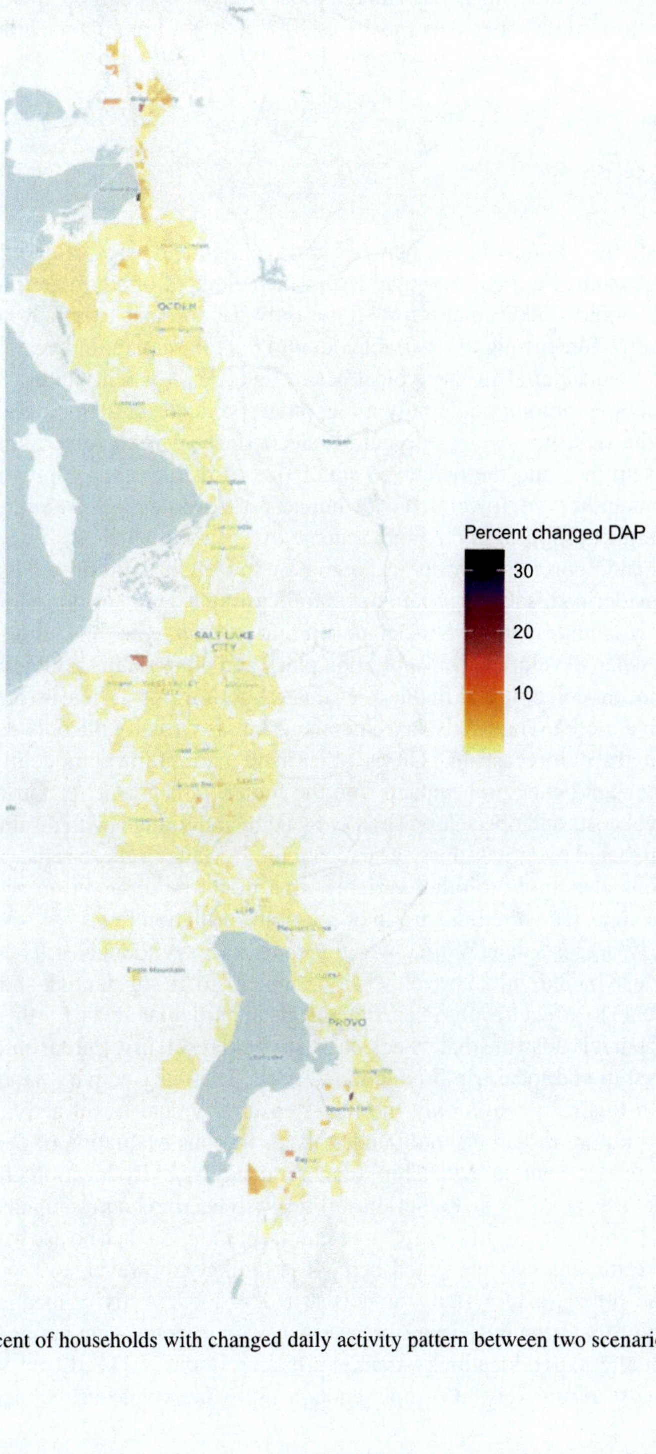

Fig. 1 Percent of households with changed daily activity pattern between two scenarios

scenarios by traffic analysis zone. Though most areas of the region experience little if any alteration, some zones see as many as 20% or 30% of their households change activity patterns.

5 Discussion

Disability status is almost never included in travel demand forecasting efforts for a number of reasons. First is a practical issue; each element of a population included in a model requires additional effort to not only catalog and calibrate in the base year, it requires forecasting that variable out into the future planning years. Limiting the variables considered to the typical set of age, employment status, household size, and income (among potentially a few others) reduces burden on planners and simplifies the data management pipeline. Second, the variety of possible travel-limiting disabilities and the perceived small size of the affected population might make understanding the travel behavior impacts of all potential disabilities a high-cost, low-impact proposal in the large scheme of a travel model.

There is an important philosophical argument to be had about the equity implications of considering disability within travel forecasting efforts. On one hand, excluding disability from travel behavior models might reinforce the "invisibility" of this population when developing transportation plans and other public policies. If a population is not modeled, it is difficult to evaluate how various policies would impact them within a model. Thus household income is left as virtually the only equity consideration in travel forecasts [4]. On the other hand, current travel modeling practice assumes constant behavioral impacts into the future. Is is correct or fair to assume that the travel costs and obstacles currently faced by individuals with disabilities will continue unabated twenty or thirty years from now? Many of these same philosophical questions also apply to other variables that might be observed to affect travel behavior, such as race or ethnic group or religious affiliation [e.g., 33].

This study suggests that omitting wheelchair use is not without cost. The impact of wheelchair use on individual travel behavior appears to be substantial, and is potentially of a scale to affect travel forecasting results in particular areas. Further research is required, but it is possible that wheelchair use could result in different outcomes for transit ridership and local trip production at least. The daily activity pattern model presented in this research is only the first step in a typical travel activity modeling process. Future research should undertake a rigorous evaluation of the effect of wheelchair use on vehicle ownership, tour frequency, activity location choice, and travel mode. It is possible that other disabilities will need to be investigated as well.

The NHTS provides a sufficiently large sample of individuals who use wheelchairs to undertake the analysis presented here. It is unlikely, however, to be useful for some of the other travel activity models just described. Activity location choice models require substantially more precise geographic detail—including detail on non-chosen alternative locations—than is provided in the NHTS. Local household travel surveys are unlikely to sample enough individuals who use wheelchairs to estimate robust effects without targeted oversampling techniques. The results of this study raise a question as to whether planning agencies ought to consider conducting this oversampling to enable future research.

The model specifications presented in Table 2 each consider wheelchair use independently from other variables. It is likely however, that there could be some interaction effects among the various individual attributes. To wit: do wheelchair users who hold bachelor's degrees behave differently with respect to activity pattern than wheelchair users with less education?

This study used information from all United States metro areas between one and three million in population. It is possible that the obstacles wheelchair users face in some of these metro areas are larger or smaller than others. It is also possible that larger or smaller metro areas would present different challenges or provide different resources to wheelchair users. The same logic extends to other countries besides the United States. Examining the role of wheelchair use on daily activity pattern in other national contexts is important, as would be further research on other categories of disability.

6 Conclusion

In the generation since the enactment of the ADA in 1990, much progress has been made in the United States in making public accommodations and transportation services available to all users regardless of their disability status. Despite this progress, the results of this research suggest that wheelchair users still face substantial friction in their daily travel after controlling for age, income, and other common sociodemographic characteristics. Though the size of the wheelchair-using population is relatively small—particularly within the context of a regional travel demand model—it may be spatially concentrated to the point where it could influence forecasts for individual transit services or local highway facilities.

As transportation planners are being asked to consider equity in more policies and contexts, the needs of wheelchair users and others with travel-limiting disabilities should be considered in more contexts. The exclusion of wheelchair users from travel models means that services and policies aimed at assisting this community may be devalued relative to highway capacity and other more traditional projects. The consequences of this exclusion should be investigated, as well as the exclusion of other variables that would enable a more holistic assessment of winners and losers in the transportation system.

Acknowledgements Figures and tables in this paper were created with a variety of R packages [2, 11, 35]. The authors would like to thank Chris Day and Christian Hunter for their help in preparing this analysis. This work was supported by the Utah Department of Transportation and the Federal Transit Administration. The authors alone are responsible for the preparation and conclusions presented herein. The contents do not necessarily reflect the views, opinions, endorsements, or policies of the Utah Department of Transportation of the US Department of Transportation. The Utah Department of Transportation makes no representation or warranty of any kind, and assumes no liability therefore.

References

1. AMPO (2020) Activitysim: an open platform for activity-based travel modeling. https://activitysim.github.io/
2. Arel-Bundock V (2021) modelsummary: summary tables and plots for statistical models and data: beautiful, customizable, and publication-ready. https://vincentarelbundock.github.io/modelsummary/. R package version 0.8.0
3. Bascom GW, Christensen KM (2017) The impacts of limited transportation access on persons with disabilities' social participation. J Transport Health 7:227–234. ISSN 2214-1405. https://doi.org/10.1016/J.JTH.2017.10.002. https://www.sciencedirect.com/science/article/pii/S2214140517300075
4. Bills TS, Sall EA, Walker JL (2012) Activity-based travel models and transportation equity analysis: research directions and exploration of model performance. Transp Res Record 2320(1):18–27
5. Bischoff JF (2019) Mobility as a service and the transition to driverless systems. PhD thesis, Technische Universität Berlin
6. Bradley M, Vovsha P (2005) A model for joint choice of daily activity pattern types of household members. Transportation 32(5):545–571. ISSN 00494488. https://doi.org/10.1007/s11116-005-5761-0
7. Brumbaugh S (2018) Issue brief travel patterns of American adults with disabilities. Technical report
8. Yves C (2020) Estimation of random utility models in R: the mlogit package. J Stat Softw 95(11):1–41. https://doi.org/10.18637/jss.v095.i11
9. Davidson W, Vovsha P, Freedman J, Donnelly R (2010) Ct-ramp family of activity-based models. In: Proceedings of the 33rd Australasian transport research forum (ATRF), vol 29. Citeseer, p 29
10. Domencich TA, McFadden D (1975) Urban travel demand: a behavioral analysis. North-Holland Pub Co. ISBN 0720431964. https://trid.trb.org/view/1175810
11. Dunnington D (2021) ggspatial: spatial data framework for ggplot2. https://CRAN.R-project.org/package=ggspatial. R package version 1.1.5
12. Erhardt G, Ory D, Sarvepalli A, Freedman J, Hood J, Stabler B (2012) Mtc's travel model one: applications of an activity-based model in its first year. In: 5th transportation research board innovations in travel modeling conference
13. Ermagun A, Hajivosough S, Samimi A, Rashidi TH (2016) A joint model for trip purpose and escorting patterns of the disabled. Travel Behav Soc 3:51–58. ISSN 2214367X. https://doi.org/10.1016/j.tbs.2015.08.002
14. Federal Transit Administration (2015) Americans with disabilities act: guidance. Circular FTA C 4710.1. https://www.transit.dot.gov/regulations-and-guidance/fta-circulars/americans-disabilities-act-guidance-pdf
15. Feeley C (2010) Evaluating the transportation needs and accessibility issues for adults on the autism spectrum in New Jersey. In: Transportation research board annual meeting, 2010
16. Frackelton A, Grossman A, Palinginis E, Castrillon F, Elango V, Guensler R (2013) Measuring walkability: development of an automated sidewalk quality assessment tool. Suburban Sustain 1(1). http://dx.doi.org/10.5038/2164-0866.1.1.4
17. Laplante M (2003) Demographics of wheeled mobility device users. In: Proceedings of the conference on space requirements for wheeled mobility, pp 1–23
18. Lubin A, Deka D (2012) Role of public transportation as job access mode. Transp Res Record (2277):90–97. ISSN 03611981. https://doi.org/10.3141/2277-11
19. Macfarlane GS, Lant NJ (2021) Estimation and simulation of daily activity patterns for individuals using wheelchairs. Report, Utah Department of Transportation Research Report No. UT-21.10. https://rosap.ntl.bts.gov/view/dot/56982
20. Macfarlane GS, Hunter C, Martinez A, Smith E (2021) Rider perceptions of an on-demand microtransit service in salt lake county, utah. Smart Cities 4(2):717–727. ISSN 2624-6511. https://www.mdpi.com/2624-6511/4/2/36

21. Paul BM, Doyle J, Stabler B, Freedman J, Bettinardi A (2018). Multi-level population synthesis using entropy maximization-based simultaneous list balancing. In: Transportation research board annual meeting
22. R Core Team (2021) R: a language and environment for statistical computing. R foundation for statistical computing, Vienna, Austria. https://www.R-project.org/
23. Soora R, Harry T (2014) Activity-based models of travel demand: promises, progress and prospects. Int J Urban Sci 18(1):31–60. https://doi.org/10.1080/12265934.2013.835118
24. Rosenbloom S (2007) Transportation patterns and problems of people with disabilities. In: Field MJ, Jette AM (eds) The future of disability in America. National Academies Press, pp 1–592. ISBN 0309668646. https://doi.org/10.17226/11898. https://www.nap.edu/catalog/11898/the-future-of-disability-in-america
25. Ruvolo M (2020) Access denied? Perceptions of new mobility services among disabled people in San Francisco. In: The SAGE encyclopedia of higher education, p 52. https://doi.org/10.4135/9781529714395.n81. https://escholarship.org/uc/item/6jv123qg
26. Schmöcker JD, Quddus MA, Noland RB, Bell MGH (2005) Estimating trip generation of elderly and disabled people: analysis of London data. Transp Res Record (1924):9–18. ISSN 03611981. https://doi.org/10.3141/1924-02
27. Shaheen S, Chan N (2016) Mobility and the sharing economy: potential to facilitate the first-and last-mile public transit connections. Built Environ 42(4):573–588. ISSN 0263-7960
28. Sweeney M (2004) Travel patterns of older americans with disabilities. Working paper 2004-001-OAS, Bureau of Transportation Statistics, pp 1–36
29. US Census Bureau (2022) Acs pums: American community survey public use microdata sample, 2022
30. U.S. Department of Transportation and Federal Highway Administration. 2017 national household travel survey, 2017. http://nhts.ornl.gov
31. Van Roosmalen L, Paquin GJ, Steinfeld AM (2010) Quality of life technology: the state of personal transportation. Phys Med Rehab Clin North America 21(1):111–125. ISSN 10479651. https://doi.org/10.1016/j.pmr.2009.07.009
32. Velho R, Holloway C, Symonds A, Balmer B (2016) The effect of transport accessibility on the social inclusion of wheelchair users: a mixed method analysis. Soc Inclus 4(3):24–35. ISSN 21832803. https://doi.org/10.17645/si.v4i3.484
33. Vyas G, Vovsha P, Paleti R, Givon D, Birotker Y (2015) Investigation of alternative methods for modeling joint activity participation. Transp Res Record 2493(1):19–28. https://doi.org/10.3141/2493-03. https://www.journals.sagepub.com/doi/abs/10.3141/2493-03
34. Wasfi R, El-geneidy A (2007) Measuring the transportation needs of people with developmental disabilities. The Lancet 369(9560):457. ISSN 01406736. https://doi.org/10.1016/S0140-6736(07)60218-9
35. Wickham H, Chang W, Henry L, Pedersen TL, Takahashi K, Wilke C, Woo K, Yutani H, Dunnington D (2020) ggplot2: create elegant data visualisations using the grammar of graphics. https://CRAN.R-project.org/package=ggplot2. R package version 3.3.3

Impacts of Inner-City Consolidation Centres on Route Distances, Delivery Times and Delivery Costs

Matthias Ribesmeier

Abstract In this study, the effects of inner-city consolidation centers on route distance, route duration and delivery costs are evaluated using the example of Äußere Neustadt a district in Dresden and the courier, express and parcel (CEP) sector as an example of inner-city freight transportation. In order to answer the research question, a multi-level model LOCAMM (**L**ogistics and **C**ity **A**rchitecture **M**ultilevel **M**odel) is presented, which forms the input data for an agent-based traffic flow simulation in the software MATSim (the Multi-Agent Transport Simulation Toolkit) in combination with JSprit. JSprit is a toolkit to solve richt Vehicle Routing Problems and Traveling Salesman Problems and is firmly integrated in MATSim. The result of the study shows that in the present case it is possible to reduce the route distance and the delivery time by using a Micro Hub. However, depending on the chosen scenario, the delivery costs are above or below the chosen comparison scenario. Furthermore, the effects of the intermodal transshipment center Bahn-City-Portal for the mentioned parameters and the link volume are described. In this study, benefits of inner-city consolidation centers on route distance, route duration and, depending on the scenario, on transportation costs were demonstrated for this case. For the Micro Hub scenarios, an increased link volume around the Micro Hub was found, but overall a lower average relative traffic volume in the delivery district.

Keywords Urban freight modelling · Impacts of freight transport · Urban-consolidation-centres · Agent-based modelling · Micro hub · Bahn-city-portal

1 Introduction

Inner-city infrastructures are increasingly reaching their performance limits. Freight traffic, which has been increasing for years, has played a major role in this situation. For example, freight vehicles account for 9–15% of total traffic in Paris

M. Ribesmeier (✉)
Technische Universität Dresden, 01069 Dresden, Germany
e-mail: matthias.ribesmeier@tu-dresden.de

C. Antoniou et al. (eds.), *Proceedings of the 12th International Scientific Conference on Mobility and Transport*, Lecture Notes in Mobility,
https://doi.org/10.1007/978-981-19-8361-0_11

[9]. However, these vehicles occupy 25% of road-lane capacity throughout Paris and 62% in the historic center Paris [9]. In general, increasing urbanization can be observed worldwide. According to a UN report, 68% of the world's population will live in cities in 2050 [22]. Increasing urbanization, in contrast, also means a further increase in freight and passenger traffic in the growing cities. This means a raised potential for conflict between the different actors of inner-city freight traffic and the population. To ensure this growth of cities and reduce the potential for conflict in the future, urban logistics must also evolve toward more sustainable logistics.

In the past, there have been some measures and ideas to make the last mile more city-friendly and environmentally friendly. These measures can include stakeholders' engagement, regulatory measures, market-bases measures, land use planning & infrastructure, new technologies or eco-logistics awareness raising [21]. One of these measures from the category land use planning and infrastructure includes inner-city consolidation centers, such as Urban-Consolidation-Centers (short: UCC) or smaller Micro Hubs. Urban-Consolidation-Centres/ Micro Hubs are facilities that are designed for district-specific logistics in order to bundle delivery traffic and also can have additional functions like parcel counters. Urban-Consolidation-Centers/ Micro Hubs also offer the opportunity to integrate new vehicles such as cargo bikes into inner-city logistics or can help to ensure a fully electrified supply chain.

In recent years, one part of inner-city logistics in particular has been growing rapidly. The courier, express, parcel service sector (short: CEP) almost doubled the volume of shipments in Germany, and the industry expects strong growth in the coming years as well [6]. The supply chain in the CEP sector generally consists of three sub-steps the first mile, the middle mile and the last mile. The first mile describes the receiving of the parcel and its transportation to the regional distribution center. From there, the middle mile begins and the parcel is transported to either a central distribution center or a regional distribution center in the target region. From the regional distribution center begins the last mile, the delivery of the parcel to the end customer. On the part of the CEP companies, the last mile causes around 41% the costs of their entire supply chain [3].

So far, the effects of inner-city consolidation centers have only been reviewed on monomodal variants with a road vehicle to road vehicle transfer (e.g [1, 10, 14, 23]). Inner-city transshipments from rail freight to road freight like the Bahn-City-Portal and their effects in comparison to the peripheral regional distributions centers have not been researched yet. This article examines the impact of a dimodal inner-city consolidation center (Bahn-City-Portal) versus peripheral delivery or peripheral delivery of an inner-city micro hub on the last mile of the CEP sector. To achieve this, we set up an freight traffic model for the CEP sector based on the LOCAMM (**L**ogistics and **C**ity **A**rchitecture **M**ultilevel **M**odel) multilevel model. We then model the last mile flows of goods using the agent-based traffic flow software MATSim in combination with the open source tool JSprit, which is a toolkit for solving rich Vehicle Routing Problems (VRP) and the Traveling Salesman Problem (TSP).

2 Methodology

2.1 Multilevel Model LOCAMM to Demonstrate the Impact of Changes in the Urban System

A multilevel model "**L**ogistics and **C**ity **A**rchitecture **M**ultilevel **M**odel" (short: LOCAMM) was developed that merges various georeferenced databases such as logistics databases, traffic databases and urban databases in three different scaling levels. These are the Micro ("Foundation Level"), Macro ("Support Level") and Meso Levels ("Data Exchange Level"). The purpose of the model is to integrate logistics into urban development and traffic planning.

Each layer has its own set of geodatabases (Urban, Logistic, Traffic), which exchanges data with the similar geodatabases of the other scale levels or receives information from other types of geodatabases in an iterative process. The structure of the model is illustrated by Fig. 1. This procedure is explained in more detail in the following by the calculation and distribution of the parcels. For this step, the Urban Databases and the Logistic Databases are used in the different scale levels. The Logistic Databases are empty at the beginning of the iterative process and are provided with values and attributes within the calculation.

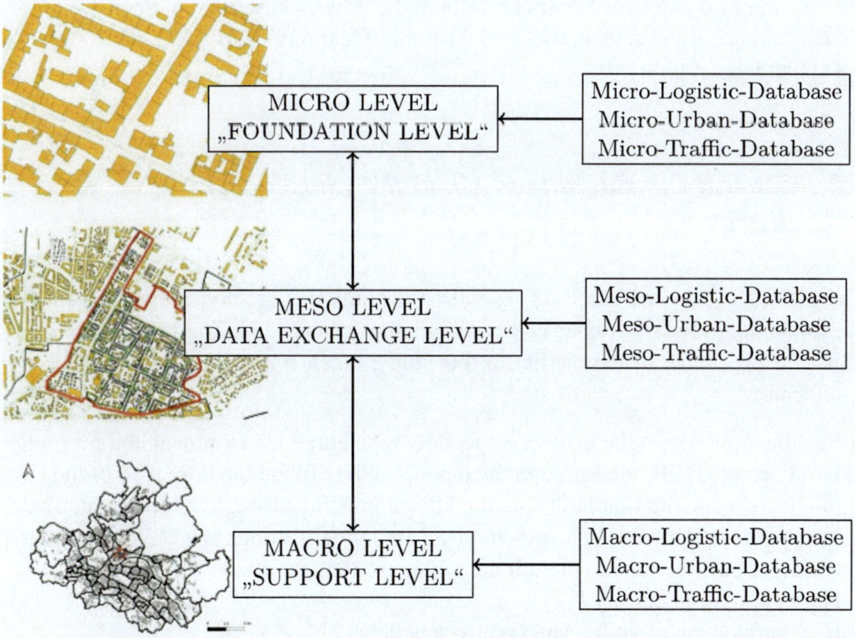

Fig. 1 Developed multilevel model as the basis for the simulation set up. *Source* own depiction

In its initial state, the Micro-Urban-Database consists of databases from Open Street Map data, which are geospatially processed. In this context, we formed centroids from the available polygons of the buildings and provide them with geospatial coordinates. From this database (Micro-Urban-Database), we selected the entirety of possible private customers and possible business customers for supply. We did this by filtering the geodatabase by the function of the buildings, whether there is a CEP-related store in that building or whether it is residential. In its initial state, the Macro-Urban database contains the districts of the statistical districts of Dresden and census data (population age, building density, number of households and population density). Household information is passed to the Meso-Urban-Database, which has the scale level of the entire district. Now the quantity of parcels per day in the delivery district can be calculated using Eq. 1 from Sect. 2.3 and this information is stored in the Meso-Logistic-Database. Then, from the totality of possible customers (Micro-Urban-Database), Monte Carlo simulation is used to select and store them in the Micro-Logistic-Database with their coordinates, their attributes (private or business customer). The Monte Carlo simulation determines the parcel quantity of the selected customers and stores it in the micro-logistic database. This database is finally transferred to MATSim. Statistics from the Micro-Logistic-Database can now be passed to the Meso-Logistic-Database or the Macro-Logistic-Database, which can be used, for example, to better calibrate the model in the future.

We use the Multi-Level Model for the following subtasks:

- Selection of the delivery district based on the criteria described in Sect. 2.2.
- Determination and selection of end customer locations (B2B, B2C and C2C)
- Demand generation

2.2 *General Considerations About the Chosen Delivery District*

The district is selected based on several constraints that are considered helpful for the implementation of a inner-city consolidation center [17]. The boundary conditions for the choice of the district in this study, taken from Richter et al. [17], are listed below:

- Specific land use in the district (especially residential development and no single family estate [17]): measured in the model by the official land use plan of the city of Dresden and the share of residential developments or single family estate share
- Narrow development [17]: measured in the model through the building density (Ratio of built-up area to unbuilt area in the settlement area)
- High population density [17]
- High proportion of young and creative residents [17]

The Richter et al. [17] study does not set a benchmark for when, for example, a population density should be considered high within a city. Therefore, we form a

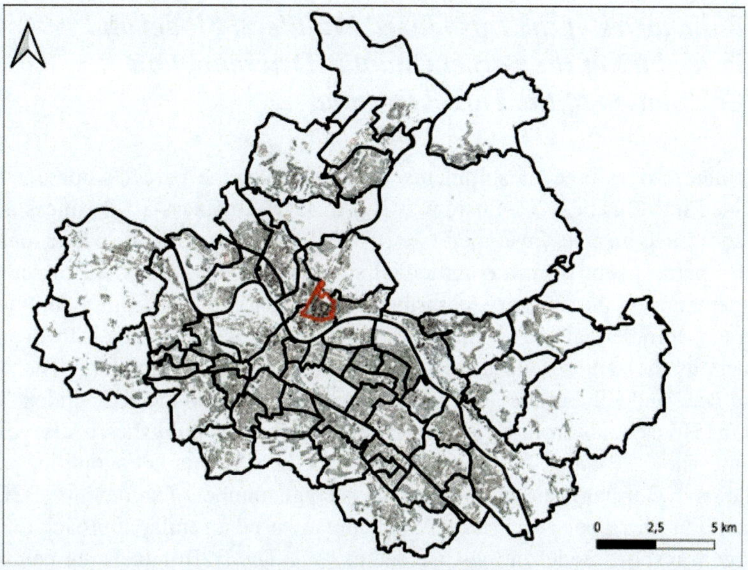

Fig. 2 Administrative borders of the city of Dresden, its districts and the selected delivery area Äußere Neustadt (outlined in red). *Source* own depiction

Dresden-specific benchmark from the values for the statistical districts in Dresden. We determined the values for the boundary conditions for all statistical districts in Dresden using Open Street Map data or official statistics from Dresden (see [8, 13]). The highest building density was in "Äußere Neustadt" with 32.4%, the "Innere Altstadt" with 30.2% and "Striesen-Ost" with 22.4%. The average building density is 11.92% in Dresden. We could not find the building density in official data of the city of Dresden, so we processed them by geospatial processing of Open Street Map databases. The lowest average age was in "Äußere Neustadt" with 33.0 years, "Südvorstadt-Ost" with 33.7 years and "Leipziger Vorstadt" with 33.8 years. Whereas the average age of the population is 43.2 years [13] in Dresden. The highest population density, measured in inhabitants per km^2, had "Äußere Neustadt" with 15981 [in/km^2], "Prohlis Süd" with 12018 [in/km^2] and "Striesen-West" with 11660 [in/km^2] [8]. The average population density in Dresden is 4528.34 inhabitants per km^2.

Ultimately, we decide on the "Äußere Neustadt" as the study area, as it has the best values in all the categories presented.

Figure 2 shows the demarcation of the Äußere Neustadt from the rest of the city of Dresden.

2.3 Boundaries of the Introduced Multilevel Model and Procedure of the Parcel Quantity Determination Calculation of the Parcel Demand

To calculate the volume of shipments in the district, the parcel shipment flows are divided into Business-To-Customer (short: B2C), Business-To-Business (short: B2B) and Customer-To-Customer (short: C2C) shipments. B2C parcel demand describes parcels sent from a commercial shipper to a private end customer. B2B parcel demand, on the contrary, describes parcels that are sent from a commercial shipper to a commercial end customer. C2C shipments describe parcels from private customers as the sender with a private end customer as the recipient. The parcel demand B2C and C2C in the area is calculated according to the KEP-Index (CEP-Index) for Households from Kurte and Esser [6], which is published every year for Germany. The CEP Index for households shows the CEP market supply of private households in Germany and is based on the average number of shipments (B2C and C2C) per household per year. Thaller 2018 already used a similar approach for their economic transport model in the CEP sector [20]. The B2B parcels are calculated by the ratio between B2C/ C2C and B2B from the reference year 2019, which is also a statistic published by Kurte and Esser. The parcel demand per B2C and C2C district M_c is calculated by Eq. 1.

$$M_c = \frac{(K_a \cdot H)}{((d_a - d_s - d_f))} \tag{1}$$

with:

M_c Parcel demand per day
K_a CEP-Index for households in the year under review
H Number of households in the observation area for the observation year
d_a Number of calendar days in the year under review
d_s Number of Sundays in the year under review
d_f Number of public holidays in the year under review, excluding public holidays that fall on Sundays

To calculate the total number of B2C and C2C parcels the following values are assumed for the variables:

K_a 47,5 [Parcels/household and year, [6]
d_a 365 [days]
d_s 52 [days]
d_f 11 [days] in Saxony

Kurte and Esser indicate the ratio B2C/ C2C and B2B with 70% B2C and C2C shipments and 30% B2B shipments for 2019 [6]. Accordingly, the B2B demand in the present delivery district can be determined as 750 parcels. As a result, the total demand for CEP services in the delivery district is 2500 parcels per delivery

Table 1 Market shares of the largest parcel service providers in Germany measured by parcel volume in 2019 and the calculated parcels for the model. *Source* own depiction with data from [11]

Delivery company	DHL	Hermes	GLS	DPD	UPS	FedEx	Others
Market share in Germany 2019	48%	16%	7%	10%	12%	6%	1%
Calculated parcels [parcels/day]	1200	400	175	250	300	150	25

day. The total number of parcels has now been randomly distributed to the specific customers as previously described. The calculated parcels that are delivered to the district per day and per delivery service provider are shown in the Table 1.

2.4 *Definition of the Scenarios*

To obtain the impact of inner-city delivery centers, we provide a baseline scenario "Scenario 0" that models the existing delivery situation. In contrast, we model the comparison scenarios, which cover the monomodal case of an inner-city delivery center (road to road transshipment) and the dimodal case (rail to road transshipment). The locations of the two supply chain starting points of the comparison scenarios are shown in Fig. 3. The locations of the Micro Hub and the Bahn-City-Portal (short: BCP, english: Rail-City-Portal) were determined by a Monte Carlo simulation. The boundary conditions for the location of the Micro Hub were the boundaries of the delivery district and the boundary conditions for the location of the Bahn-City-Portal were the city boundaries and the location on rail infrastructure, at a level of development suitable for rail freight. We simulated coordinates via Monte Carlo simulation that met these boundary conditions for the Bahn-City-Portal and the coordinates of the Micro Hub. The Micro Hub and the Bahn-City-Portal are considered white label solutions in the scenarios presented. This implies that from this point on, a delivery service provider delivers to the end customer. This allows the advantages or disadvantages of cooperation in the CEP sector on the last mile to be examined under the condition of a mutually independent middle mile.

The scenarios set up are described in detail below:
Scenario 0: "peripheral delivery"
In this scenario, the delivery service providers each separately deliver their parcels to the end customers in the delivery district. The optimized vehicle fleet of the delivery service providers consists of vans and they only deliver the parcels in the delivery district. After completing the optimized delivery routes, the fleets drive directly back to the individual regional distribution center of the respective delivery service provider. The responsible regional distribution centers of the delivery service providers were determined based on the data from Source [16]. The last mile supply chain thus contains the following sequence:

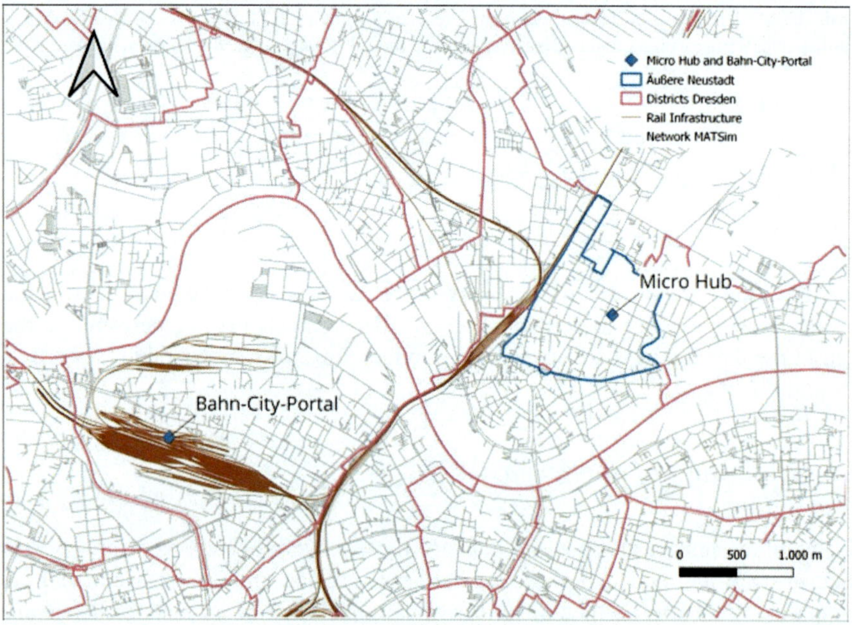

Fig. 3 Location of the Micro Hub and the Bahn-City-Portal. *Source* own depiction

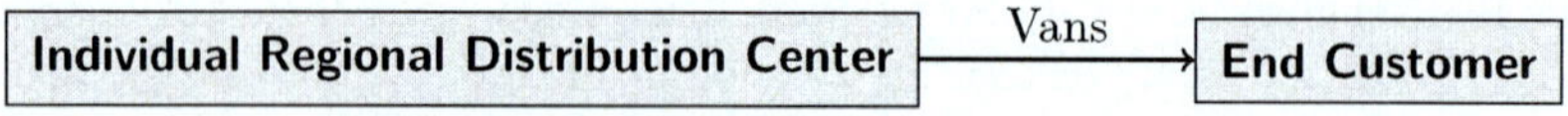

Scenario 1: "peripheral delivery centers in combination with Micro Hub"
In this scenario, the delivery service providers first deliver to the Micro Hub in the delivery district. The location of the Micro Hub is shown in the Fig. 3. Then, a white label fleet drives the parcels to the end customers, as previously explained. Delivery to the Micro Hub is optimized for each delivery service provider from a vehicle fleet consisting of trucks or vans. The delivery of end customers from the Micro Hub is carried out by a delivery fleet consisting of vans. In order to investigate the effects of time inconsistencies in the delivery, arising from the delivery to the Micro Hub on the investigated parameters, the unloading time of the vehicles is varied in the scenarios. The last mile supply chain thus contains the following sequence:

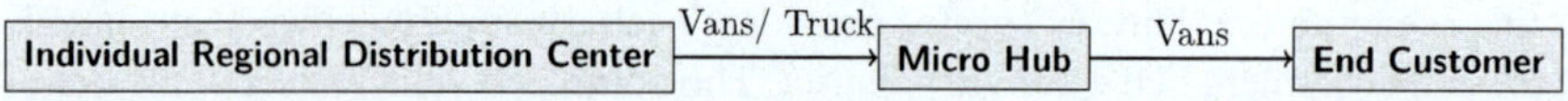

Scenario 2: "Bahn-City-Portal"
The Bahn-City-Portal is an inner-city intermodal goods handling center that ensures a middle route by rail and fine distribution in the last mile with delivery solutions such as vans with alternative drive concepts or cargo bikes. Additional information about the Bahn-City-Portal can be found in the publication by König et al. [7]. In this scenario, a delivery is made with the starting point of the Bahn-City-Portal. This

means that the middle mile is made by rail. From the Bahn-City-Portal, a white label fleet delivers to the end customers in the delivery district. The last mile supply chain thus contains the following sequence:

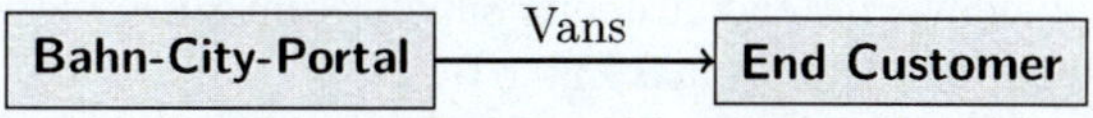

Scenario 3: "Bahn-City-Portal in combination with Micro Hubs"
The Bahn-City-Portal supplies the Micro Hub with one delivery fleet and no distinction is made between different delivery service provider. The delivery fleet consists out of trucks or vans when it comes to the delivery to the Micro Hub. As the starting point of the Micro Hub, the end customers are supplied with a delivery fleet consisting of vans. The last mile supply chain thus contains the following sequence:

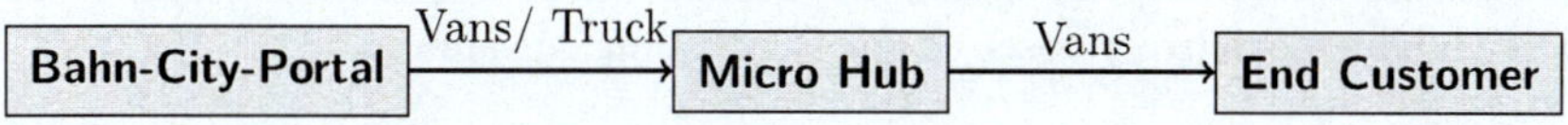

2.5 *Simulation Boundaries*

The capacity of the delivery vehicles and their specific parameters, which significantly influence the optimization of the delivery vehicle fleet, are summarized in Table 2. The capacity of the vehicle "Van" represents an average value from the analysis of the capacity by Bogdanski [2]. The capacity of the vehicle type "Truck" is a back calculation based on the volume of the vehicle, its interior volume and the space requirement per parcel calculated from the vehicle type "Van".

Table 2 Assumed costs of the vehicles (fixed costs, variable costs per kilometer and the time-dependent costs or personnel costs), weights of the vehicles and the capacities of the vehicles. *Source* own depiction

	Van[a]	Truck[a]
Permissible total weight [t]	3.5	12.0
Maximum capacity [PU/v]	165	695[b]
Fixed cost [€/d]	76	84
Variable cost per kilometer [€/km]	0,28	0,47
Time dependent cost or personnel cost per second [€/s]	0,0077778	0,0077778

[a]With data from [18]
[b]Own calculation

Summary of the boundary conditions used for the scenarios:

- Capacity of vehicles from Table 2
- Costs of vehicles and labour costs per time from Table 2
- Delivery time windows for end customer supply from 8 a.m until 3 p.m and for the supply of the Micro Hub from 3 a.m to 6 a.m
- No reloading of the vehicle in the depot (single route for each vehicle)
- No pickup of parcels on the route
- The waiting time due to delivery is set to 6 min for every stop (taken from [19]), as no better data is currently available
- The unloading time at the Micro Hub is set at 30 min; no values can be found in the literature; this time is used in Scenario 1 and Scenario 3
- We only simulate the traffic caused by the CEP sector in the district

In the following, we also simulate the influence of the unloading time at the Micro Hub on the delivery costs, as this is an estimated value.

2.6 Simulation in MATSim Using JSprit

The agent-based software solution MATSim (the Multi-Agent Transport Simulation Toolkit, www.matsim.org) is used in combination with JSprit for traffic flow simulation. JSprit is a proprietary open source software solution designed to solve Vehicle Routing Problems (VRP) and the Traveling Salesman Problem (TSP). JSprit is generally a proprietary software solution, but firmly integrated into MATSim through the Freight Distribution. In order to be able to integrate multi-link logistics chains (such as UCCs or Micro Hubs), the supply chain must be separated into two different parts [12]. The first part represents the delivery of the Micro Hub and the second part is the distribution of the goods starting from the Micro Hub to the end customer. The above mentioned boundary conditions and the Micro-Logistic-Database with the customer data were handed over to MATSim/ JSprit. JSprit optimizes the delivery fleet and delivery routes of the vehicles in the model for each scenario. The combination of the two software solutions can also be used for an overall traffic simulation, which, however, is not applied in this study.

3 Results of the Simulation Study

To compare the scenarios, we introduce three variables. These are the last mile supply chain costs (referred to as costs in the further course), the route distance and the delivery time. Last mile supply chain costs are the costs incurred by delivery. Costs arising from the regional distribution centers, the Micro Hub or the Bahn-City-Portal are not taken into account. The delivery time consists of the time required for transport and the unloading times during delivery.

Table 3 Results of the different simulation scenarios regarding delivering distances, delivering costs and delivering times seen relative to the base line scenario 0 for the whole last-mile delivery chain with 30 min unloading time in the Micro Hub scenarios with V for Vans and T for Trucks. *Source* own depiction

Scenario	Scenario 0	Scenario 1	Scenario 2	Scenario 3
Route distance [%]	0	–59	–80	–90
Delivery time [%]	0	–1	–60	–14
Waiting time [%]	0	+6.7	0	+1.4
Costs [%]	0	+16	–26	–23
Used vehicles	19 V	24[a] V, 2 T	15 V	15[b] V, 2 T

[a]From which 9 deliver to the Micro Hub and 15 deliver to the end customers
[b]From which 15 deliver to the end customer

As shown in Table 3 the simulation scenarios differ significantly from one another. It was shown that a Micro Hub supplied from the regional distribution centers without a change in the delivery fleet cause an increase in delivery costs (see Scenario 0 and Scenario 1). This can be explained by the increasing number of vehicles. The vehicles of the individual providers are needed to supply the Micro Hub and, in the second step, the vehicles of the Micro Hub operator are needed to supply the end customers (see 19 Vans for Scenario 0 and 24 Vans in Scenario 1 of which 15 Vans supply the end customers in Table 3). In Scenarios 2 and Scenarios 3, supply costs decreased by 26% and 23%, respectively. We observed a lower reduction in transportation costs in Scenario 3, which we attribute to the costs arising from the delivery of the Micro Hub. Subsequently, we were able to identify the waiting time in the Micro Hub as a factor affecting the cost of the scenario.

To elicit the effect of the unloading time in the Micro Hub on the simulation, we ran the simulation with three different unloading times. We used 30 min, 60 min and 90 min and show the effects of the iterated unloading time on the total cost structure of the supply and on the delivery time. The results are presented in Table 4. Table 4 shows that Scenario 1 is more dependent on the applied unloading times of the unloading process in the Micro Hub than Scenario 3. This can be explained by the additional number of vehicles required and the further distance to the Micro Hub. With an increase of 30 min in unloading time, the total cost also increased by 5% in the analysis. The total costs compared to Scenario 0 increased from 16% before to 21% with an unloading time of 60 min (see Table 4). Overall, Table 4 shows that for Scenario 1 there is a higher dependence of the two variables Cost and Transport Time on the unloading times at the Micro Hub than compared to Scenario 3.

In all comparison Scenarios (Scenarios 1, 2 and 3), we were able to determine a significant reduction in the route distance. Route distance in Scenario 1 decreased by 59% compared to Scenario 0, in Scenario 2 decreased by 60% compared to Scenario 0, and in Scenario 3 decreased by 90% compared to Scenario 0 (see Table 3. We attribute the reduction in route distance in Scenarios 2 and 3 to the inner-city loca-

Table 4 Influence of the unloading times (30 min, 60 min and 90 min) for Scenario 3 (BCP 30, BCP 60, BCP 90) and Scenario 1 (MH 30, MH 60, MH 90) on the cost structure and Transport Time of the Micro Hub supply. *Source* own depiction

	Summed Waiting Time for the used vehicles [min]	Share of the Micro Hub's waiting time costs in the total costs %	Development of total costs compared to Scenario 0 [%][a]	Transporting Times relative to Scenario 0 [%]
BCP 30	60	8.7	–23	–14
BCP 60	120	9.3	–22	–13
BCP 90	180	9.8	–21	–12
MH 30	330	38.7	+16	–1
MH 60	660	41.2	+21	6
MH 90	990	58.2	+26	13

[a] Sum of the costs from waiting and delivery last mile costs (delivery last mile costs: Fixed Costs, Variable Costs per kilometer and time dependent cost or personnel cost per second)

tion of the Bahn-City-Portal, and in Scenario 3 to the double consolidation effect of goods in the Bahn-City-Portal and the Micro Hub. In Scenario 1, the lower route distance is mainly due to the consolidation effect in the Micro Hub, the close location of the Micro Hub to the end customers, and ultimately the better routing options due to the consolidation. The reduction in route distance also means a reduction in emissions caused by delivery traffic.

The key figures of the supply chain evaluated so far should not be considered without an analysis of the link volumes caused by the CEP traffic. For this analysis, we consider only the traffic volumes generated by CEP traffic in the district. We examine the absolute traffic volumes per day (crossings of a link by CEP traffic per day) and the relative traffic volumes per day (crossings of a link by CEP traffic compared to link capacity).

The absolute traffic volume is used as an indicator in Figs. 4 and 5. The different magnitudes of the numbers for the colour scales in Figs. 4, 5, 6 and 7 should be noted. It is evident that the two scenarios (Scenario 0 and Scenario 1 in Figs. 4 and 5) differ in terms of absolute link volumes. One difference is the area of the Micro Hub. In this area, a significantly higher absolute link volumes can be observed in Scenario 1. A relief of some streets in the delivery district is noticeable in Scenario 1. This is evident from the results of Table 5 and Fig. 5.

Similar applies to the two Scenarios 2 and 3. The introduction of the Micro Hub in Scenario 3 has seen a reduction in traffic on the feeder roads to the delivery district in comparison to Scenario 2. Compared to the other scenarios, some streets have more traffic volume with the introduction of a Micro Hub in the delivery district. In general, the two scenarios with the Bahn-City-Portal put less strain on the streets in the delivery district than the two comparison Scenarios Scenario 0 and Scenario 1. This can be explained by the fact that the delivery in both scenarios is a white label solution. Therefore, Scenario 3 could show better values in terms of relative

Fig. 4 Absolute traffic volume resulting from CEP-services in Scenario 0. *Source* own depiction

Fig. 5 Absolute traffic volume resulting from CEP-services in Scenario 1. *Source* own depiction

Fig. 6 Absolute traffic volume resulting from CEP-services in Scenario 2. *Source* own depiction

Fig. 7 Absolute traffic volume resulting from CEP-services in Scenario 3. *Source* own depiction

traffic volume per day compared to the other Micro Hub Scenario Scenario 1 (also see Table 5). This applies especially in the area around the Micro Hub, which we identified as a critical area.

Table 5 Comparison of the four scenarios based on the different link volumes within the delivery district caused by the CEP-sector. *Source* own depiction

	Scenario 0	Scenario 1	Scenario 2	Scenario 3
Max. absolute link volume per day	16	52	15	34
Max. relative link volume per day	2,7%	8,7%	2,0%	5,7%
Avg. relative link volume per day	0,66%	0,56%	0,41%	0,48%
Avg. Absolute link volume per day	3,46	2,96	2,49	2,31

In addition to the visual analysis of the results, the key figures average relative volumes, average absolute volumes and the maximum absolute and relative volumes within the delivery district are evaluated. These analyzed metrics for the link volume analysis are shown in Table 5. The absolute and relative volumes per street in the delivery district determined by MATSim were taken. Access roads are not included in the calculations. The introduction of a Micro Hub (Scenario 1) decreases the overall mean link volume on the streets in the district from 0,66% (Scenario 0) to 0,56%. Also an increase from 2,7–8,7% in maximum relative volumes can be observed. This is due to the additional traffic generated around the Micro Hub (see Figs. 4 and 5). A decrease of the maximum absolute volume was achieved by combining a Bahn-City-Portal with a Micro Hub. We attribute this effect to the coupling of the two white label solutions and thus a better possibility to consolidate shipments. The maximum absolute volume value of this scenario is 34 vehicles per delivery day. In comparison, the maximum absolute volume value of the other Micro Hub scenario (Scenario 1) is 52. In addition, by combining the Bahn-City-Portal with the Micro Hub, a reduction in maximum and average relative volumes per day could be achieved compared to Scenario 1. However, the maximum relative volume could not be reduced by Scenario 3. We attribute this primarily to the concentration of CEP traffic in the Micro Hub area.

4 Conclusion

The limited data basis in commercial traffic models makes it difficult to model traffic flows accurately [4, 15]. There is a lack of more detailed information that are necessary for modelling like parcel sizes, age dependent parcel distribution, or parcel volume distribution in different districts of the city. These data are available to the delivery service providers but are considered company secrets and often not available for research purposes [10]. We present an experimental environment consisting of a multi-level model LOCAMM and the traffic flow software MATSim to evaluate delivery strategies in the logistics industry. In the presence of a total traffic model, this approach also offers the possibility of modeling commercial traffic and the interaction with individual traffic and public transport. At the time the study was conducted, we did not have a calibrated traffic model of Dresden that included individual traffic and public transport. To circumvent this limitation, we tried to restrict deliveries to off-peak hours of urban traffic in Dresden to still obtain valid results. Future research will be needed to compare the simulation with a full traffic model.

Nevertheless, an indication of the impact on costs, delivery times, delivery distances could be provided. We were able to prove the influence of cooperations in the CEP sector or the necessity to carry out traffic simulations before placing an inner-city consolidation center. In particular, we were able to show the impact of inner-city intermodal consolidation centers such as the Bahn-City-Portal or multi-link supply chains starting from the Bahn-City-Portal. In that an unloading time of 30 min is an estimate when the Micro Hubs are supplied, we modeled the two scenarios with Micro Hub (Scenario 1 and Scenario 3) with three alternative unloading times (30 min, 60 min and 90 min duration). We were able to show that the change in waiting times during the delivery of the Micro Hub has significant effects on the cost structure of the supply chain and on the time duration. The influence was higher the more vehicles were in use in total. Accordingly, the Bahn-City-Portal with Micro Hub (Scenario 3) was able to achieve better results than the comparison case in Scenario 1. This proved the influence of cooperation in the last mile area in the CEP sector and the opportunities created by an inner-city intermodal logistics center. It has also been demonstrated that the use of the Bahn-City-Portal can enable more cost-efficient delivery of goods in the CEP sector. For other areas of freight traffic, the analyses have to be carried out again, as these partly have different supply chain compositions (e.g [5]).

In order to work out further differences between the intermodal and monomodal inner-city distribution centers, it will also be necessary to model the middle mile in the future and the connections between more Bahn-City-Portals in different cities.

In the future, it will be also necessary to incorporate other sustainable logistics concepts such as parcel stations into the experimental environment in addition to the inner-city consolidation centers. However, this simulation would have to be carried out with an overall traffic simulation. This is due to the fact that savings potentials through parcel stations only arise if the parcels are picked up on foot, by bike or similar vehicle and not if the car is used for this purpose. In addition, it will be of particular interest to be able to load vehicles several times and to send them off from a depot several times a day. This has an impact on the fixed costs of the delivery fleet, which should be correspondingly lower with a smaller number of vehicles.

References

1. Browne M, Allen J, Leonardi J (2011) Evaluating the use of an urban consolidation centre and electric vehicles in central london. IATSS Res 35(1):1–6. https://doi.org/10.1016/j.iatssr.2011.06.002
2. Bogdanski R (2014) Nachhaltige stadtlogistik durch kep
3. Capgemini: The last-mile delivery challenge (2019)

4. Chow JYJ, Yang CH, Regan AC (2010) State-of-the art of freight forecast modeling: lessons learned and the road ahead. Transportation 37(6):1011–1030. https://doi.org/10.1007/s11116-010-9281-1
5. Dreischerf AJ, Buijs P (2022) How urban consolidation centres affect distribution networks: an empirical investigation from the perspective of suppliers. Case Stud Transp Policy 10(1):518–528. https://doi.org/10.1016/j.cstp.2022.01.012
6. KE-Consult Kurte & Esser GbR: Kep-studie 2020 – analyse des marktes in deutschland (2021)
7. König R, Schmidt M, Noennig JR (2014) Bahn-city portale - teil 2. In: Störungsmanagement (PRODUCTIVITY Management 1/2014). Gito-Verlag, pp 55–58
8. Landeshauptstadt Dresden-Kommunale Statistikstelle: Stadtteilkatalog 2019 (2019)
9. Laura Ramoneda Cuenca: Le problème du dernier kilomètre à paris intra-muros. introduction du paramètre environnemental dans la réservation d'aires de livraison (2008)
10. Llorca C, Moeckel R (2021) Assesment of the potential of cargo bikes and electrification for last-mile parcel delivery by means of simulation of urban freight flows. Europ Transp Res Rev 13(1) (2021). https://doi.org/10.1186/s12544-021-00491-5
11. Logistik Heute: Marktanteile der größten paketdienste in deutschland gemessen am paketvolumen im jahr 2019 [graph] (2021)
12. Martins-Turner K, Nagel K, Zilske M (2019) Agent-based modelling and simulation of tour planning in urban freight traffic. Transp Res Proc 41:328–332. https://doi.org/10.1016/j.trpro.2019.09.054
13. Melderegister der LH Dresden: Bevölkerungsbestand (2021)
14. Moutaoukil A, Neubert G, Derrouiche R (2015) Urban freight distribution: the impact of delivery time on sustainability. IFAC-PapersOnLine 48(3):2368–2373. https://doi.org/10.1016/j.ifacol.2015.06.442
15. Outwater M, Smith C, Wies K, Yoder S, Sana B, Chen J (2013) Tour based and supply chain modeling for freight: integrated model demonstration in chicago. Transp Lett 5(2):55–66. https://doi.org/10.1179/1942786713z.0000000009
16. Paketda!: Alle paketzentren in deutschland (2021). https://www.paketda.de/paketdepot-alle.html
17. Richter R, Söding M, Christmann GB (2020) Logistik und mobilität in der stadt von morgen: Eine expert*innenstudie über letzte meile, sharing-konzepte und urbane produktion: Irs dialog. http://hdl.handle.net/10419/219451
18. Schroder S, Dabidian P, Liedtke G (2015) A conceptual proposal for an expert system to analyze smart policy options for urban cep transports. In: 2015 smart cities symposium prague (SCSP). IEEE (2015), pp 1–6. https://doi.org/10.1109/SCSP.2015.7181555
19. Schäfer PK, Quitta A, Blume S, Schocke K, Höhl S, Kämmer A, Brandt J (2017) Wirtschaftsverkehr 2.0: analyse und empfehlungen für belieferungsstrategien der kep-branche im innerstädtischen bereich. Technical report, Frankfurt University of Applied Sciences: Fachbereich 1: Architektur Bauingenieurwesen Geomatik und Fachbereich 3: Wirtschaft und Recht
20. Thaller C (2018) Strategische verkehrsprognose. PhD thesis, Technische Universität Dortmund (2018). https://doi.org/10.17877/DE290R-19348
21. Stefanelli T, Di Bartolo C, Galli G, Pastori E, Quak H, Smart choices for cities-making urban freight logistics more sustainable
22. United Nations: World urbanization prospects-the 2018 revision (2019). https://population.un.org/wup/Publications/Files/WUP2018-Report.pdf
23. Veličković M, Stojanović Đ, Nikoličić S (2018) Maslarić M (2018) Different urban consolidation centre scenarios: impact on external costs of last-mile deliveries. Transport 33(4):948–958. https://doi.org/10.3846/16484142.2017.1350995

Forecasting Parking Search Times Using Big Data

Kleio Milia, Magnus Duus Hedengran, Thomas Jansson, Filipe Rodrigues, and Carlos Lima Azevedo

Abstract Searching for parking (also known as cruising) is one of the key contributors to urban congestion and consequently air pollution. Providing information about parking availability is already present in many cities, efficiently guiding drivers to vacant parking spaces, either based on past data or using technologies for real-time information. Informing drivers about the expected cruising time at their destination is an additional pathway, which can impact their departure time and mode choices, and consequently improve the overall mobility system performances. One of the main challenge of cruising time estimation lies within how cruising itself is detected. This study examines the case of detecting cruising using GPS traces from trajectories. A new parametric detection method is proposed establishing speed and acceleration/deceleration related conditions. A sensitivity analysis to the method is presented along with a comparison against existing and similar GPS-based cruising detection methods and validation against labelled data. After the cruising detection, about 800,000 GPS trips were used to estimate and validate an offline machine learning algorithm to forecast the cruising time in three different urban areas in the City of Copenhagen, Denmark, with clear distinct parking conditions. Neighbor-

K. Milia (✉) · M. D. Hedengran · F. Rodrigues · C. L. Azevedo
Management Department, Transport Division, Technical University of Denmark, Bygningstorvet 116B, 2800 Kgs. Lyngby, Denmark
e-mail: kleiomil@gmail.com

M. D. Hedengran
e-mail: magnus.duus@gmail.com

F. Rodrigues
e-mail: rodr@dtu.dk

C. L. Azevedo
e-mail: climaz@dtu.dk

T. Jansson
Data Science, Connected Cars A/S, Banemarksvej 2B, 2600 Glostrup, Denmark
e-mail: tj@connectedcars.dk

C. Antoniou et al. (eds.), *Proceedings of the 12th International Scientific Conference on Mobility and Transport*, Lecture Notes in Mobility,
https://doi.org/10.1007/978-981-19-8361-0_12

hoods were divided into spatial cells for which hourly cruising times were estimated. Feed-Forward Neural Network (FFNN) and eXtreme Gradient Boosting (XGBoost) architectures were tested as machine learning algorithms and outperformed a simple moving average (RMSE gains from 62.01 to 52.57 s). The present study paves the ground for the exploration of large datasets with GPS trajectories in urban areas for tackling the lack of information on parking search. Despite the improved overall prediction power, the potential errors from the cruising detection method, lack of data needed to capture patterns when cruising time is high or the existence of many missing values due to aggregation of data could be the reason for the observed algorithm's inability to predict the larger values of cruising time.

Keywords Parking search · GPS · Cruising · Big data · Forecasting · Machine learning

1 Introduction

Today, congestion continues to be a serious problem in many cities of the world because of excess time spent in traffic resulting in substantial greenhouse gas emissions. Searching for parking (cruising) can lead to both excess travel by itself, but it also contributes to congestion during peak hours [1]. The problem is magnified in megacities where people are likely to park illegally which increases traffic congestion, especially during peak hours, and consequently affects the economy and the environment [2, 3]. Therefore, insufficient parking availability and/or information thereof can lead to motorists driving in excess near their destination in search of vacant parking spaces.

The extend of cruising in cities remains unclear. Some studies have given their suggestions for methods and estimates of the extend of cruising in certain cities [4]. But even though, the conceptual definition of cruising is agreed upon, it cannot be easily translated to detection methods.

There are different ways to reduce the extend of cruising in cities, for example by increasing the supply of parking spaces or by revising the pricing scheme for parking (and its enforcement). Often the means are expensive to be implemented or may have other undesirable outcomes (e.g. increased car demand). If the time spent on cruising can be detected, an inexpensive and flow-regulating way to reduce cruising time could be simply to provide drivers with the information of the current (and future) expected cruising time near their destinations. This smart solution can potentially facilitate improvements of the parking conditions in cities accompanying the already existing parking technologies.

In this study we extend the existing literature with the exploration of big data to: (1) design and estimate a method for detecting and quantifying cruising, and (2) forecast expected cruising times in a relative high resolution of subareas within a city using a machine learning algorithm. These two objectives rely on an extensive data set of GPS traces for three months in 2019 and 2020 for Copenhagen, Denmark.

2 Cruising for Parking Background

2.1 Parking Search Behavior

It is generally accepted that parking search occurs when drivers approach their destination and gradually increase their awareness in search of vacant parking spaces. Willingness to park is complex and is based on the cost of parking, the level of rule enforcement, potential parking guidance systems and any prior knowledge of the spatial and temporal distribution of parking opportunities [5]. Parking search can be described as the process by which drivers are able to match their preferences for parking with the present parking availability [6].

2.1.1 Cruising Definition

Parking search is often referred in literature as *cruising*. The cruising part of a trip can be described in a temporal (time spent on cruising) and spatial dimension (chosen route while cruising) [7]. Kaplan and Bekhor identified a sequence of decisions the driver has to take during his parking process [8] which are: (1) decide when to start parking search, (2) determine the type of parking space, (3) choice of parking facility (if type is off-street) and (4) route choice. Laurier defines the decision process of parking as a 3-step process [9]. Firstly, the driver enters the area or block for parking search and quickly tries to understand the parking conditions in the area by observing the already parked cars (e.g. double parking) and traffic conditions. Afterwards, in the first priority street she begins decreasing her speed in order to increase her awareness in search for parking availability. Lastly, there is a circling close to the destination and the cruising process in this area stops when the driver exceeds her limit of times traversing the street of main interest and eventually end ups in a further location.

Shoup stated that cruising is "*to wait in a queue of unknown length, where the next person called to the window is determined by lottery*" [4]. This statement reveals the uncertainty of cruising time duration, the randomness of the next vacant parking space and the difficulty the driver has to face on estimating the cruising time at his destination prior to his trip. Weinberger et al. [10] concludes that only a subset of all trips include cruising and that some of them can never be cruising, such as when the end position of the car is a reserved parking spot or when the trip purpose is to drop off someone.

There are many factors that have an influence on the duration and length of cruising for parking. Longer stays in on-street parking spaces leads to high occupancy rates which increases cruising time of other drivers. This also affects walking distance since drivers looking for parking in a specific area will potentially move further from it in the case of many occupied spaces [11]. This means that motorists can change their initial preferences for parking close to their destination and end up in a parking position which was not an option when cruising started [9]. Parking prices also raise the incentive of drivers to cruise for parking hoping to find lower parking charges according to their preferences. However, even in the case when the level of parking prices meets the needs of the driver, he may also cruise due to unfamiliarity with the area and the lack of knowledge of usual vacant spaces or off-street parking areas [12]. Some other studies have examined sociodemographic characteristics of drivers, such as sex, age, income and education, to investigate potential correlations to cruising time [12, 13].

Through paper, the term *cruising* is widely used, meaning the act of driving while searching for a parking space.

2.1.2 Cruising Detection and Estimation

Despite a consistent cruising definition, the exact point at which the driver starts cruising is a rather controversial issue. In literature, there are two dominant used methods to analyze and quantify cruising for parking: GPS traces and videotaping driver's behavior. In each method, after collecting the needed data, the cruising part was quantified based on different criteria: driver's body movement, speed profile, comparison between traversed and shortest path to the destination and distance from the final destination when the driver starts searching for parking, as shown in Table 1.

Once a detection method and criteria for cruising has been defined and applied, the resulting data can be used to estimate cruising time of trips. Of course, the estimations of cruising time are directly affected by the detection method. More recent studies have showed that the average cruising time in cities is generally lower than the estimates of 3.5 to 14 min by Shoup [4]. Newer estimates of average cruising time in cities range between about 2 to 3 min [10, 20]. However, since cruising time can vary over time and location, neither of the results can easily be rejected.

Table 1 Quantifying cruising

	Driver's behavior	Average cruising speed (km/h)	Shortest path	Distance to destination (m)
Zhu et al. [14]	Yes	13.53		
Martens et al. [15]		12		
Hampshire et al. [16]	Yes		300 m (70%)	100
Jones et al. [17]		15		400
Chen et al. [18]		<10		
Belloche [19]		9.87		
van der Waerden et al. [7]		<23		
Weinberger et al. [10]			>200 m	400

3 Study Areas

The City of Copenhagen, the capital of Denmark, consists of 4 municipalities: Copenhagen, Frederiksberg, Tårnby and Dragør [21]. Copenhagen and Frederiksberg Municipalities are the most densely populated in Denmark [22]. A significant percentage of car trips ending in these two areas may therefore be cruising and both have different parking conditions.

Copenhagen Municipality (CPH) comprises a zone system for the pricing of on-street parking spaces. The general idea is that parking is more expensive closer to the city center and during working hours (8 a.m. to 6 p.m.). The parking prices per hour vary between 3 DKK per hour to 39 DKK per hour (while being free during weekends) [23]. Frederiksberg has different rules and prices for parking. It is free to park for the first 2 h and then parking is priced around 20 DKK per hour. A 24 h ticket with a fixed price of 90 DKK is available for a parking duration of more than 6 h [24].

Within Copenhagen Municipality, the neighborhoods of Vesterbro and Indre Nørrebro had the highest parking occupancy rates in the city in 2018 with 90-108% of on-street parking spaces being occupied in three different time slots during a weekday (12 a.m., 5 p.m., 10 p.m.) [25]. On the other hand, Islands Brygge had the lowest occupancy rates ranging from 71% to 79% during the three time slots.

In another study, the time spent on searching for parking was estimated [26] through a sample of 2,441 trips. Workers and residents in Copenhagen were recruited, asked to use a tailor made smartphone application and registered the moment at which they would start search for parking space while driving. The study showed that Vesterbro area had the largest expected cruising time with 250 s between 8 p.m. and 12 p.m. during workdays.

Three different neighborhoods were chosen for the study: Frederiksberg, Vesterbro (CPH) and Islands Brygge (CPH). Vesterbro was chosen due to its high occupancy rates and expected cruising time. Since Islands Brygge has comparably low occupancy rates, it was also selected in order to compare its cruising time values to Vesterbro. Frederiksberg was chosen due to its significant differences in parking policy. The extend of the study areas is shown on Fig. 1.

4 The Data

The data set for this study was provided by *Connected Cars A/S* (https://connectedcars.dk/). The company collects and stores a broad range of data in large volumes from smart devices installed in cars. The data can be applied for efficient car and fleet management to keep track of vehicle trajectories and performance. The Connected Cars A/S data contains trip trajectories for the connected car fleet of more than 100,000 cars in Denmark (more than 3% of the total Danish car fleet) [27, 28]. The company covers data mainly from the Volkswagen Group (e.g. Volkswagen,

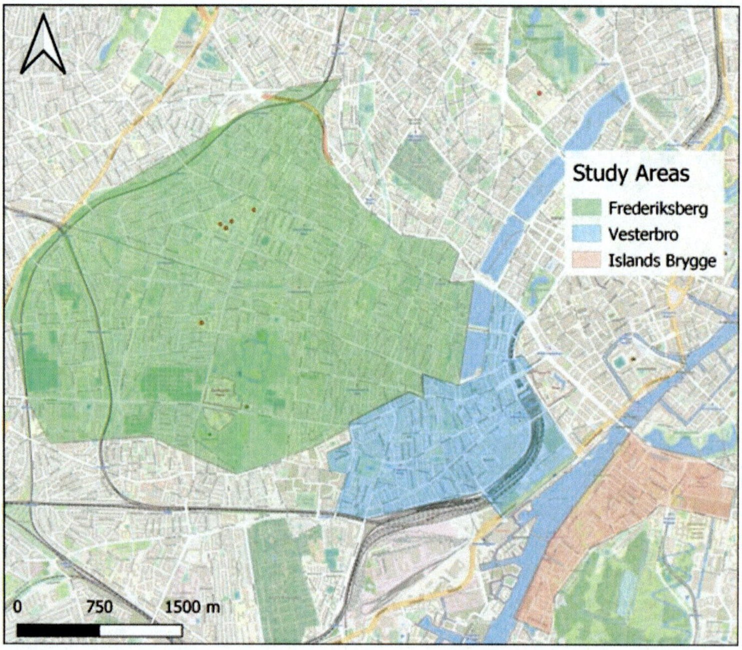

Fig. 1 Selected neighborhoods

Skoda, Audi) and the distribution of car segments is representative of the Danish car fleet [27].

For the present project, Connected Cars A/S provided a data set containing pseudo-anonymized information of trip trajectories and vehicle types. The data set contains GPS positions of trips having their final destination in either of the three study areas, Frederiksberg, Vesterbro or Islands Brygge. The chosen time period was selected to be September to November in 2019 and 2020 respectively. The idea was to focus on days outside vacation periods, e.g. summer or Christmas holidays, in order to reflect better the normal traffic conditions during workdays. In addition, investigating potential differences in cruising time before and during the COVID-19 pandemic was deemed as an interesting topic worth examining. During the autumn period of 2020 there was no lockdown in Denmark and travel patterns were likely more normalized compared to the initial lockdown in spring of 2020.

Two separate data sets were provided: one of 577,349 trips from 2019 and another one of 674,762 trips from 2020. The total number of unique registered cars in the data set was 32,105 and 37,422 in 2019 and 2020 accordingly. For GPS traces, the available information is shown in Table 2. To preserve the anonymity of the users the exact starting location was not provided to the authors and there was no intention of driver recognition. Additionally, the identifier for unique cars was a sequential number with no relation to the vehicle identification number (VIN). Given the data

Table 2 Data set columns

Data column	Format	Description
CarId	Number	Sequential identifier for car
TripId	String	Sequential identifier for trip
FuelType	Categorical	Type of fuel (diesel, gasoline, hybrid, electric)
CarSegment	Categorical	AF classification (A—mini car, B—small car, C—medium car, D—large car, E—executive car, F—luxury car, M—multi purpose car, J—sport utility car, S—sports car, V—van) [29]
Start GPS position	(Latitude, longitude, timestamp)	This GPS position is the same in every row of the corresponding trip
End GPS position	(Latitude, longitude, timestamp)	This GPS position is the same in every row of the corresponding trip
Recorded GPS position	(latitude, longitude, timestamp)	Each recorded GPS point corresponds to one row
Drive duration	Minutes	Duration of the whole trip
Consumed fuel	Liters	Estimated used fuel in liters of the trip
Idle time	Seconds	Idle time of a trip
GPS direction	Degrees	Bearing from the North
GPS speed	km/h	Instantaneous speed recording from GPS unit

set, it was not possible to determine the drivers' parking preferences and trip purpose, which would have assisted the project.

Trips having their end point outside the boundaries of Frederiksberg, Vesterbro or Islands Brygge, were excluded from the analysis. The number of the final studied trips is shown in Fig. 2. There are two reasons explaining the increase in the number of trips in 2020 compared to 2019. Foremost, Connected Cars A/S increased its car fleet in 2020 (16.2% more cars in 2020 data set), hence there are more observed trips (17.5%) during this period. A second potential reason is the COVID-19 pandemic. During the study period of 2020, the number of cases started to increase again in Denmark [30] as a consequence of the second wave of COVID-19 spread in Europe starting around mid October [31]. Furthermore, since March 2020, recommendations to avoid unnecessary public transportation were published in Denmark, which could have led to a shift to private cars to occur [32].

The analysis revolves around only the cruising part of trips and the precise trajectory far from the end position of each trip is unimportant. Therefore, all rows of GPS

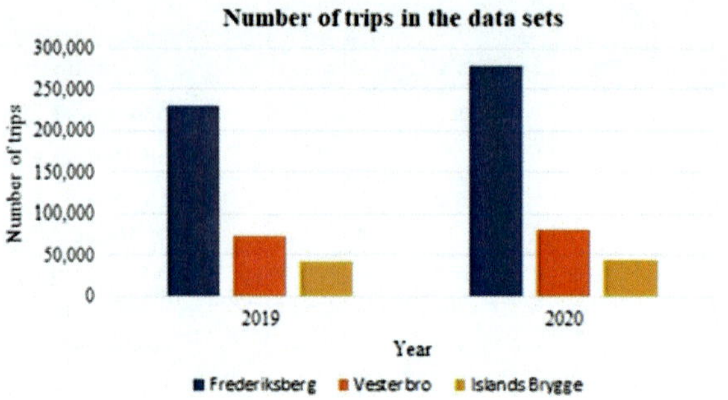

Fig. 2 Number of trips in the data sets

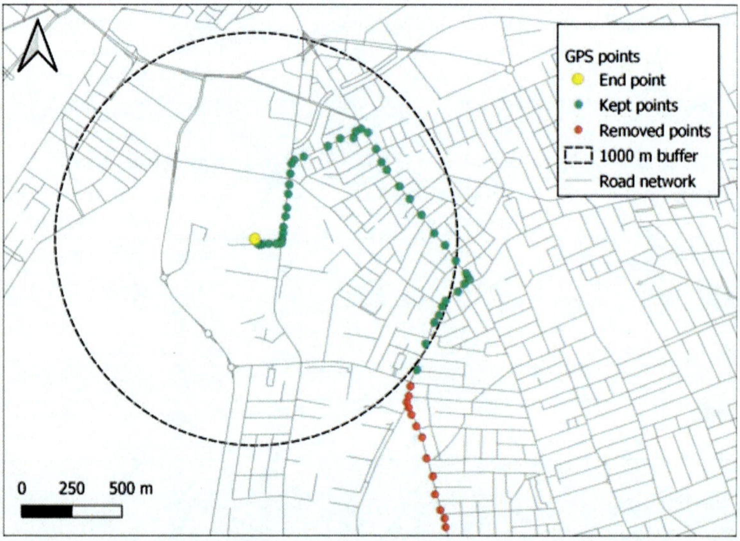

Fig. 3 Example of GPS traces removal in a trip

traces before the vehicle entered the area with a radial distance of 1,000 meters from the end point of the corresponding trip were excluded. After this point, the whole trajectory of each trip was kept, i.e. if the vehicle again drove out of the 1,000 meter circle, it was still captured in the data set. One GPS point before entering the circle was also kept in order to ensure at least 1,000 meters of travel. Figure 3 shows an example of the GPS points removed for a specific trip.

The speed feature was recalculated due to missing values in GPS speed column. The haversine formula was used for computing the distance between GPS points [33].

5 Detecting Cruising

Given the provided data set two pathways were possible to be examined: (1) investigating the speed profile of the trips (2) comparison between the traversed and shortest path. However, these criteria come with their disadvantages. Driving speed depends on many parameters, such as traffic volume, type of road and capacity. Hence, the speed might decrease for other reasons than cruising. On the same principle, drivers might choose a different route from the shortest path because of road work, missed turns, preference over the fastest and not the shortest path, area unfamiliarity or they may find a vacant parking space along the shortest path to their destination.

Considering the above, the initial approach of detecting cruising was to use the method given by van der Waerden et al. [7]. In their analysis using GPS traces from the city of Turnhout in Belgium, they concluded that the starting point of cruising is when the average speed of five consecutive time periods falls under 23 km/h and the speed difference between each pair of GPS points is below 5 km/h. The detailed analysis about how they ended up with these speed thresholds is not publicly available. However, it is mentioned that they adjusted the thresholds after several iterations to numbers that provided the least misinterpreted trips. In essence, it was perceived as a good starting point for investigating cruising on GPS traces.

After the first time of applying these criteria to the data set, small changes took place to enhance its performance for this case study. Firstly, weighted average speed was calculated, having the travel time between each pair of consecutive points as the weight, instead of average speed as used in the referred study [7]. This comes from the fact that the time resolution in the GPS data set is not constant. Furthermore, it was observed that in most cases speeds close to 0 km/h were the reason of having an average speed less than 23 km/h because drivers were not cruising but just starting and stopping before and after a traffic light. This is why a condition which only accepts speeds > 1 km/h was added to the condition for starting of cruising as well as a 5 km/h lower bound for weighted average speed.

Another issue had to do with *all* the GPS points after the starting point of cruising in the trip. In some cases after this point, the rest of the speed observations were below 10 km/h and the distances to the end position were lower than 10 m which was interpreted as the parking part of the trip where the driver tries to fit his car in a parking space and that was the reason why the weighted average speed was reduced. Therefore, an extra condition to the algorithm was to avoid stating as starting of cruising points where in the whole remaining part of the trip, the speeds and distances to end position are under 10 km/h and 10 m respectively. All these changes led to the following results comprising the so-called Base Scenario (see Figs. 4 and 5; note that the neighborhoods are labelled using the acronyms FB, VB and IB for Frederiksberg, Vesterbro and Islands Brygge, respectively).

After adjusting the cruising criteria, a set of changes to the thresholds was run to examine the sensitivity of the output (number of cruising trips) to generate new pathways of cruising investigation and additional modifications to improve and adapt the criteria to the study areas. These threshold changes refer to the weighted average

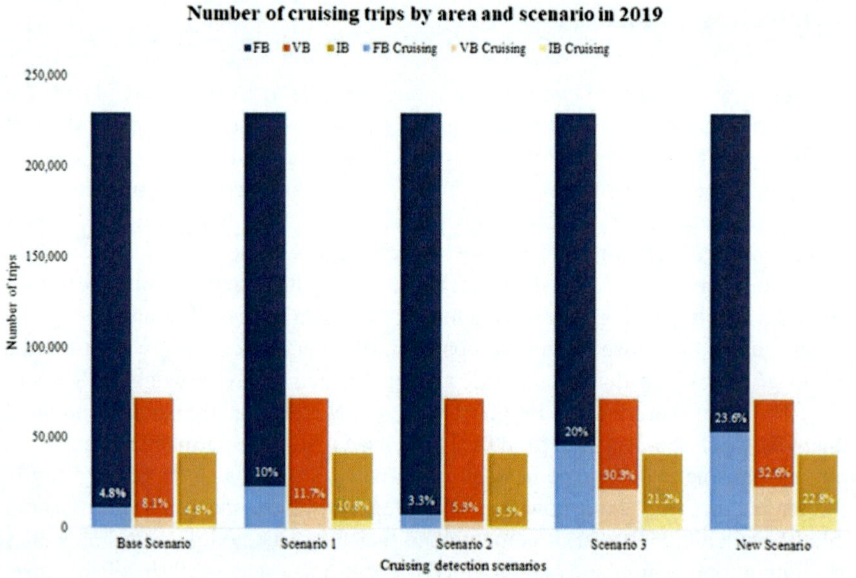

Fig. 4 Number of cruising trips in 2019

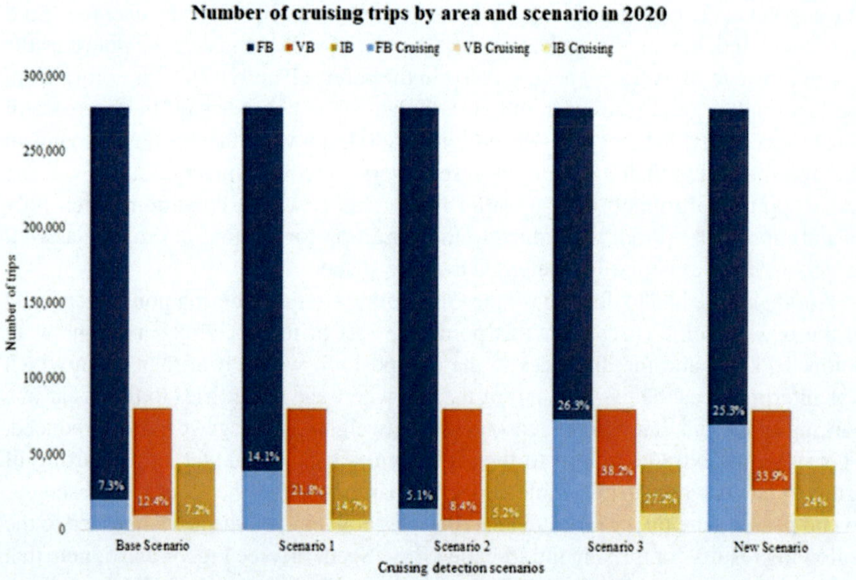

Fig. 5 Number of cruising trips in 2020

Table 3 Parameters of scenarios

Parameter name	Base scenario	Scenario 1	Scenario 2	Scenario 3	New scenario
Weighted average speed (km/h)	23	23	17	23	21
Time periods	5	4	5	5	4
Speed of time periods (km/h)	>1	>1	>1	>1	>5
Speed difference of time periods (km/h)	≤5	≤5	≤5	≤10	–
Acceleration/deceleration rates of time periods (m/s^2)	–	–	–	–	<0.51
Weighted average speed—rest of the trip (km/h)	≤23	≤23	≤17	≤23	≤21
Speeds to end position (km/h)[a]	<10	<10	<10	<10	–
Distances to end position (m) [b]	<10	<10	<10	<10	–

[a] Not *ALL* the speeds to end position should follow the condition
[b] Not *ALL* the distances to end position should follow the condition

speed, the number of consecutive GPS points—time periods to detect the starting point of cruising and the speed difference. Thus, three scenarios were created for the analysis: (1) change the number of examined consecutive GPS points from 6 to 5 (4 time periods) (2) decrease the weighted average speed threshold from 23 km/h to 17 km/h (3) increase the allowed speed difference between points from 5 km/h to 10 km/h (see Table 3). The new thresholds in Scenarios 1 and 3 are related to the variable GPS time resolution. In Scenario 1, the number of consecutive GPS points being examined did not increase, because the algorithm might end up examining the start of cruising based on the speed values of 1 or more minutes ago. In Scenario 3, considering that GPS traces are received in high resolution especially when the car turns, small errors in the GPS location may have a significant effect on the calculated speed, thus the speed difference threshold increased. Lastly, in Scenario 2, the weighted average threshold was decreased to be closer to other studies shown in Table 1.

At first, the low percentages in Scenario 2 compared to the others were noticed which are even lower than in the Base Scenario. This is because the condition for the weighted average speed being less than 17 km/h in the whole cruising part of the trip eliminates many trips. By observing the trips that were not perceived as *cruising* in the Base Scenario, it was derived that Scenario 1 captures more “*real*” *cruising* trips as perceived by the authors than the Base Scenario. The Base Scenario is stricter in cases where there are a few GPS points along a road before the driver turns and thus the speed difference condition does not stand. In Scenario 1, more trips are perceived as *cruising* than before which appears to be more correct. The same issue applies in Scenario 3. The lost cruising trips due to the restrictions in the Base Scenario are captured also by Scenario 3 when the speed difference increases to 10 km/h. However, this scenario considers as *cruising* also trips that end up in loosely populated residential areas (mostly in Frederiksberg).

5.1 New Cruising Detection Method

Taking into account the results in the Base Scenario and the sensitive features of time periods and speed difference, it was decided to examine more thoroughly the trips and try to understand the driving patterns during cruising in the given data set. The aim was to build a perceived truth, if possible, for the cruising part of the trips as perceived and derived by the authors.

150 randomly selected trips from 2019 data set were analyzed. Three tools were used in order to have a better understanding of each trip: (1) the profile of the trip, mainly considering the speed, distance to end position, travel time between points and weighted average speed of 5 and 6 consecutive GPS points (4 and 5 time periods respectively), (2) the map matched route to visualize the trajectory and (3) Google Maps to see the type of the end position of each car (e.g. private or public parking space) as well as to understand the type of roads that were used. The matching of each trip was done using the readily available algorithm *pgMapMatch*, specifically designed to match trajectories with parking behavior to an underlying road network [34].

After completing this process, the most important insight was the wrong interpretation of the starting point of cruising in the Base Scenario. In that case, cruising was starting when the latest 5 consecutive GPS points together with the starting point of cruising (so in total 6 consecutive points) fulfilled the cruising conditions. That is to say, these 5 latest points were not included in the cruising part of the trip. That definition implies that the cruising starts after the driver has reduced his speed to the corresponding threshold. However, this statement is not in line with the authors' perception of cruising, which also includes the part of the trip when the driver starts decreasing his speed in order to look for a parking space. Cruising makes for a change in the driving behavior and it is important to capture this shift from the beginning and not a few seconds or minutes later.

After manually labelling the cruising trips, several iterations were executed to find the appropriate criteria and thresholds for cruising. It was noticed that the speed does not change significantly when searching for parking, which means that its rate may remain under a certain limit. Considering that, it was observed that the cruising starts when the weighted average speed of 4 time periods falls under 21 km/h, where speeds and acceleration/deceleration rates along these periods are greater than 5 km/h and less than 0.51 m/s^2 respectively (which value is close to [14] study). The weighted average speed should remain below 21 km/h in the rest of the trip. The percentages of cruising trips using these criteria are shown in Figs. 4 and 5.

Figures 6 and 8 illustrate the distribution of cruising time in 2019 and 2020 respectively. Both graphs are positively skewed with a few outlier values reaching up to approximately 5500 s meaning that most drivers experienced a relatively short cruising time, approximately less than 5 min. After going through some of the trips with cruising time more than 3000 s, it was observed that after the driving part, the cars remained idled at the end position and their engines were still turned on. Thus, the GPS detector kept receiving GPS data of the car resulting in more GPS points for

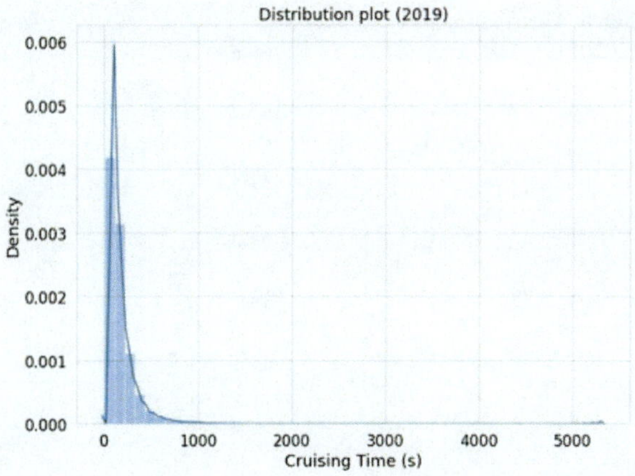

Fig. 6 Distribution plot (2019)

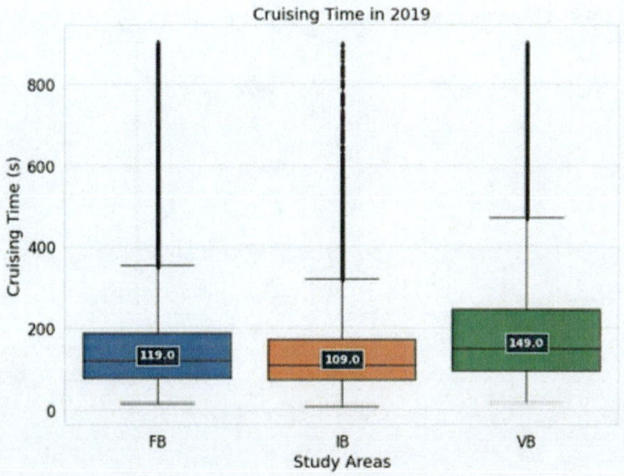

Fig. 7 Cruising time in 2019

these trips. For better visualization of the results these outliers were excluded from Figs. 7 and 9.

The median value of cruising time in Vesterbro is larger compared to Frederiksberg and Islands Brygge (see Figs. 7 and 9). In 2019 in Vesterbro it was 25% larger than in Frederiksberg and 37% larger than in Islands Brygge. This relation almost holds in 2020, where they were 25% and 31% respectively. Moreover, a decrease in median values can be easily observed from 2019 to 2020 in all study areas. Since more people likely preferred the car to move around the city, there was potentially a circulation of cars during the day resulting in parking spaces becoming vacant more frequently than before. The sizes of the boxplots indicate that cruising time in Vesterbro and Frederiksberg fluctuate more in comparison to Islands Brygge.

Another computed feature was the distance from the end position when drivers start cruising. The median distance to end position was 203 m, 168 m and 209 m in Frederiksberg, Islands Brygge and Vesterbro respectively. There are also some trips starting cruising in a distance little higher than 1 km before reaching the end

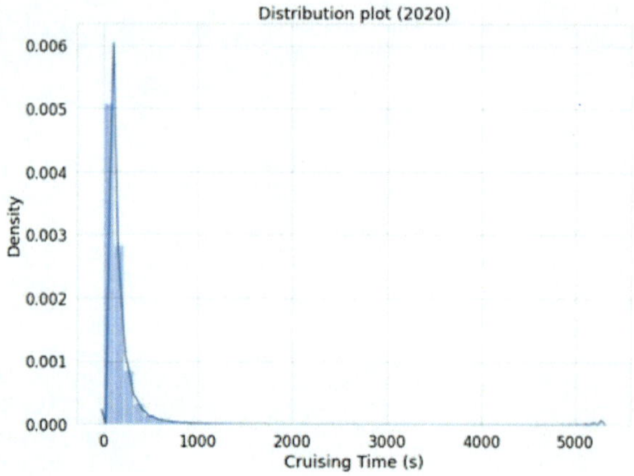

Fig. 8 Distribution plot (2020)

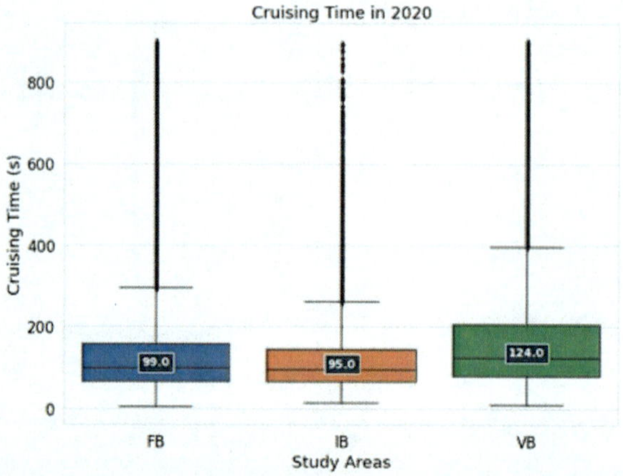

Fig. 9 Cruising time in 2020

position but they account for 1% of the whole data set. This means that there are cases where the driver exits the circle with a 1 km radius from the end position after starting cruising or the first time that the driver was within a 1 km distance from the end position was at the starting of his trip, which means that on the way back to the same area he started cruising outside the circle. In 2020 the median distances from the end position were shorter than before (160 m, 138 m and 167 m keeping the same order as before) whereas the percentage of trips starting cruising in a distance more than 1000 m from the end position fell to 0.35%.

In order to validate the new criteria, several trips from the sample of 150 trips were investigated before and after applying the cruising conditions. An example is illustrated in Fig. 10, where only the suggested criteria identified it as *cruising*. Along the road after the black dot, there are many on-street parking spaces, which was the main reason why this trip was perceived as *cruising* during the investigation process, and the new criteria detected the starting point of cruising the point when the car entered this road. The reason why the previous criteria did not capture this trip

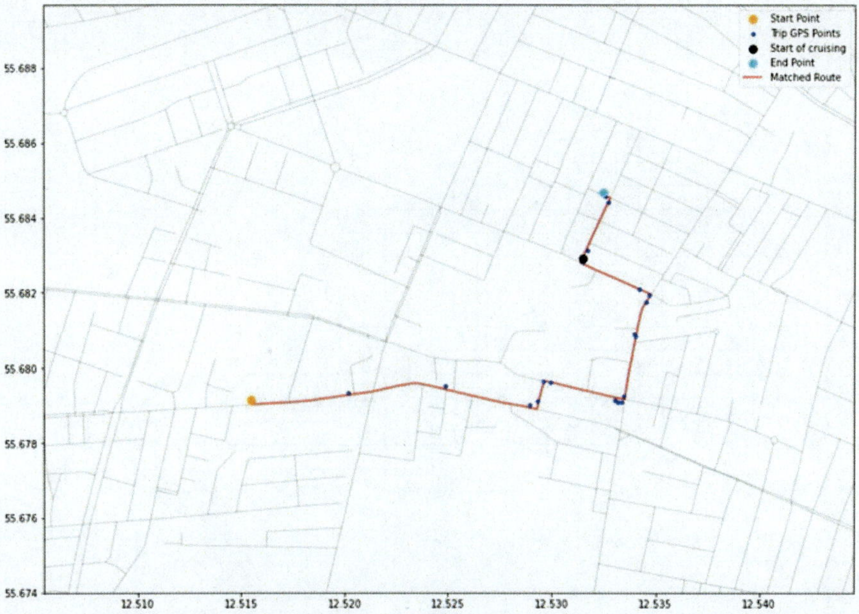

Fig. 10 Trip interpreted as "cruising" by the New Scenario

was the speed difference threshold where the speeds before and after the turn were computed to be 35 km/h and 7 km/h respectively. In addition to that, the next speed after the starting point also outreached the speed difference threshold. Therefore, it was not possible for the Base Scenario to identify the cruising part of this trip. In the study on which the construction of the Base Scenario was based, there was a standard resolution of 3s [7] which probably is a requirement to apply the speed difference condition.

In general, most of the examined trips were correctly classified as trips including *cruising* or *not cruising* using the new criteria. There were also a few cases which correspond to likely misinterpreted trips. False-positives cannot be completely avoided given a method of strict thresholds on speed and acceleration. All in all, the New Scenario seems to perform better compared to the other scenarios since it has a higher success rate and it was decided to proceed with this. Since the detection of cruising is rather controversial and subjective, in this study the New Scenario was considered as the best method to the authors to identify the trips that are cruising.

6 Forecasting Cruising Time

After applying the cruising criteria to the data set, each trip is split into a cruising part and a non-cruising part. The objective is to apply different ML methods in order to forecast cruising time. In order to do so, the data set was first modified from

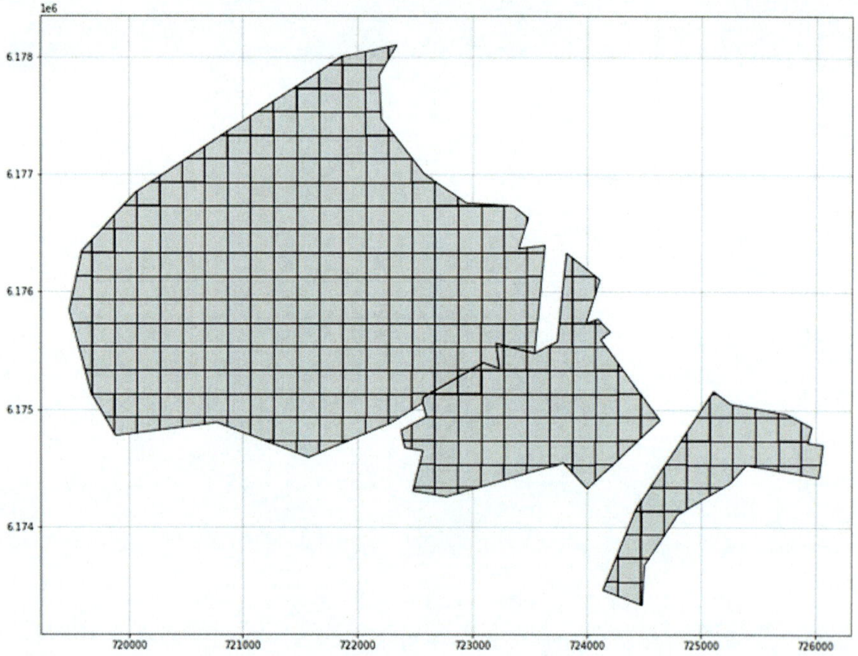

Fig. 11 Grid network for spatial aggregation

a GPS-point based data set into a trip-based data set, by grouping and calculating the time spent on cruising for each trip. Then a grid network was set up for spatial aggregation into square cells with sides of 200 meters (see Fig. 11).

The data set was also set up with a temporal aggregation by aggregating into 1 h intervals. The resulting dataframe was a table with one row for each combination of cell and hour. The average cruising time for a specific cell in a given hour was obtained by calculating the average cruising time spent for all trips (either cruising or not) having their end point in that cell and within the time period of interest.

The dataframe was enriched with a number of explanatory variables from different sources to estimate the expected cruising time (see Table 4). A number of explanatory variables was also derived directly from the GPS based data from Connected Cars A/S.

In some cases, no trips were conducted within a cell for a specific hour (in a row of the dataframe), therefore the expected cruising time was unknown. Since the prediction algorithms depend on previous values of cruising time, these unknown values had to be handled in some way. This was done by first reducing the data set and second filling the unknown values.

Looking at the distribution of unknown values by hours and cells respectively, it was apparent that certain hours and cells had many rows of unknown cruising time. For example, hours during night and cells covering water or parks had very few trips ending there. For this reason, rows with hours between 7 p.m. and 6 a.m., cells with

Table 4 Explanatory variables

Data column	Format	Description
Grid ID	Numerical	Sequential identifier for grid
Grid centroid	(Latitude, longitude)	Coordinates of grid centroid
Parking supply	Numerical	Number of public on-street parking spaces (Source: Open Data DK [25, 35])
Parking price	DKK	Price of parking per hour (Source: Open Data DK [25, 35])
Time parking restriction	Hour	Upper limit of parking time in hours (Source: Open Data DK [25, 35])
Temperature	Celsius	Average hourly temperature (Source: WorldWeatherOnline [36])
Precipitation	mm/h	Average hourly precipitation (Source: WorldWeatherOnline [36])
Number of non-cruising trips	Numerical	
Number of cruising trips	Numerical	
POIs	Numerical	One column for each category (Leisure facilities, University, Bus stop, Libraries, Schools, Shopping Center, Sport facility, Train station) (Source: Open Data DK [25, 35])
Familiarity	Numerical	Percentage of familiar trips
Study area	Categorical	Frederiksberg, Vesterbro, Islands Brygge
Historical cruising time	Numerical	Average cruising time in the same weekday/hour in the past 4 weeks
Cruising time of nearby cells	Numerical	Average cruising time of nearby cells in the past 4 weeks
Land use	Categorical	Continuous urban fabric, Discontinuous dense urban fabric, Discontinuous low-medium density urban fabric, Green urban areas, Industrial/commercial/public, military and private units, Sports and leisure facilities (Source: Urban Atlas [37])
Hour of day	Categorical	
Weekday	Categorical	
Year	Categorical	
Autumn break	Binary	0—for non autumn holiday, 1—for autumn holiday

Table 5 Parameter settings

XGBoost hyperparameters	Values	FFNN hyperparameters	Values
max_depth	6	Optimizer, O	Adamax
min_child_weight	8	Activation function, A	ReLU
gamma	0.2	Dropout rate, dr	0.25
subsample	0.7	Weight initialization, W_{in}	Glorot normal
colsample_bytree	0.7	Learning rate, lr	0.001
reg_alpha	100	Batch size, β	120
reg_lambda	2	Number of epochs, E	15
eta	0.02	Number of layers	3
n_estimators	635	Number of neurons in each hidden layer	200,75,20

zero supply of parking spaces and cells with less than 100 total trips were removed from the data set. This operation reduced the share of rows with missing values from 74.3% to 52.6%.

The remaining missing values had to be filled in some way. A number of different methods was applied to do this. Two simple and naive methods: linear interpolation (a method to fill intermediate missing values in a data set), monthly average, and two more advanced methods, which allow non-linear correlations between the dependent and explanatory variables: XGBoost (eXtreme Gradient Boosting) and FFNN (Feed-Forward Neural Network). The two advanced methods utilized the explanatory variables as shown in Table 4 to estimate the expected cruising time for a row after tuning their hyper parameters. The parameter settings of these advanced methods are shown in Table 5.

The XGBoost and FFNN performed significantly better with RMSE 50.8 and 52.3 respectively, compared to the naive methods with RMSE 95.2 (linear interpolation) and 95.6 (monthly average).

Since the RMSE of XGBoost and FFNN were close, the unknown values of cruising time were filled using both methods. Looking at the distribution of the filled values (Figs. 12 and 13), FFNN seemed more reasonable. FFNN was the only method that filled some rows with 0 cruising time (meaning no cars were cruising). Therefore, the missing values were filled using FFNN, which enabled a prediction algorithm for cruising time.

Once the missing values of cruising time were filled, methods for predicting the expected cruising time were set up. A naive method was the simple moving average, where the cruising time at a certain time slot was estimated by taking the average cruising time of the same hour, weekday and cell from the past four weeks. This was compared to a more advanced method, using FFNN with the explanatory variables from Table 4. However, for the stochastic variables that are varying over time, the

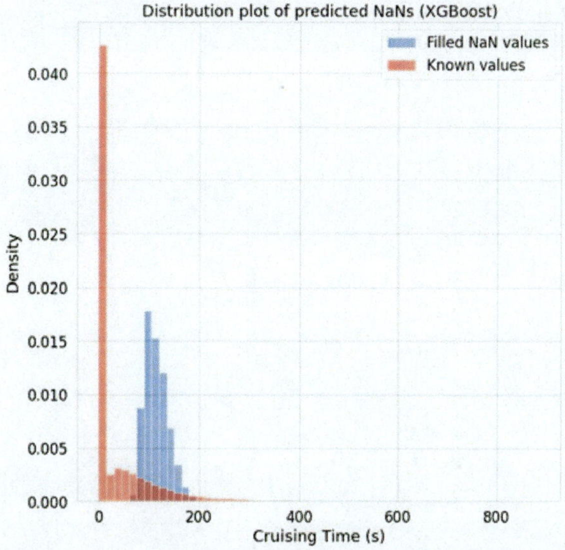

Fig. 12 Distribution plot (XGBoost)

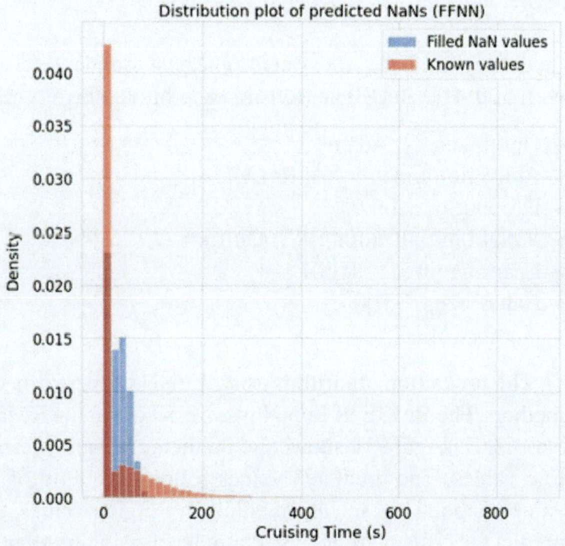

Fig. 13 Distribution plot (FFNN)

values of the current time cannot be used. Instead, lagged values of cruising time at the same cell, cruising time in nearby cells and number of trips were introduced. By examining autocorrelation plots, the largest correlation between current cruising time and previous values could be found. From this, the lagged values from 24 h (1 day) and 168 h (1 week) before were chosen and included as explanatory variables into the FFNN model. Again, the hyperparameters of the FFNN model were tuned using pairwise grid search and a 3-fold cross validation was implemented for final

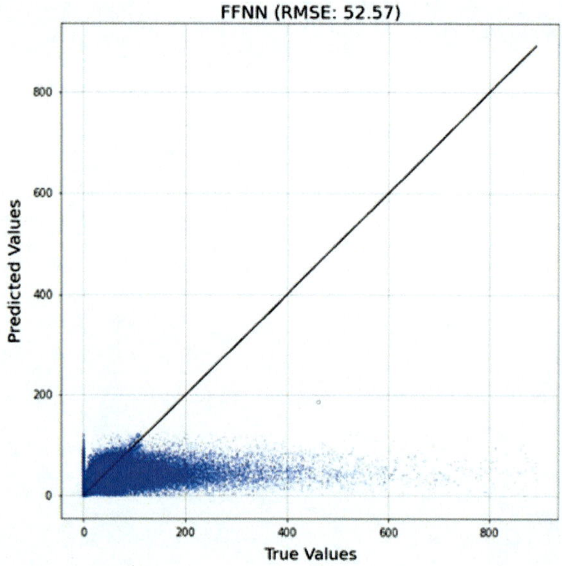

Fig. 14 Prediction errors for FFNN (forecasting)

model performance assessment. The best found model is a three layer neural network with 300, 100 and 30 neurons in each hidden layer with:

- Optimizer, O: Adam
- Activation function, A: ReLU
- Dropout rate dr: 0.0
- Weight initialization, W_{in}: Uniform
- Learning rate, lr: 0.001
- Batch size, β: 100
- Number of epochs, E: 5

The prediction algorithm using FFNN outperformed the simple moving average method. The RMSE of FFNN was 52.57 while the RMSE of simple moving average was 62.01. Figure 14 shows the predicted values of cruising time compared to their true values. The predicted values follow the trend of $y = x$ up to a cruising time value of about 150 s. Unfortunately, for higher values, the prediction method fails to predict the cruising time. For example, cruising times beyond 400 s make for 5.7% of 2019 trips and 3.9% of 2020 ones. There are three main explanations for this: (1) the detection criteria applied for cruising were incorrect or insufficient, (2) explanatory variables that could explain large values of cruising time were missing, and/or (3) too few observations.

7 Conclusions

The present study revolves around two related parts: detection of cruising and prediction of cruising time. A cruising detection method containing only speed-related criteria was found in literature, modified and applied to the whole data set. The average cruising time in Vesterbro was 149 s in 2019 and decreased by 25 s during COVID-19 pandemic in 2020. These values were 119 and 109 s for Frederiksberg and Islands Brygge in 2019 and 99 and 95 s in 2020 respectively.

When trying to predict the cruising time using an aggregated approach, there were many hours where no trips were observed to subgrids in the network. This resulted in the need to reduce and fill in the number of unknown values. The filled data set was then used to predict cruising time based on lagged values from 24 h and 168 h, due to high autocorrelation before the predicted value. For cruising time values of more than 150 s, the value could not be predicted due to either errors from the cruising detection method, from explanatory variables that were not included in the study or from too few observations. Examples of possible explanatory variables that potentially could explain these outliers include traffic accidents and road work. However, big data for parking search detection and estimation is promising, enabling extensive and analytical research exploration. Having a large amount of data leads to good predictions, as it was shown for cruising time values below 150 s, where the prediction was quite accurate.

To validate the cruising detection criteria before applying it for prediction in practice, one should obtain a similar extensive GPS data set with a sample of ground truth cruising data. This enables validation of the criteria for cruising, and thereby reduces the possible errors that are passed on to the prediction of cruising part. Such a ground truth data set could be obtained by employing a group of drivers to behave as they usually would during a defined period. Then, the drivers will locate ex-post the parts of the trip where they have been cruising (if any).

Acknowledgements The authors would like to thank Carlos M. Lima Azevedo and Filipe Rodrigues, associate professors at the Technical University of Denmark, for their assistance and guidance. Additionally, a big thanks to Thomas Jansson from Connected Cars A/S for providing the needed data and for his overall contribution to the project. The data was initially provided during the thesis study of the authors at the Technical University of Denmark in 2021.

References

1. Wijayaratna S (2015) Impacts of on-street parking on road capacity. Australasian transport research forum. Australia, Sydney, pp 1–15
2. Li D, Ma J, Cheng T, van Genderen JL, Shao Z (2018) Challenges and opportunities for the development of megacities. Int J Digit Earth
3. Stepanov M, Skrininkova V (2021) Features of parking in megacities. In: E3S web of conferences, EDP sciences, vol 281
4. Shoup DC (2006) Cruising for parking. Transp. Policy 13(6):479–486

5. Polak J, Axhausen KW (1990) Parking search behaviour: a review of current research and future prospects. Transport studies unit. University of Oxford
6. Salomon I (1986) Towards a behavioural approach to city centre parking: the case of Jerusalem's CBD. Cities 3(3):200–208
7. van der Waerden P, Timmermans H, Van Hove L (2015) GPS Data and car drivers' parking search behavior in the city of Turnhout. In: Geoinformatics for intelligent transportation. Springer, Belgium, pp 247–256
8. Kaplan S, Bekhor S (2011) Exploring en-route parking type and parking-search route choice: decision making framework and survey design. In: 2nd international choice modelling conference, Leeds. Citeseer
9. Laurier E (2005) Searching for a parking space
10. Weinberger RR, Millard-Ball A, Hampshire RC (2020) Parking search caused congestion: where's all the fuss? Transp Res Part C: Emerg Technol 120:102781
11. van Ommeren J, McIvor M, Mulalic I, Inci E (2021) A novel methodology to estimate cruising for parking and related external costs. Transp Res Part B: Methodol 145:247–269
12. Van Ommeren JN, Wentink D, Rietveld P (2012) Empirical evidence on cruising for parking. Transp Res Part A: Policy Pract 46(1):123–130
13. Brooke S, Ison S, Quddus M (2014) On-street parking search: review and future research direction. Transp Res Rec 2469(1):65–75
14. Zhu Y, Ye X, Chen J, Yan X, Wang T (2020) Impact of cruising for parking on travel time of traffic flow. Sustainability 12(8):3079
15. Martens K, Benenson I, Levy N (2010) The dilemma of on-street parking policy: exploring cruising for parking using an agent-based model. In: Geospatial analysis and modelling of urban structure and dynamics. Springer, pp 121–138
16. Hampshire RC, Jordon D, Akinbola O, Richardson K, Weinberger R, Millard-Ball A, Karlin-Resnik J (2016) Analysis of parking search behavior with video from naturalistic driving. Transp Res Rec 2543(1):152–158
17. Jones M, Khan A, Kulkarni P, Carnelli P, Sooriyabandara M (2017) Parkus 2.0: automated cruise detection for parking availability inference. In: Proceedings of the 14th EAI international conference on mobile and ubiquitous systems: computing, networking and services, pp 242–251
18. Chen X, Santos-Neto E, Ripeanu M (2012) Crowdsourcing for on-street smart parking. In: Proceedings of the second ACM international symposium on design and analysis of intelligent vehicular networks and applications, pp 1–8
19. Belloche S (2015) On-street parking search time modelling and validation with survey-based data. Transp Res Proc 6:313–324
20. Mantouka EG, Fafoutellis P, Vlahogianni EI (2021) Deep survival analysis of searching for on-street parking in urban areas. Transp Res Part C: Emerg Technol 128:103173
21. Statistics Denmark (2021) Regions, provinces and municipalities, v1:2007-
22. Statistics Denmark (2019) Befolkning 1. Jan 2019
23. Copenhagen Municipality (2021) Parkering i København
24. Frederiksberg Municipality (2021) Parkering i Frederiksberg
25. Copenhagen Municipality (2018) Open data DK - Københavns Kommune)
26. Anders Rody Hansen PE, Wulff C (2018) Parkeringssøgende trafik i København
27. Rantorp C (2020) 100.000 personbiler hjælper Cowi med at tage temperaturen på trafikken efter corona
28. Moric K (2021) 61 pct. af dagens biler kører stadig i 2030
29. Sommerlund E (2014) Kender du dine biltyper?
30. Worldometer (2021) Coronavirus cases Denmark
31. Cohen J (2020) Second wave of coronavirus intensifies across Europe
32. Local T (2020) Coronavirus in Denmark: these are the recommendations for using public transport
33. Jansson T (2021) Vectorized GPS distance/speed calculation for pandas

34. Millard-Ball A, Hampshire RC, Weinberger RR (2019) Map-matching poor-quality GPS data in urban environments: the pgMapMatch package. Transp Plan Technol 42(6):539–553
35. Municipality F (2021) Open data DK - Frederiksberg Kommune
36. Viriyakovithya E (2020) WorldWeatherOnline historical weather data API wrapper
37. Agency EE (2018) Copernicus land monitoring service - Urban atlas

Real-Time And Robust 3D Object Detection with Roadside LiDARs

Walter Zimmer, Jialong Wu, Xingcheng Zhou, and Alois C. Knoll

Abstract This work aims to address the challenges in autonomous driving by focusing on the 3D perception of the environment using roadside LiDARs. We design a 3D object detection model that can detect traffic participants in roadside LiDARs in real-time. Our model uses an existing 3D detector as a baseline and improves its accuracy. To prove the effectiveness of our proposed modules, we train and evaluate the model on three different vehicle and infrastructure datasets. To show the domain adaptation ability of our detector, we train it on an infrastructure dataset from China and perform transfer learning on a different dataset recorded in Germany. We do several sets of experiments and ablation studies for each module in the detector that show that our model outperforms the baseline by a significant margin, while the inference speed is at 45 Hz (22 ms). We make a significant contribution with our LiDAR-based 3D detector that can be used for smart city applications to provide connected and automated vehicles with a far-reaching view. Vehicles that are connected to the roadside sensors can get information about other vehicles around the corner to improve their path and maneuver planning and to increase road traffic safety.

Keywords Autonomous and connected vehicles · Traffic perception · 3D object detection · Roadside LiDAR · Point clouds · Deep learning · Transfer learning · Infrastructure · Testfield A9

W. Zimmer (✉) · J. Wu · X. Zhou · A. C. Knoll
Department of Computer Engineering, School of Computation, Information and Technology, Technical University of Munich (TUM), 85748 Garching, Germany
e-mail: walter.zimmer@tum.de

J. Wu
e-mail: jialong.wu@tum.de

X. Zhou
e-mail: xingcheng.zhou@tum.de

A. C. Knoll
e-mail: knoll@in.tum.de
URL: https://www.ce.cit.tum.de/air

C. Antoniou et al. (eds.), *Proceedings of the 12th International Scientific Conference on Mobility and Transport*, Lecture Notes in Mobility,
https://doi.org/10.1007/978-981-19-8361-0_13

1 Introduction

1.1 Problem Statement

High quality and balanced data is crucial to achieve high accuracy in deep learning applications. We analyze and split the task into four challenges. The first challenge is the lack of labeled frames of roadside LiDARs. Considering the high labor cost of manual labeling for 3D bounding boxes in LiDAR point clouds, we need to find a solution to deal with small datasets and use only few labeled frames for training. Publicly available LiDAR datasets were recorded and labeled from a driver perspective which makes is difficult to apply these datasets for roadside LiDARs. The second research question this work is dealing with lies in the area of domain adaptation. How can a neural network that was trained in one operational design domain (ODD), e.g. on data recorded by vehicle sensors like in the *KITTI* dataset [8, 9], be adapted to a different domain (e.g. to data recorded by roadside sensors on a highway or a rural area? Much research has been done in the area of domain adaptation and transfer learning [29, 35]—training a model on a large dataset (source domain) and fine-tuning it on a smaller dataset (target domain). Another challenge is real-time 3D object detection on roadside LiDARs, i.e. to detect objects with a high enough frame rate to prevent accidents. This highly depends on the LiDAR type and the number of points per scan that need to be processed. The final challenge this work is dealing with is a robust 3D detection of all traffic participants. Detecting small and occluded objects in different weather conditions and rare traffic scenarios is a highly important research area to increase safety of automated vehicles (Fig. 1).

1.2 Objectives

The first objective is to create a large infrastructure dataset with sufficient labeled point cloud frames. This dataset should be balanced in terms of object classes and contain a high variety so that objects can be detected in different scenarios and different environment conditions. Another objective is to analyze whether transfer learning from a larger roadside LiDAR dataset, such as the recently released *IPS300+* dataset [30], can improve the model performance. The first batch of the published IPS300+ dataset includes 1,246 labeled point cloud frames and contains on average 319.84 labels per frame. In this work the aim is to design a single-stage 3D object detector that can detect objects in real-time using roadside LiDARs. An example of an intersection that is equipped with LiDARs is shown in Fig. 2. The goal is to reach an inference rate of at least 25 Hz, which leads to a maximum inference time of 40 ms per point cloud frame. In terms of accuracy the target is to achieve at least 90% mean average precision (mAP) on the Car class within the test set. In the end the designed model needs to reach a reasonable trade-off between the inference time and the model performance. The final target is to evaluate the model performance

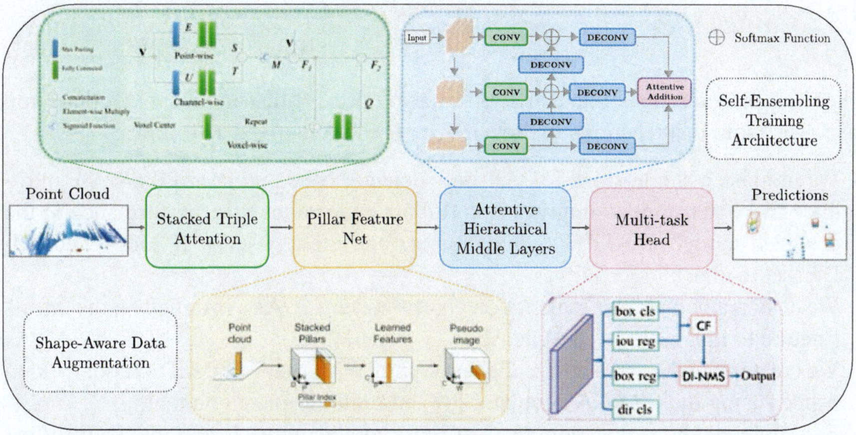

Fig. 1 Overview architecture of SE-ProPillars, a LiDAR-only single-stage pillar-based 3D object detector. The detector is based on PointPillars [17], with the following five additional extensions. (1) Shape-Aware Data Augmentation [37], a training technique to improve the accuracy without adding any additional costs in the inference time. (2) The Stacked Triple Attention Mechanism [21] enhances the learned features from the raw point cloud using the triple attention mechanism, including channel-wise, point-wise, and voxel-vise attention. (3) The pillar feature net turns point-wise features into pillar features and scatters the pillar features into a pseudo image. (4) We propose the Attentive Hierarchical Middle Layers to perform 2D convolution operations on the pseudo image. Hierarchical feature maps are concatenated with an attentive addition operation. (5) The Multi-task detection head [36] is used for the final prediction, that includes an IoU prediction to alleviate the misalignment between the localization accuracy and classification confidence. (6) Finally, a Self-Ensembling Training Architecture (a teacher and student training framework) [37] is used as a training technique in the training process that leads to a substantial increase in the precision without affecting the inference time

Fig. 2 Left: Intersection where 3D Object Detection with roadside LiDARs is performed. Right: Roadside LiDAR mounted above the Autobahn A9

and speed on the first released batch of the manually labeled *A9-Dataset* [2] and on publicly available infrastructure datasets. The *A9-Dataset* was recorded on the Providentia++ Test Stretch in Munich, Germany [16].

1.3 Contribution

The work provides a complete analysis and solution for the single LiDAR detection task. The main contributions in this work can be summarized as follows:

- We propose a single-stage LiDAR-only detector based on PointPillars. We introduce five extensions to improve PointPillars and evaluate the performance of the designed model on KITTI's validation set and the test set of IPS300+ and A9-Dataset to check its validity.
- We design a point cloud registration algorithm for LiDARs installed on the infrastructure to increase the point density.
- We create a synthetic dataset, called proSynth, with 2k labeled LiDAR point cloud frames using the CARLA simulator [5] and train our model on that.
- Experiments show that our *SE-ProPillars* model outperforms the *PointPillars* model by (+4.20%, +1.98%, +0.10%) 3D mAP on easy, moderate and hard difficulties respectively, while the inference speed reaches 45 Hz (22 ms).
- We train on the IPS300+ dataset, perform transfer learning on A9-Dataset, and achieve a $3D\ mAP@0.5$ of 50.09% on the Car class of the A9 test set using 40 recall positions and a NMS IoU of 0.1.

2 Related Work

According to the form of feature representation, LiDAR-only 3D object detectors can be divided into four different types, i.e. point-based, voxel-based, hybrid and projection-based methods.

2.1 Point-Based Methods

In point-based methods, features maintain the form of point-wise features, either by a sampled subset or derived virtual points. PointRCNN [28] uses a PointNet++ backbone [26] to extract point-wise features from the raw point cloud, and performs foreground segmentation. Then for each foreground point, it generates a 3D proposal followed by a point cloud ROI pooling and a canonical transformation-based bounding box refinement process. Point-based methods usually have to deal with a huge amount of point-wise features, which leads to a relatively lower inference speed. To accelerate point-based methods, 3DSSD [32] introduces a feature farthest-point-sampling (F-FPS), which computes the feature distance for sampling, instead of Euclidean distance in traditional distance farthest-point-sampling (D-FPS). The inference speed of 3DSSD is competitive with voxel-based methods.

2.2 Voxel-Based Methods

VoxelNet [39] divides the 3D space into equally spaced voxels and encodes point-wise features into voxel-wise features. Then 3D convolutional middle layers operate on these encoded voxel features. 3D convolution on sparse point cloud space brings too much unnecessary computational cost. SECOND [31] proposes to use a sparse convolutional middle extractor [10, 20], which greatly speeds up the inference time. In PointPillars [17], the point cloud is divided into pillars (vertical columns), which are special voxels without any partition along the z-direction. The feature map of pillars can be treated as a pseudo-image, and therefore the expensive 3D convolution is replaced by 2D convolution. PointPillars achieves the fastest speed with the help of TensorRT acceleration. SA-SSD [11] adds a detachable auxiliary network to the sparse convolutional middle layers to predict a point-wise foreground segmentation and a center estimation task, which can provide additional point-level supervision. SA-SSD also proposes a part-sensitive warping (PS-Warp) operation as an extra detection head. It can alleviate the misalignment between the predicted bounding boxes and classification confidence maps, since they are generated by two different convolutional layers in the detection head. CIA-SSD [36] also notices the misalignment issue. It designs an IoU-aware confidence rectification module, using an additional convolutional layer in the detection head to make IoU predictions. The predicted IoU value is used to rectify the classification score. By introducing only one additional convolutional layer, CIA-SSD is more lightweight than SA-SSD. SE-SSD [37] proposes a self-ensembling post-training framework, where a pre-trained teacher model produces predictions that serve as soft targets in addition to the hard targets from the label. These predictions are matched with student's predictions by their IoU and supervised by the consistency loss. Soft targets are closer to the predictions from the student model and therefore can help the student model to fine-tune its predictions. The Orientation-Aware Distance-IoU Loss is proposed to replace the traditional smooth-L_1 loss of bounding box regression in the post training, in order to provide a fresh supervisory signal. SE-SSD also designs a shape-aware data augmentation module to improve the generalization ability of the student model.

2.3 Hybrid Methods

Hybrid methods aim to take advantage of both point-based and voxel-based methods. Point-based methods have a higher spatial resolution but involve higher computational cost, while voxel-based methods can efficiently use CNN layers for feature extraction but lose local point-wise information. Hybrid methods try to strike a balance between them.

HVPR [23] is a single-stage detector. It has two feature encoder streams extracting point-wise and voxel-wise features. Extracted features are integrated together and scattered into a pseudo image as hybrid features. An attentive convolutional middle module is performed on the hybrid feature map, followed by a single-stage detection head. STD [33] is a two-stage detector that uses PointNet to extract point-wise features. A point-based proposal generation module with spherical anchors is designed to achieve high recall. Then a PointsPool module voxelizes each proposal, followed by a VFE layer. In the box refinement module, CNNs are applied on those voxels for final prediction. PV-RCNN [27] uses the 3D sparse convolution for voxel feature extraction. A Voxel Set Abstraction (Voxel-SA) module is added to each convolutional layer to encode voxel features into a small set of key points, which are sampled by farthest point sampling. Key point features are then re-weighted by foreground segmentation score. Finally, they are used to enhance the ROI grid points for refinement. $H^2$3D R-CNN [4] extracts point-wise features from multi-view projection. It projects the point cloud to a bird's eye view under Cartesian coordinates and a perspective view under the cylindrical coordinates, separately. BEV feature and point-voxel (PV) features are concatenated together for proposal generation in BEV and fused together as point-wise hollow-3D (H3D) features. Then voxelization is performed on the 3D space, and point-wise H3D features are aggregated as voxel-wise H3D features for the refinement process.

2.4 Projection-Based Methods

RangeDet [7] is an anchor-free single-stage LiDAR-based 3D object detector that is purely based on the range view representation. It is compact and has no quantization errors. The inference speed is 12 Hz using a RTX 2080 TI GPU. The runtime is not affected by the expansion of the detection distance, whereas a BEV representation will slow down the inference time as the detection range increases. RangeRCNN [18] is another 3D object detector that uses a range image, point view and bird's eye view (BEV). The anchor is defined in the BEV to avoid scale variation and occlusion. In addition, a two-stage regional convolutional neural network (RCNN) is used to improve the 3D detection performance.

3 Data Generation

3.1 Real Data Generation

The first step is to record point cloud data from roadside LiDARs. These recordings should cover a large variety of different scenarios (e.g. traffic jam, overtaking, lane change, lane merge, tail-gate events, accidents, etc.). In the next step (data selection), scenarios of high importance are selected for labeling. The point clouds are then converted into the right format (.pcd) and the ground is removed using the *RANSAC* algorithm. Dynamic objects are then labeled using a custom 3D point cloud labeling

Fig. 3 Left: Example of the intersection that is part of the A9 Test Stretch that was modelled in the CARLA simulator. Right: Synthetic point cloud frames with labeled ground truth vehicles extracted from CARLA simulator

tool called *proAnno*, that is based on the open source 3D bounding box annotation tool *3D-BAT* [40]. 456 LiDAR frames were manually labeled to form the first batch of the A9-Dataset.

3.2 Synthetic Data Generation

We created a synthetic dataset (proSynth) with 2000 point cloud frames using the CARLA simulator and train our *SE-ProPillars* model on it. Figure 3 shows an example of an intersection with generated traffic in the CARLA simulator. A simulated LiDAR sensor represents a real Ouster OS1-64 (gen. 2) LiDAR sensor with 64 channels and a range of 120 m. In the simulation a noise model is used with a standard deviation of 0.1 to disturb each point along the vector of its raycast. The LiDAR emits 1.31 million points per second and runs at 10 Hz (131,072 points per frame). The extracted labels were stored in .json files according to the OpenLABEL standard [3].

4 Approach

We design a real-time LiDAR-only 3D object detector (*SE-ProPillars*) that could be applied to real-world scenarios. The architecture of the designed *SE-ProPillars* model is shown in Fig. 1. We describe the each part of the detector in the following sections.

4.1 Point Cloud Registration

We first design a point cloud registration algorithm for LiDARs installed on the infrastructure to increase the point density and facilitate the detection task. The two Ouster OS1-64 (gen. 2) LiDARs are mounted side by side (about 13 m apart from each other) on the gantry bridge. Both LiDARs are time synchronized using the ROS time that itself is synchronized with an NTP time server. One LiDAR is treated as the source (L_s) and the other one as target (L_t). Both LiDAR sensors capture N point cloud scans and each scan is marked with a Unix timestamp: $A = \{a_i\}_{i=1,\ldots,N}$ and $B = \{b_i\}_{i=1,\ldots,N}$. The goal is to put both point clouds into the same coordinate system, by transforming the source point cloud into the target point cloud's coordinate system. To achieve this a set of correspondence pairs $S = \{s_i\}_{i=1,\ldots,n}$ and $T = \{t_i\}_{i=1,\ldots,n}$ need to be found, where s_i corresponds to t_i. Then the rigid transformation T that includes the rotation R and translation t is estimated by minimizing the root-mean-squared error (RMSE) between correspondences. Inspired by [15], we provide an initial transformation matrix to help the registration algorithms better overcome local optima. The initial transformation was acquired with a Real-time Kinematic (RTK) GPS device. With this initial transformation the continuous registration is less likely to be trapped in a local optimum. The continuous registration is done by point-to-point ICP. The registration process of two Ouster LiDARs, running at 10 Hz, takes 18.36 ms on an Intel Core i7-9750H CPU. A RMSE of 0.52164 could be achieved using a voxel size of 2 m. Figure 4 shows the point cloud scans before and after registration.

4.2 Voxelization

We divide the raw point cloud into vertical pillars before feeding them into a neural network. These are special voxels that are not split along the vertical axis. Pillars have several advantages over voxels. A pillar-based backbone is faster than a voxel-based backbone due to fewer grid cells. Time consuming 3D convolutional middle layers are also being eliminated and instead 2D convolutions are being used. We also do

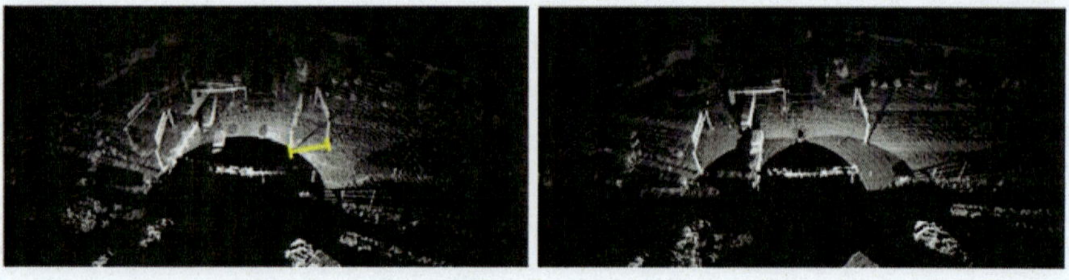

Fig. 4 Left: Two point cloud scans before registration. The displacement error is marked in yellow. Right: Two point cloud scans after registration

not need to manually tune the bin size along the z-direction hyperparameter. If a pillar contains more points than specified in the threshold, then the points are being subsampled to the threshold using farthest point sampling [6]. If a pillar contains fewer points than the threshold, then it is padded with zeros to make the dimensions consistent. Due to the sparsity issue most of the pillars are empty. We record the coordinates of non-empty pillars according to the pillar's center. Empty pillars are not being considered during the feature extraction until all pillars are being scattered back to a pseudo image for 2D convolution.

4.3 Stacked Triple Attention

The *Stacked Triple Attention* module is used for a more robust and discriminative feature representation. Originally introduced in *TANet* [21] by Liu et. al., this module enhances the learning of objects that are hard to detect and better deals with noisy point clouds. The TA module extracts features inside each pillar, using point-wise, channel-wise, and voxel-wise attention. The attention mechanism in this module follows the Squeeze-and-Excitation pattern [13].

The structure of the triple attention module is shown in Fig. 5. The input V to the module is a $(P \times N \times C)$ tensor, where P is the number of non-empty pillars, N is the maximum number of points in each pillar, and C is the dimension of the input point-wise feature. In the upper branch point-wise attention, following the Squeeze-and-Excitation pattern, we firstly perform max pooling to aggregate point-wise features across the channel-wise dimensions, and then we compute the point-wise attention score S using two fully connected layers. Similarly, the middle branch channel-wise attention aggregates channel-wise features across their point-wise dimensions, to get the channel-wise attention score T. Then S and T are combined by element-wise multiplication, followed with a sigmoid function to get the attention scale matrix M, $M = \sigma(S \times T)$. M is then multiplied with input V, to get the feature tensor $F1$. In the bottom branch voxel-wise attention, the C-dim channel-wise feature in $F1$ is enlarged by the voxel center (arithmetic mean of all points inside the pillar) to $C + 3$-dim for better voxel-awareness. Then the enlarged $F1$ is fed into two fully connected layers. The two FC layers respectively compress the point-wise and the

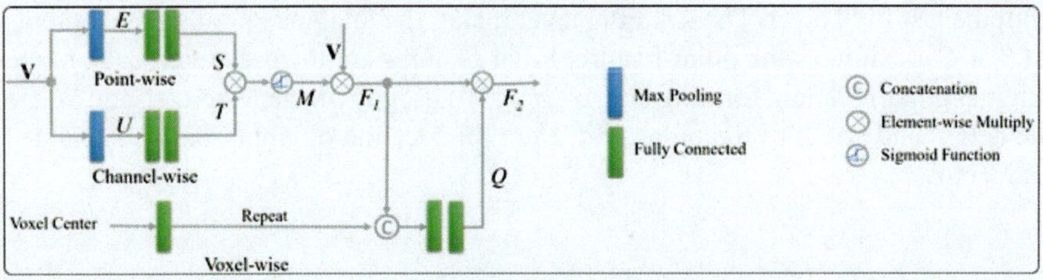

Fig. 5 Structure of the triple attention module [21]

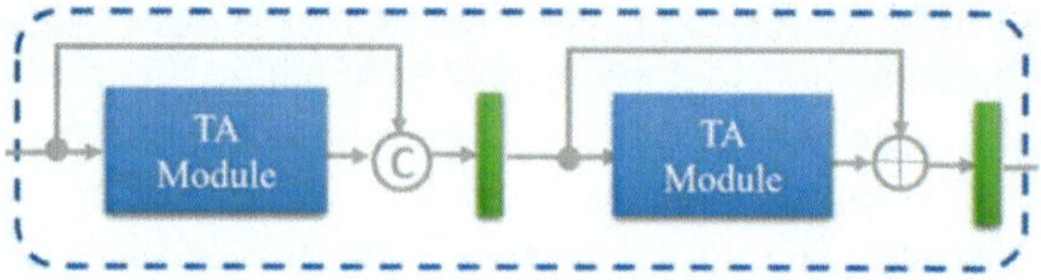

Fig. 6 Structure of the stacked triple attention module

channel-wise dimensions to 1, to get the voxel-wise attention score. Finally, a sigmoid function generates the voxel-wise attention scale Q, multiplied with the original $F1$ to generate the final output of the TA module $F2$.

To further exploit the multi-level feature attention, two triple attention modules are stacked with a structure similar to the skip connections in ResNet [12] (see Fig. 6). The first module takes the raw point cloud as input, while the second one works on the extracted high dimensional features. For each TA module the input is concatenated or summed to the output to fuse more feature information. Each TA module is followed by a fully connected layer to increase the feature dimension. Inside the TA modules, the attention mechanism only re-weights the features, but does not increase their dimensions.

4.4 Pillar Feature Net

We choose *PointPillars* [17] as our baseline and improve its 3D detection performance at the expense of inference time. The inference speed of *PointPillars* is 42 Hz without the acceleration of *TensorRT*. Since there is a trade-off between speed and accuracy, we can further boost the accuracy by incorporating additional modules without sacrificing the inference speed too much. The pillar feature net (PFN) shown in Fig. 1, introduced by Lang et. al, takes pillars as input, extracts pillar features, and scatters pillars back to a pseudo image for 2D convolution operations in the middle layers. The pillar feature net acts as an additional feature extractor to the stacked triple attention module. The point-wise pillar-organized features from the stacked TA module with shape $(P \times N \times C)$ are fed to a set of PFN layers. Each PFN layer is simplified PointNet [25], which consists of a linear layer, Batch-Norm [14], ReLU [22], and max pooling. The max-pooled features are concatenated back to the ReLU's output to keep the point-wise feature dimension inside each pillar, until the last FPN layer. The last FPN layer makes the final max pooling and outputs a $(P \times C)$ feature as the pillar feature. Pillar features are then scattered back to the original pillar location, forming a $(C \times H \times W)$ pseudo image, where H and W are the height and width of the pillar grid. Here the location of empty pillars is padded with zeros.

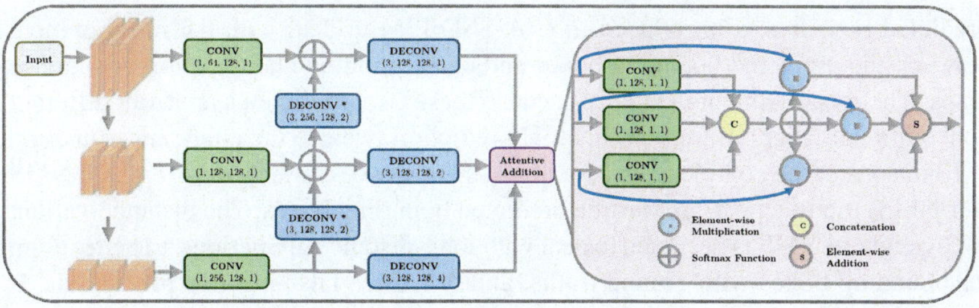

Fig. 7 Left: Structure of the attentive hierarchical middle layers. Right: Structure of the attentive addition operation

4.5 Attentive Hierarchical Middle Layers

We exchange the default backbone of *PointPillars* with an *Attentive Hierarchical Backbone* to perform 2D convolution on the pseudo image from the pillar feature net. Figure 7 depicts the structure of the attentive hierarchical middle layers. In the first stage, the spatial resolution of the pseudo image is gradually downsampled by three groups of convolutions. Each group contains three convolutional layers, where the first one has a stride of two for downsampling, and the two subsequent layers act only for feature extraction. After downsampling, deconvolution operations are applied to recover the spatial resolution. Deconvolutional layers (marked with an asterix) recover the size of feature maps with stride 2 and element-wise add them to upper branches. The remaining three deconvolutional layers make all three branches have the same size (half of the original feature map). Then the final three feature maps are combined by an attentive addition to fuse both, spatial and semantic features. The attentive addition uses the plain attention mechanism. All three feature maps are being passed through a convolutional operation and are channel-wise concatenated as attention scores. The softmax function generates the attention distribution and feature maps are multiplied with the corresponding distribution weight. The element-wise addition in the end gives the final attention output, a $(C \times H/2 \times W/2)$ feature map.

4.6 Multi-task Head

The multi-task head outputs the final class (based on a confidence score), the 3D box position (x, y, z), dimensions (l, w, h), rotation (θ) and the direction of the detected object. The direction (front/back) is being classified to solve the problem that the sine-error loss [31] cannot distinguish flipped boxes. Four convolutional layers operate on the feature map separately. Figure 1 shows the brief structure of the multi-task head in the bottom right corner. One of the four heads is the IoU prediction head that predicts an IoU between the ground truth bounding box and the predicted box. It was introduced in CIA-SSD [36] to deal with the misalignment between the predicted bounding boxes and corresponding classification confidence maps. The misalignment is mainly because these two predictions are from different convolutional layers. Based on this IoU prediction, we use the confidence function (CF) to correct the confidence map and use the distance-variant IoU-weighted NMS (DI-NMS) module post-process the predicted bounding boxes. The distance-variant IoU-weighted NMS is designed to deal with long-distance predictions, to better align far bounding boxes with ground truths, and to reduce false-positive predictions. If the predicted box is close to the origin of perspective, we give higher weights to those box predictions with high IoU. If the predicted box is far, we give relatively uniform weights, to get a more smooth final box.

4.7 Shape-Aware Data Augmentation

Data augmentation has proven to be an efficient way to better exploit the training dataset and help the model to be more generalized. We use the shape-aware data augmentation method proposed by SE-SSD [37] (see Fig. 8). This module simplifies the handling of partial occlusions, sparsity and different shapes of objects in the same class. The ground truth box is divided into six pyramidal subsets. Then we independently augment each subset using three operations, random dropout, random swap and random sparsifying.

Some traditional augmentation methods are also applied before the shape-aware augmentation, e.g. rotation, flipping, and scaling.

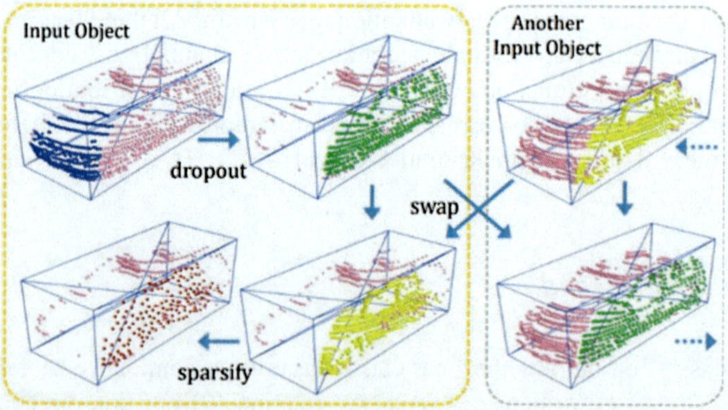

Fig. 8 Illustration of the shape-aware data augmentation module [37]. It includes random dropout in the blue subset, random swap between green and yellow subsets, and random sparsifying from the yellow to the red subset

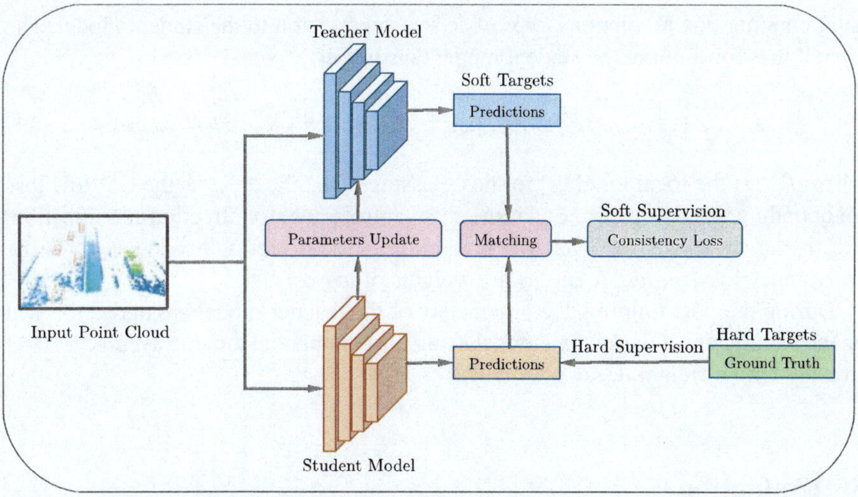

Fig. 9 Self-ensembling training architecture

4.8 Self-ensembling Training Framework

In addition, we introduce the self-ensembling training framework [37] to do a post training: We first train the model shown in Fig. 9 but without the self-ensembling module, and then we take the pre-trained model as a teacher model to train the student model that has the same network structure.

The self-ensembling training architecture is show in Fig. 9. The teacher model is trained with the following parameters: An adapted learning rate for each epoch was implemented with a maximum of 0.003 and a OneCycle learning rate scheduler. The model is trained for 80 epochs with a batch size of 2 for pre-training and for 60 epochs with a batch size of 4 for post-training to accelerate the training process. Adam is used as optimizer. The predictions of the teacher model can be used as soft targets. Combined with the hard targets from the ground truth, we can provide more information to the student model. The student model and the teacher model are initialized with the same pre-trained parameters. During training, we firstly feed the raw point cloud to the teacher model and get teacher predictions. Then we apply global transformations to the teacher predictions as soft targets. For the hard targets (ground truths), we apply the same global transformations and additionally the shape-aware data augmentation. After that we feed the augmented point cloud to the student model and get student predictions. The orientation-aware distance-IoU (OD-IoU) loss is introduced in the hard supervision, to better align the box centers and orientations between the student predictions and hard targets. Comparing to the plain IoU loss [34], OD-IoU loss also considers the distance and the orientation difference between two boxes. Finally, we use an IoU-based matching to match student and teacher predictions. We use the consistency loss over classification scores and bounding box predictions to provide soft supervision to the student model. The overall loss for training the student model consists of:

$$\mathcal{L}_{student} = \mathcal{L}^{s}_{cls} + \omega_1 \mathcal{L}^{s}_{OD-IoU} + \omega_2 \mathcal{L}^{s}_{dir} + \lambda \mathcal{L}_{iou} + \mu_t \mathcal{L}_{consist}, \quad (1)$$

where $\mathcal{L}^{s}_{cls}$ is the focal loss [19] for box classification, $\mathcal{L}^{s}_{OD-IoU}$ is the OD-IoU loss for bounding box regression, $\mathcal{L}^{s}_{dir}$ is the cross-entropy loss for direction classification, $\mathcal{L}_{iou}$ is the smooth-L_1 loss for IoU prediction in the detection head, $\mathcal{L}_{consist}$ is the consistency loss, ω_1, ω_2, λ and μ_t are weights of losses.

During the post-training, the parameters of the teacher model are updated based on the parameters of the student model using the exponential moving average (EMA) strategy with a weight decay of 0.999.

5 Evaluation

To prove the effectiveness of our proposed modules in *SE-ProPillars*, we perform some ablation studies in that we stepwise include some modules into our training pipeline. We train and evaluate the model on the *KITTI* dataset which is one of the most popular datasets in the autonomous driving domain. Furthermore, we train and evaluate the model on two infrastructure datasets with roadside LiDAR sensors: the *IPS300+* dataset and the recently released *A9* dataset. The evaluation was performed on a Nvidia GeForce RTX 3090 GPU.

5.1 Ablation Studies

By adding the *Stacked Triple Attention* module (*S-TA* module) the mAP_{3D} can be increased by 1.3% mAP_{3D}. This module adds 2 ms of inference time to the baseline model and runs with 35 ms per frame (28.57 FPS). Considering only the nearby area of the LiDAR point cloud (0–30 m), the mAP_{3D} can even be increased by 8.21% compared to 1.3% when taking the whole area of 70 m into account. The second module (*Attentive Hierarchical Backbone*) increases the mAP_{3D} again by 2.26% compared to the baseline model.

The result of single-class pre-training is shown in Table 1 (second last row). After applying the confidence function (CF), we call this model *naive SE-ProPillars* (no training tricks). We also apply the distance-variant IoU-weighted NMS (DI-NMS) and the shape-aware data augmentation (SA-DA) module in the pre-training. After enabling the multi-task head but without applying the CF, although the precision drops, the inference speed gets a large increase. The runtime is reduced from 32 to 19.4 ms. The CF in the lightweight multi-task head is trying to solve the misalignment problem between the predicted bounding boxes and corresponding classification confidence maps. After applying the confidence function, an increase is visible in

Table 1 Ablation studies of the *SE-ProPillars* 3D object detector after the pre and post training (last two rows). We report the 3D and BEV mAP of the Car and Van category (similar type enabled) on the KITTI validation set under 0.7 IoU threshold with 40 recall positions. Post-training especially improves the accuracy on the hard examples (+2.12 3D mAP)

Method	3D mAP			BEV mAP			Time (ms)
	Easy	Moderate	Hard	Easy	Moderate	Hard	
ProPillars	92.46	80.53	75.50	94.25	89.80	84.98	32.0
Naive without CF	89.64	79.59	74.43	93.88	89.29	86.29	**19.4**
Naive	89.71	79.95	75.03	93.85	89.55	86.75	20.1
Naive+DI-NMS	90.25	80.48	75.60	94.18	89.84	85.11	21.4
Naive+DI-NMS+SA-DA	92.38	80.93	76.05	96.30	89.85	87.21	22.0
SE-ProPillars (ours)	**93.10**	**81.26**	**78.17**	**96.44**	**89.88**	**87.21**	22.0
Improvement	*+0.64*	*+0.33*	*+2.12*	*+0.14*	*+0.03*	*+0.00*	*+0.00*

all metrics, except the BEV easy difficulty keeps at the same level. After adding the DI-NMS module, the precision becomes competitive with *ProPillars* considering the balance of precision and speed, although there is a relatively large gap in the 3D easy mode. The shape-aware data augmentation leads to a further increase in precision. After the pre-training our model outperforms the baseline in almost all metrics, except a tiny 0.1% gap in the 3D easy difficulty. The speed of our model is much faster compared to the baseline. There is a small latency in the runtime after enabling the shape-aware data augmentation module, because as the precision increases, the computation of the DI-NMS increases.

5.2 *KITTI Dataset*

We evaluate our *SE-ProPillars* detector on the *KITTI* validation set using the conventional metric of KITTI, i.e. the BEV and 3D mAP result under 0.7 IoU threshold with 40 recall positions. All three levels of difficulty are taken into account. We follow the NMS parameters in [37], set the NMS IoU threshold to 0.3, and the confidence score threshold to 0.3. We evaluate the performance on both single-class detection - Car category, and multi-class detection—Car, Pedestrian, and Cyclist. In the single-class detection, we enable the similar type, which means we train on both Car and Van categories, but treat them all as Cars. *ProPillars* [38] is used as our baseline.

We use the pre-trained model as the initial teacher model in the self-ensembling training architecture. Results are shown in Table 2. In the post-training, the consistency loss is enabled to provide supervision for the student model, as well as the replacement from the traditional smooth-L_1 loss to the orientation-aware distance-

Table 2 Comparison on the KITTI validation set. We show 3D multi-class detection results of SE-ProPillars and report the 3D mAP of the Car category under 0.7 IoU threshold with 40 recall positions

Method	Year	3D mAP (Car)		
		Easy	Medium	Hard
VoxelNet [28]	2018	82.00	65.50	62.90
PointPillars [17]	2019	87.75	78.39	75.18
F-PointPillars [24]	2021	88.90	79.28	78.07
SE-ProPillars (ours)	2022	**93.10**	**81.26**	**78.17**
Improvement		*+4.20*	*+1.98*	*+0.10*

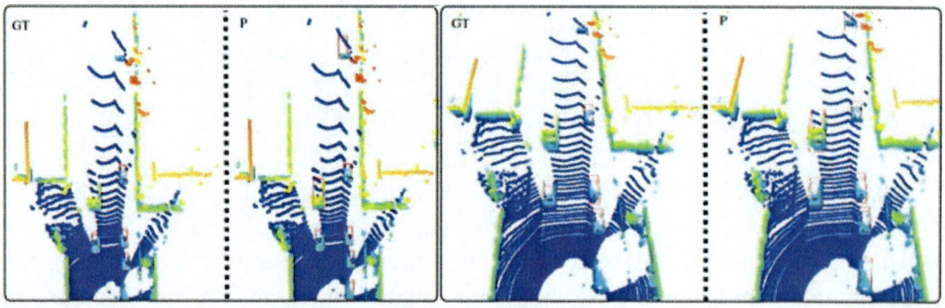

Fig. 10 Left: BEV visualization of single-class detection results on the KITTI validation set. Ground truth bounding boxes are marked with GT, predicted bounding boxes with P. In KITTI, some ground truth boxes are filtered out because they are smaller than 25 pixels in height in the corresponding captured image, or completely out of the image. Some of these filtered objects can still be detected by our *SE-ProPillars* which shows the good performance for detecting small objects. Right: 3D front view visualization of single-class detection results on the KITTI validation set. In this example a larger vehicle is detected as a Car

IoU loss for the bounding box regression. The SA-DA module is compulsory for providing more noisy samples to the student model. After 60 epochs of post-training, all metrics gain further improvement. Our model outperforms the baseline by (0.64%, 0.73%, 2.67%) 3D mAP and (2.19%, 0.08%, 2.23%) BEV mAP on easy, moderate, and hard difficulties respectively. At the same time, we have a much lower inference time of 22 ms, compared to our baseline of 32 ms. The inference time includes 5.7 ms for data pre-processing, 16.3 ms for network forwarding and post-processing (NMS). The hard disk used to load the model is relatively slow, which leads to a higher data loading time. The self-ensembling architecture is a training technique and therefore has no additional cost in the inference time. Some qualitative results are shown in Fig. 10. We also compare our *SE-ProPillars* model with other state-of-the-art detectors on the KITTI validation set (see Table 2).

Table 3 3D object detection results of the naive SE-ProPillars model with SA-DA. We report the 3D and BEV mAP of the Car class on a single LiDAR and two registered LiDARs on the split IPS300+ test set under 0.7 and 0.5 IoU thresholds with 40 recall positions. Using the full unsplit (original) IPS300+ dataset, we noticed a performance improvement of +13.99 3D mAP at an IoU threshold of 0.7. This shows the effect of splitting the IPS300+ point clouds into four sub-areas. Objects at the split boundary get truncated and become difficult to detect. Hence the 3D mAP values on the split datasets are lower

	3D mAP@0.7	3D mAP@0.5	BEV mAP@0.7	BEV mAP@0.5
1 LiDAR, split-IPS300+	54.91	83.76	74.57	85.97
2 LiDARs, split-IPS300+	63.58	87.52	78.34	89.42
Full (unsplit) IPS300+	**68.90**	**97.82**	**95.53**	**98.16**
Improvement	*+13.99*	*+14.06*	*+20.96*	*+12.19*

Fig. 11 Left: Visualization of detection results of a single LiDAR on the split IPS300+ validation set. Ground truth bounding boxes are marked with GT and predicted bounding boxes with P. Here only 18 out of 26 vehicles were detected because of the sparse point cloud from a single LiDAR. Right: Visualization of detection results of two registered IPS300+ LiDAR point cloud scans of the IPS300+ validation set. Here 31 out of 32 vehicles could be detected and only one vehicle was missed because of the far range

5.3 IPS300+ Roadside Dataset

Table 3 shows the evaluation results on the full unsplit (original) IPS300+ roadside test set as well as two modified test sets: A single-LiDAR on a split IPS300+ test set and two registered LiDAR sensors on the split IPS300+ test set. The IPS300+ roadside dataset contains full-surround 3D labels whereas the KITTI dataset covers only the frontal view of the vehicle. That is why we split the IPS300+ dataset into four sub areas that represent the four different driving directions at the intersection where the dataset was recorded (Fig. 11).

Table 4 Results of naive SE-ProPillars with SA-DA. We report the 3D and BEV mAP of Car on the A9 test set under 0.5 and 0.25 IoU threshold, with 40 recall positions

Metric	3D mAP@0.5	3D mAP@0.25	BEV mAP@0.5	BEV mAP@0.25
ProPillars	N/A	N/A	N/A	47.56
Ours (NMS IoU = 0.2)	31.03	48.88	39.91	49.68
Ours (NMS IoU = 0.1)	30.13	50.09	40.21	**51.53**
Improvement	–	–	–	*+3.97*

5.4 A9-Dataset

We use the *naive SE-ProPillars* with SA-DA module trained on a single-LiDAR split-IPS300+ as our model, fine-tune on the training set of the A9 dataset for 100 epochs, and test on the testing set. We report our result using 3D and BEV mAP under an IoU threshold of 0.5 and 0.25 like in [38]. The confidence score is set to a threshold of 0.1, and we again decrease the NMS IoU threshold to 0.2. This is exactly the same setting as the convention in the nuScenes [1] dataset defines. Since the LiDAR frames of the first batch of the A9 dataset are captured on the Highway A9, the distance between cars is further and there are no parked cars. We also test the case of using 0.1 as the NMS threshold. The result is shown in Table 4.

6 Conclusion

In this work we present our *SE-ProPillars* 3D object detector that is an improved version of the *PointPillars* model. We replace the detection head from [11] with a multi-task head for the final prediction, that includes an IoU prediction to alleviate the misalignment between the localization accuracy and classification confidence. Attentive hierarchical middle layers were proposed to perform 2D convolution operations on the pseudo image. The attentive addition operation concatenates the hierarchical feature maps. We add two training techniques to our baseline: the shape-aware data augmentation module and the self-ensembling training architecture to improve the accuracy without adding additional costs in the inference time. We show the generalization ability of our model and make it more robust to noise by applying multiple data augmentation methods. Sufficient data collection is key to achieve a good accuracy. That is why we generated a synthetic dataset proSynth with 2k labeled LiDAR point cloud frames and trained our model on that. We do several sets of experiments for each module to prove its accuracy and runtime performance in different ablation studies. Finally, we evaluate our 3D detector on different datasets that shows that our model significantly ourperforms the baseline model.

7 Future Outlook

Our *SE-ProPillars* 3D object detector is a significant contribution within the area of roadside LiDAR-based 3D perception and can be used for smart city applications in the future. A far-reaching view can be provided to connected and automated vehicles and detected vehicles around the corner can be communicated to all traffic participants to improve road traffic safety. LiDAR sensors can be used for anonymous people and vehicle counting to calculate the traffic density in real-time. Traffic monitoring will improve traffic flow and increase public safety. Anonymous 3D detections will be sent to connected vehicles to improve path and maneuver planning. Detecting edge cases and rare events like accidents or breakdowns is part of the future work to notify emergency vehicles immediately. The used LiDAR sensors are part of an Cooperative Intelligent Transport System (C-ITS), that can share dangers on the road (e.g. vehicle breakdowns or falling objects) with all connected vehicles in real-time. Using these LiDAR sensors in combination with smart traffic lights will improve the traffic flow in the future by enabling a green wave for traffic participants (e.g. cyclists). Connecting roadside LiDARs with vehicles will lead to a safe and efficient driving and accidents will be prevented before they happen. Connected and cooperative vehicles will be able to make safer and more coordinated decisions and improve their path and maneuver planning in real-time. Finally, an online and anonymized database of traffic participants can be created and shared with connected traffic participants to automatically react to potential incidents. This platform will share all traffic data and stream live sensor data to improve public safety.

Acknowledgements This work was funded by the Federal Ministry of Transport and Digital Infrastructure, Germany as part of the research project Providentia++ (Grant Number: 01MM19008A). The authors would like to express their gratitude to the funding agency and to the numerous students at TUM who have contributed to the creation of the first batch of the A9-Dataset.

References

1. Caesar H, Bankiti V, Lang AH, Vora S, Liong VE, Xu Q, Krishnan A, Pan Y, Baldan G, Beijbom O (2020) Nuscenes: a multimodal dataset for autonomous driving. In: Proceedings of the IEEE/CVF conference on computer vision and pattern recognition, pp 11,621–11,631
2. Cress C, Zimmer W, Strand L, Fortkord M, Dai S, Lakshminarasimhan V, Knoll A (2022) A9-dataset: multi-sensor infrastructure-based dataset for mobility research. arXiv preprint
3. Croce N (2021) Openlabel version 1.0.0 standardization project by asam association for standardization of automation and measuring systems. https://www.asam.net/standards/detail/openlabel/. Accessed 12 Nov 2021
4. Deng J, Zhou W, Zhang Y, Li H (2021) From multi-view to hollow-3d: Hallucinated hollow-3d r-cnn for 3d object detection. IEEE Trans Circuits Syst Video Technol
5. Dosovitskiy A, Ros G, Codevilla F, Lopez A, Koltun V (2017) Carla: an open urban driving simulator. In: Conference on robot learning. PMLR, pp 1–16
6. Eldar Y, Lindenbaum M, Porat M, Zeevi YY (1997) The farthest point strategy for progressive image sampling. IEEE Trans Image Process 6(9):1305–1315

7. Fan L, Xiong X, Wang F, Wang N, Zhang Z (2021) Rangedet: in defense of range view for lidar-based 3d object detection. In: Proceedings of the IEEE/CVF international conference on computer vision, pp 2918–2927
8. Geiger A, Lenz P, Stiller C, Urtasun R (2013) Vision meets robotics: the kitti dataset. Int J Robot Res 32(11):1231–1237
9. Geiger A, Lenz P, Urtasun R (2012) Are we ready for autonomous driving? the kitti vision benchmark suite. In: 2012 IEEE conference on computer vision and pattern recognition. IEEE, pp 3354–3361
10. Graham B, Engelcke M, Van Der Maaten L (2018) 3d semantic segmentation with submanifold sparse convolutional networks. In: Proceedings of the IEEE conference on computer vision and pattern recognition, pp 9224–9232
11. He C, Zeng H, Huang J, Hua X.S, Zhang L (2020) Structure aware single-stage 3d object detection from point cloud. In: Proceedings of the IEEE/CVF conference on computer vision and pattern recognition, pp 11,873–11,882
12. He K, Zhang X, Ren S, Sun J (2016) Deep residual learning for image recognition. In: Proceedings of the IEEE conference on computer vision and pattern recognition, pp 770–778
13. Hu J, Shen L, Sun G (2018) Squeeze-and-excitation networks. In: Proceedings of the IEEE conference on computer vision and pattern recognition, pp 7132–7141
14. Ioffe S, Szegedy C (2015) Batch normalization: accelerating deep network training by reducing internal covariate shift. In: International conference on machine learning. PMLR, pp 448–456
15. Kloeker L, Kotulla C, Eckstein L (2020) Real-time point cloud fusion of multi-lidar infrastructure sensor setups with unknown spatial location and orientation. In: 2020 IEEE 23rd international conference on intelligent transportation systems (ITSC). IEEE, pp 1–8
16. Krämmer A, Schöller C, Gulati D, Lakshminarasimhan V, Kurz F, Rosenbaum D, Lenz C, Knoll A (2019) Providentia-a large-scale sensor system for the assistance of autonomous vehicles and its evaluation. arXiv preprint arXiv:1906.06789
17. Lang AH, Vora S, Caesar H, Zhou L, Yang J, Beijbom O (2019) Pointpillars: fast encoders for object detection from point clouds. In: Proceedings of the IEEE/CVF conference on computer vision and pattern recognition, pp 12,697–12,705
18. Liang Z, Zhang M, Zhang Z, Zhao X, Pu S (2020) Rangercnn: towards fast and accurate 3d object detection with range image representation. arXiv preprint arXiv:2009.00206
19. Lin TY, Goyal P, Girshick R, He K, Dollár P (2017) Focal loss for dense object detection. In: Proceedings of the IEEE international conference on computer vision, pp 2980–2988
20. Liu B, Wang M, Foroosh H, Tappen M, Pensky M (2015) Sparse convolutional neural networks. In: Proceedings of the IEEE conference on computer vision and pattern recognition, pp 806–814
21. Liu Z, Zhao X, Huang T, Hu R, Zhou Y, Bai X (2020) Tanet: robust 3d object detection from point clouds with triple attention. Proceedings of the AAAI conference on artificial intelligence 34:11677–11684
22. Nair V, Hinton GE (2010) Rectified linear units improve restricted boltzmann machines. In: Icml
23. Noh J, Lee S, Ham B (2021) Hvpr: hybrid voxel-point representation for single-stage 3d object detection. In: Proceedings of the IEEE/CVF conference on computer vision and pattern recognition, pp 14,605–14,614
24. Paigwar A, Sierra-Gonzalez D, Erkent Ö, Laugier C (2021) Frustum-pointpillars: a multi-stage approach for 3d object detection using rgb camera and lidar. In: Proceedings of the IEEE/CVF international conference on computer vision, pp 2926–2933
25. Qi CR, Su H, Mo K, Guibas LJ (2017) Pointnet: deep learning on point sets for 3d classification and segmentation. In: Proceedings of the IEEE conference on computer vision and pattern recognition, pp 652–660
26. Qi CR, Yi L, Su H, Guibas LJ (2017) Pointnet++ deep hierarchical feature learning on point sets in a metric space. In: Proceedings of the 31st international conference on neural information processing systems, pp 5105–5114
27. Shi S, Guo C, Jiang L, Wang Z, Shi J, Wang X, Li H (2020) Pv-rcnn: point-voxel feature set abstraction for 3d object detection. In: Proceedings of the IEEE/CVF conference on computer vision and pattern recognition, pp 10,529–10,538

28. Shi S, Wang X, Li H (2019) Pointrcnn: 3d object proposal generation and detection from point cloud. In: Proceedings of the IEEE/CVF conference on computer vision and pattern recognition, pp 770–779
29. Triess LT, Dreissig M, Rist CB, Zöllner JM (2021) A survey on deep domain adaptation for lidar perception. In: 2021 IEEE intelligent vehicles symposium workshops (IV workshops), pp 350–357. IEEE
30. Wang H, Zhang X, Li J, Li Z, Yang L, Pan S, Deng Y (2021) Ips300+: a challenging multimodal dataset for intersection perception system. arXiv preprint arXiv:2106.02781
31. Yan Y, Mao Y, Li B (2018) Second: sparsely embedded convolutional detection. Sensors 18(10):3337
32. Yang Z, Sun Y, Liu S, Jia J (2020) 3dssd: point-based 3d single stage object detector. In: Proceedings of the IEEE/CVF conference on computer vision and pattern recognition, pp 11,040–11,048
33. Yang Z, Sun Y, Liu S, Shen X, Jia J (2019) Std: sparse-to-dense 3d object detector for point cloud. In: Proceedings of the IEEE/CVF international conference on computer vision, pp 1951–1960
34. Yu J, Jiang Y, Wang Z, Cao Z, Huang T (2016) Unitbox: an advanced object detection network. In: Proceedings of the 24th ACM international conference on multimedia, pp 516–520
35. Zhang W, Li W, Xu D (2021) Srdan: scale-aware and range-aware domain adaptation network for cross-dataset 3d object detection. In: Proceedings of the IEEE/CVF conference on computer vision and pattern recognition, pp 6769–6779
36. Zheng W, Tang W, Chen S, Jiang L, Fu CW (2020) Cia-ssd: confident iou-aware single-stage object detector from point cloud. arXiv preprint arXiv:2012.03015
37. Zheng W, Tang W, Jiang L, Fu C.W (2021) Se-ssd: self-ensembling single-stage object detector from point cloud. In: Proceedings of the IEEE/CVF conference on computer vision and pattern recognition, pp 14,494–14,503
38. Zhou X, Zimmer W, Erçelik E, Knoll A (2021) Real-time lidar-based 3d object detection on the highway. Master's thesis, Technische Universität München. Unpublished thesis
39. Zhou Y, Tuzel O (2018) Voxelnet: end-to-end learning for point cloud based 3d object detection. In: Proceedings of the IEEE conference on computer vision and pattern recognition, pp 4490–4499
40. Zimmer W, Rangesh A, Trivedi M (2019) 3d bat: a semi-automatic, web-based 3d annotation toolbox for full-surround, multi-modal data streams. In: 2019 IEEE intelligent vehicles symposium (IV). IEEE, pp 1816–1821

Estimating the Number of Tourists in Kyoto Based on GPS Traces and Aggregate Mobile Statistics

Tomoki Nishigaki, Jan-Dirk Schmöcker, Tadashi Yamada, and Satoshi Nakao

Abstract A clear understanding of the demand patterns, is one of the key contributors to laying a firm foundation for tourist planning. In pursuit of that aim, we estimated the number of tourists at specific areas and times in Kyoto City using regression analysis and hierarchical linear models (HLM). We first discuss how to extract the tourists' data from a "mesh population" obtained from aggregate mobile network operational data. We then propose that a relatively small sample of GPS tracking data for a population that has been monitored over a longer time than the mesh population can be used as a surrogate. To distinguish tourists from other persons, we find that a specified threshold of visiting a certain number of tourist attractions per day is useful. We also examine the effect of months and time of days by HLM on the model fit and number of tourists. Finally, we show that the accessibility of information such as the level of the attractiveness of particular Points of Interests (POIs) measured in terms of "Google ratings", in conjunction with the GPS records significantly contributes to a better estimation of the number of tourists at specific areas and times in Kyoto City.

Keywords Tourism · Hierarchical linear model · Mesh population · GPS data · Population estimation

T. Nishigaki (✉) · J.-D. Schmöcker · S. Nakao
Graduate School of Engineering, Kyoto University, Kyoto, Japan
e-mail: nishigaki@trans.kuciv.kyoto-u.ac.jp

J.-D. Schmöcker
e-mail: schmoecker@trans.kuciv.kyoto-u.ac.jp

S. Nakao
e-mail: nakao@trans.kuciv.kyoto-u.ac.jp

T. Yamada
Graduate School of Management, Kyoto University, Kyoto, Japan
e-mail: yamada.tadashi.2x@kyoto-u.ac.jp

C. Antoniou et al. (eds.), *Proceedings of the 12th International Scientific Conference on Mobility and Transport*, Lecture Notes in Mobility,
https://doi.org/10.1007/978-981-19-8361-0_14

1 Introduction

Over the past few years, Japan's popularity as a tourist destination has been gradually increasing with exception of the sudden interruption by the COVID-19 crisis. Consequently, problems such as traffic congestion and crowding at and around touristic places have become increasingly become more serious issues, causing dissatisfaction amongst both the tourists and residents alike [1]. This study focuses on Kyoto, Japan's old capital, one of the most significant tourist destinations. Before the COVID-19 crisis, the annual number of tourists continuously increased for two decades, reaching nearly 60 million per year, and the tourism consumption reached 1.2 trillion JPY, as illustrated in Fig. 1 [2]. However, along with the rapid rise in tourism, dissatisfaction amongst the tourists with their touristic experience has also gradually increased over the years, congestion being one of the major contributing factors. For instance, between 2011 and 2019, the mention of "crowding" as the main reason for tourist dissatisfaction increased from just over 10 to 20% [2].

For Kyoto City and Japan at large, tourism is an essential stimulant of economic prosperity, and as such improvement of tourists' satisfaction is considered both as a priority and a common fundamental policy objective. Furthermore, in the wake of the COVID crisis and its aftermath, adequate prevention of crowding in touristic areas, after the end of travel restriction policies has become an additional concern. One of the recommended approaches to solving this problem is obtaining a clear understanding of the tourism demand. We aim at contributing to that objective by estimating the number of tourists in specific areas, at specific points in time in Kyoto city. The paper explores if two data sets and transport accessibility measures are suitable to extract and predict tourist numbers.

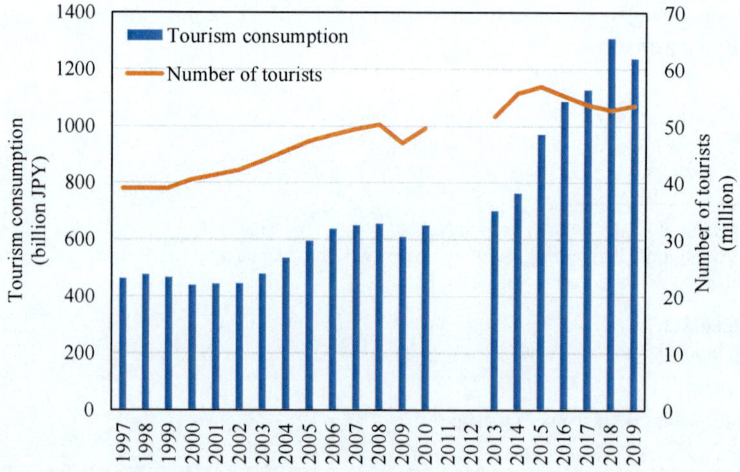

Fig. 1 The trend of tourism consumption and the number of tourists in Kyoto, Japan

Since there is far less data on tourist behavior than residential travel behavior, most travel surveys mainly target residents because obtaining representative samples from tourist surveys is quite difficult. One of the novel approaches to addressing this challenge is the use and reliance on "big data" for tourist travel patterns as discussed by Schmöcker [3]. Though there are several kinds of data such as mobile network operational data, GPS data, Wi-Fi access point data, traffic IC card data, probe car data…etc., all of these require additional analysis to distinguish tourists from other travelers or residents. Furthermore, the spatial and temporal units of a majority of these data sources do not often match those of interest for planning purposes. For instance, in most cities, there could be several areas where tourists tend to gather, at which the planner intends to have a clear estimate of the visitors and crowding.

In this study, the two main data sources used are mobile network operational data and GPS tracking data. The mobile network data provides us with information about the population within predefined mesh areas. This means that for our interest, touristic areas, an estimation based on "interpolation" or other statistical methods is needed. To note is that the data are based on a very large number of mobile phones so that the total number of persons in a mesh can be considered to be a fairly accurate estimate. The GPS data is the location data of users of a travel planning app with a time stamp. This data provides us with detailed individual data of those who agreed to share their location information. Because of this, the sample size is significantly smaller as well as biased towards public transport users. This means that it is more difficult to obtain an accurate population estimate from this data. In addition to these two data sets, we used "Point of Interest" (POI) data and public transportation network data to consider the touristic features of each mesh and its accessibility.

Both, mobile mesh data and GPS data, are hence not the "ground truth data" of tourists in touristic areas. In this paper, we discuss their limitations and show their correlation. In particular, we aim to understand how well the GPS data can be used to estimate tourist numbers. The reasons are twofold. One is that this smaller sample data set is often more available for researchers (or affordable at a lower price). Secondly, if the GPS data can be used for our total tourist number estimates and if the biases in the dataset are understood, this also provides us with more confidence for further analysis of tourist characteristics, such as which places are visited in conjunction and what the typical stay times are, at those touristic areas. Such information is not available from the aggregate mobile phone mesh data.

This motivates the establishment of a model where the mobile mesh data is the dependent variable and the GPS data, accessibility information from POI and public transportation accessibility data are independent variables. We first establish linear regression (LR) models and then also establish models which consider the effect of month or time of day with hierarchical linear models (HLM). Lastly, we apply our model to estimate the population within each area as defined by the Kyoto city government.

2 Literature Review

A relatively rich body of literature exists about the estimation of the static number of persons in specific areas. Some researchers have proposed the use of multiple existing population maps i.e., GPW (Gridded Population of the World), LandScan, WorldPop, GRUMP (Global Rural–Urban Mapping Project), GHS-POP (Global Human Settlement-Population), HYDE (History Database of the Global Environment), census data, etc., to estimate the population within a square of any size as so called "mesh populations" [4, 5], and to evaluate their accuracy [6–10]. Each of these databases is based on census data or information from scanning data of the earth by satellite, etc. and are in general highly reliable.

Notable is that, we found little research on other "shape regions" such as touristic areas because most of them focus on only mesh populations. There are, however, some significant contributions on other population group estimates. Balakrishnan [11] estimated a residential density with a 30 m resolution using street density, building heights, and ward-level data on car ownership. Bakillah et al. [12] estimated the building level population using building footprints and POI data. Shimosaka et al. [13] estimated the population within 100 m square meshes using POI and anonymized large-scale GPS data. There is also research on refining the census population [14, 15]. These focused on the specific areas used in census data, but these only refined the population obtained from census data based on multiple spatial resolutions, optical imagery, or telecommunications data. Kikuchi et al. [16] instead used mesh data to calculate an "expansion factor" of a population estimate obtained from census data. This expansion factor describes the ratio of the mesh data to the census data, so that a population estimate can be obtained from the sample. Otake and Kikuchi [17] used mesh data as the actual value of a population and then refined the origin–destination specific traffic volume based on data assimilation. Similar to what will be done in this study, they redistributed the mesh population into the area used in the census based on the sizes of each area. These contributed to estimating the population within the fine-grained mesh or regions which do not fit the mesh patterns.

There is further research on estimating the number of persons at a specific time in a spatial area. Khodabandelou et al. [18] and Cecaj et al. [19] estimated the urban scale dynamic population densities with mobile network traffic data. Bachir et al. [20] and Aasa et al. [21] estimated the dynamic population densities within an area used in the census. All of the results make methodological contributions, but none is related to the problem of tourism estimation. For tourism applications, we note the work of Ahas et al. [22] who used the anonymized GPS data, extracted foreigners' data based on the information on the mobile phone as the tourists' data from foreign countries, and analyzed the difference in behavior among nationalities. Ahas et al. [23] also analyzed the data with accommodation statistics and showed the distribution of the bias of the data. They emphasized that their data was a sensitive issue due to privacy concerns. If we could access data similar to theirs, our research objective would have been much more quickly accomplished, however, we are unable to access such data due to privacy restrictions in Japan, as they emphasized.

Recent research about Japanese tourism, has mainly focused upon the use of other forms of data other than that from the traditional surveys. Ubukata et al. [24] estimated the features of tourists' behavior using GPS data, i.e., the number of those visiting a specific area, the staying duration, the origin and the traffic mode to visit the site(s). Their data was collected with the smartphone users' consent every 5 min. They reported that the number of users who consented was only about 5% of the Japanese population, implying that there could be a bias in their findings. They extracted the tourists' data by following a 2-step criteria; First, the number of visits to the objective area per year is less than 12. Secondly, the users walked around at least two touristic areas. Kobayashi et al. [25] also used GPS traces to estimate tourist flows but only used the simple criteria of how many days a person is observed as a criterion to distinguish tourists from other persons in their sample. Dantsuji et al. [26] used Wi-Fi packet data and estimated the stay time of tourists. Nakanishi et al. [27] used Wi-Fi packet data and counted the number of persons visiting the facilities with Wi-Fi packet sensors. Kawakami et al. [28] used the same population mesh data that will be used in our subsequent study and OD traffic volume that is published from the same mobile phone provider to estimate the OD traffic volumes. They showed that the data could contribute to understanding the tourists' behavior and also pointed out the need to better estimate the total tourist population. However, they focused on the predefined meshes or areas created by combining the meshes, so their study did not consider the non-uniformly shaped touristic areas. Gao and Schmöcker [29], also suggested that Wi-Fi packet data is a good source to estimate point specific tourist numbers only near the Wi-Fi sensor as well as flows of tourists between specific parts of the city, however, a large number of sensor installations is required which may be nonviable to obtain the total number of tourists.

In conclusion, the majority of the past research on tourist estimation studies is based on data from the traditional surveys, except for some recent contributions [30–32]. There is relatively little research on estimating the number of tourists using other sources of data The main novelty of this study is the estimation of the population from mesh data using the population from small sample GPS data and accessibility information. In addition to this, we also estimate the population within the areas defined by the Kyoto city government. As shown by the literature, land-use and other "map data" also appear to be promising to partly overcome these problems like estimating the number of tourists. The subsequent study aims to explore this further within the touristic areas.

3 Tourism in Kyoto and Data Overview

3.1 Tourism in Kyoto

Tourism behavior in Kyoto is widely varied and dispersed, and the tourists use various transport modes to travel between touristic areas. Figure 2 shows the map of Kyoto city and the 37 tourist areas, and Table 1 indicates the name and size of each area. Shen et al. [33] used the same map's definitions to estimate tourist flows between these areas based on a survey of tourists at public transport stations. The density of public transportation is higher in the south region of Kyoto city than in the north. Kyoto city government strongly recommends tourists to travel without their cars. This policy is called "Arukumachi Kyoto."; in English, "Kyoto, the town for walking around." Thanks to this policy, many tourists use a variety of other transportation modes. These areas shown in Fig. 2 have multiple sizes and features: some include only one famous touristic point, some include a few touristic points, and some cover huge areas like hiking trails. As Fig. 2 and Table 1 illustrate, our challenge is to estimate the number of tourists within the various size areas and characterize them.

Ishigami et al. [34] mentioned that mobile network operational data and GPS data focus on all traffic modes, Wi-Fi access point data mainly targets pedestrians within the vicinity of Wi-Fi spots, traffic IC card data targets only public transportation users, whereas probe car data clearly records only car users. From this perspective, mobile network operational data and GPS data are most suitable for our study, which is why we sought access to these two types of data sources.

3.2 Mesh Data

First, our mobile network operational data are "mobile spatial statistics" from a major Japanese mobile phone service provider [35]. This data is generated based on the following criteria; counting the number of mobile phones around the cellular phone base station, expanding the counted value based on the diffusion rate of the mobile phone provider, and using these values to provide estimates within standardized 1 km square mesh population. This mesh data is only published based on predetermined meshes defined for all of Japan. Adjusting the data to non-uniform meshes is not trivial. Consider a case where half the mesh is covered by inaccessible mountains and the other half of the mesh contains touristic POIs. (A scenario that is common in Kyoto as several temples are located at the gateways to mountains.) Then presuming that half of the mesh population is in the touristic part of the mesh is clearly an underestimation. As will be discussed in Sect. 5–3, we, therefore, use the GPS data to account for such cases.

The total number of persons in a mesh can be considered accurate due to the significant market share of the mobile phone provider. The mesh data also allows us to distinguish between the people from Kyoto, other provinces of Japan and

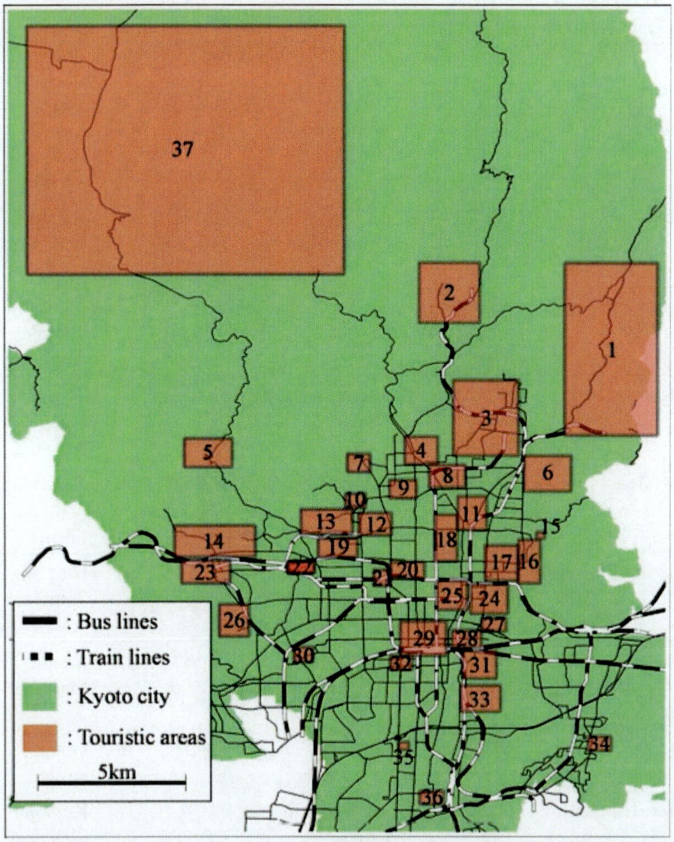

Fig. 2 Touristic areas in Kyoto city (*Source* Survey by Kyoto city government)

foreigners. The data provider is able to do so according to the registered address of the mobile phone. However, for privacy reasons, only aggregated data is available, and the number of persons within a mesh is not published if too small. This particularly implies that, the number of foreign tourists can only be obtained for large space and/or time intervals. In this study, therefore, we focus on Japanese tourists who make up about 90% of the total tourists in Kyoto.

Data was obtained for some Wednesdays and weekends/public holidays for the period (October 2018–January 2019) i.e., from the period before COVID-19. We considered averages for Wednesdays as representative of weekdays and averages for weekends and public holidays as representative of holidays.

Table 1 The name and size of each touristic area

No	Name	Size [km^2]	No	Name	Size [km^2]	No	Name	Size [km^2]
1	Ohara/Yase	21.9	15	Ginkaku-Ji Temple	0.0742	29	Kyoto station Vicinity	2.20
2	Kurama Area	5.03	16	The path of philosophy	1.34	30	Katsura imperial villa	0.396
3	Takaragaike	6.83	17	Heian Jingu Shrine	1.59	31	Tofuku-ji Temple Area	1.31
4	Kamigamo Shrine	1.28	18	Kyoto imperial palace	1.45	32	To-ji Temple Area	0.423
5	Takao Area	1.89	19	Hanazono Area	1.03	33	Fushimi inari Shrine	1.45
6	Shugakuin/Shisendo	2.35	20	Nijo castle Area	0.765	34	Daigo-ji Temple Area	0.476
7	Koetsu-Ji Temple	0.579	21	Nijo station Vicinity	0.382	35	Jonan-gu Shrine Area	0.111
8	Kitayama-dori street	1.17	22	Uzumasa Area	0.431	36	Fushimi Area	0.448
9	Daitoku-Ji Temple	0.685	23	Arashiyama Area	1.54	37	Keihoku direction	108
10	Kinkaku-ji Temple	0.408	24	Gion Area	1.53			
11	Shimogamo Shrine	1.36	25	Kawaramachi	1.25			
12	Kitano Temmangu Shrine	1.03	26	Matsuo Taisha Area	1.30			
13	Kinugasa / Omuro	1.77	27	Kiyomizu-dera Temple	0.433			
14	Sagano Area	3.48	28	Sanjusangendo	0.702			

3.3 GPS Data

Secondly, we have access to GPS data from a public transport planning mobile phone application called "Arukumachi Kyoto." Some users have given their consent to being tracked, and their locations and timestamps are stored mostly while the app is in use. With this individual data, user ID, and using language in users' OS, it is possible to distinguish between Japanese and foreigners. To match our analysis with the mesh data, we use only data from those presumed as Japanese based on the language settings. Mainly due to the necessity of the users' consent to obtain the data, the sample size of this GPS data is much smaller than the mesh data. The data will be further biased towards those using public transport as car users have less need to use this app. Because of this, even if we aggregate the data, it does not provide the actual value of the number of persons within an area and we need to consider it in conjunction with other data. Our data covers all days for the period between October 2018 and January 2019.

We note, that GPS data also has accuracy issues. However, we suggest this is a minor issue in our case, since we used GPS data only for counting the number of users per hour within each mesh. Specifically, we consider a traveler was in the mesh at each hour when at least one GPS record is inside the area. Hence errors can be made only if all records of a user over an hour (if there are multiple ones) are continuously outside the touristic area which the traveler visited. That case can be made only when a user stayed or walked around near the mesh edge for over one hour. This is, however, not likely the case because almost all tourists walk around in various directions for tourism. Clearly, the aforementioned problem, of missing records, is a more significant one.

3.4 *Point of Interest (POI) Data*

We also used "POI data" collected from Google maps API. We collected the information on the objects labeled "tourist_attraction." The information includes the latitude, longitude as well as the average rating by visitors of the POI and the number of ratings. Figure 3 shows the number of POIs per mesh, the average rating, and the total number of ratings within each mesh. As can be seen, 165 meshes have no POI. Their distribution is shown in Fig. 3. The figure shows that most meshes do not have POIs and the ratings are concentrated within a few meshes. Furthermore, most of the POIs are rated highly.

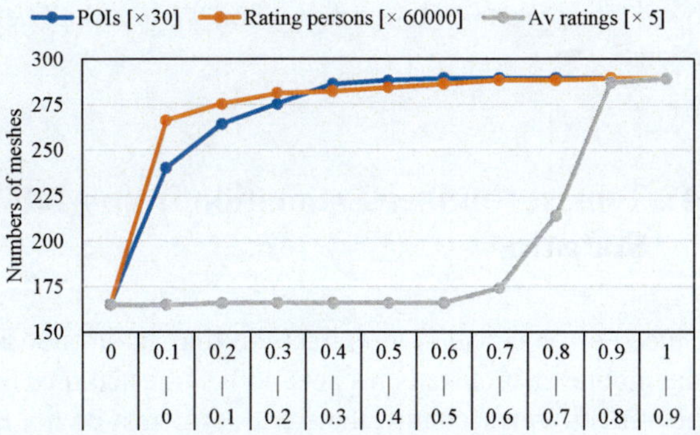

Fig. 3 The distribution of the three types of POI values per mesh

3.5 Public Transportation Data

We also used public transportation data to explain in how far the accessibility of an area explains the tourist presence. We estimated this accessibility by the number of stations or bus stops and travel costs. Since GTFS data was not available for Kyoto, the information on routes, their frequencies and fare information was gathered from the operators' web pages. The average waiting time of each line at its terminal during 3 h time intervals was used as the frequency of the line as a representative. We used the fare table for Kyoto city subway [36] as the fare for all links on the train as fares for each train link are not published and since there is no fixed fare per km. The fare for all buses was set to 230 JPY as this is the Kyoto city bus flat fare, though some operators charge in some parts of the city slightly different (distance-depending) fares that are not published on web pages. The generalized travel cost, therefore, includes travel time, waiting time, and fare. We used the time value of 29.8 JPY/min suggested in the VOT meta-analysis of Kato and Hashimoto [37]. Frequency is converted into waiting time, assuming regular service arrivals and random passenger arrival.

We used the above public transport information as direct indicators of the accessibility in conjunction with the POI data as shown by Eq. (1).

$$W_{it} = \sum_j \mathrm{w}_j exp(-C_{ijt}) \tag{1}$$

where C_{ijt} is the generalized cost from station or bus stop i to mesh j in time of day t. This includes the travel time, waiting time, and fare. w_j is the weight of mesh j based on the POIs in the mesh. We consider four types of POI weights to reflect their attractiveness to tourists: w_j^p is the average number of ratings (number of rating persons) and w_j^r the average rate within a mesh. The product of ratings and the number of ratings can be considered as a more comprehensive measure of attractiveness, so that we further define $\mathrm{w}_j^a = \mathrm{w}_j^p \mathrm{w}_j^r$ as well as a logarithmic version of $ln(\mathrm{w}_j^a)$. Our rationale for testing the logarithmic value of w_j^a was that it is closer to a normal distribution than w_j^a. In the regression we tested all versions as will be described in Sect. 5.

4 Tourist Number Estimation from Mobile Spatial Statistics

Since our objective is to extract the tourists in the various areas shown in Fig. 2, the mobile spatial statistics need to be adjusted. For one, not all non-residents will be tourists and, secondly, the areas of interest do not match those for which data is available.

To address these problems, we tested two approaches. First, we extracted the visitors' data within the mesh, overlapping with the touristic areas. There are 890 meshes within Kyoto city out of which 289 overlap with the touristic areas shown in Fig. 2. The number of persons for each month within the 890 meshes is shown in Fig. 4, and the number of persons within the 289 "touristic meshes" in each month is shown in Fig. 5.

The figures illustrate the concentration of visitors on the touristic meshes, but clearly there are also non-tourists amongst the visiting population. We refer to this estimate of tourists as $\hat{P}$, noting that it will overestimate the number of tourists. The figures also illustrate that November is the busiest tourist month and December the least busy month. November is the month of autumn foliage in Kyoto, usually attracting large numbers of tourists. Instead, December, due to weather conditions and working and school schedules, is generally a month with relatively little domestic tourism. Hence as a conservative (underestimate) of the tourist population, we extracted the part of the tourists' data in November by removing the number of visitors in December and refer to this as $\check{P}$ and show it in Fig. 6

Instead of taking a mean of the two estimates, we aim at obtaining evidence as to which approach is more appropriate. First, we note that Kawakami et al. [28] used the same data as ours for tourist flow estimation and compared their approach with survey data and suggested that results from $\hat{P}$ matched well with the survey result.

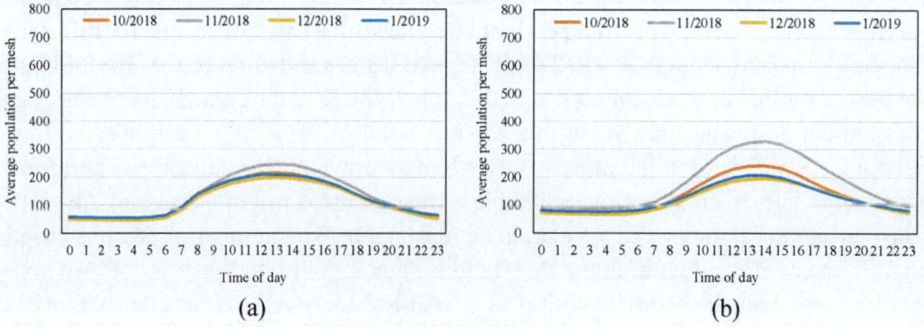

Fig. 4 Number of visitors in all Kyoto meshes **a** on weekdays (left) **b** on holidays (right)

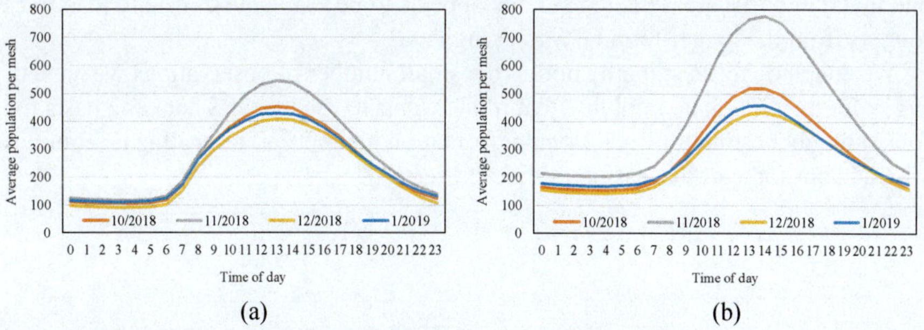

Fig. 5 $\hat{P}$ in each time of day **a** on weekdays (left) **b** on holidays (right)

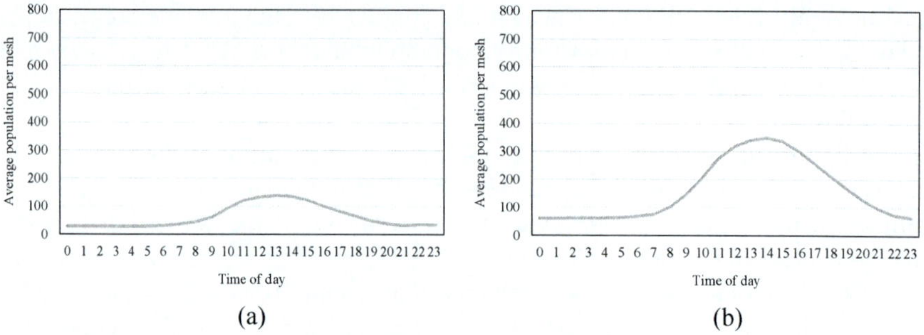

Fig. 6 $\check{P}$ in each time of day **a** on weekdays (left) **b** on holidays (right) in November

To obtain further evidence, we return to our second data set, the GPS traces. Also here, the problem of distinguishing visitors from the tourists remains. As an indicator of whether a person is likely a visitor, we consider the number of touristic places visited by the respondents over the period for which we have (infrequent) tracking records. Our hypothesis is that the number of recorded tourist areas per day referred to as μ, will tend to be larger for tourists. To find a suitable threshold, we conducted a sensitivity analysis comparing mesh data in the form $\hat{P}$ and $\check{P}$ as the dependent variable and GPS data as the independent variable with a LR. The results for different thresholds ranging from $\mu = 0$ to 2 with steps of 0.1 are shown in Fig. 7. We find that the best fit with $\hat{P}$ is achieved with $\mu = 0.3$. For $\check{P}$ the R^2 continuously increased, but the gradient of this became negligible when $\mu > 0.4$. A value of 0.3 might seem low, but our GPS data is sampled mostly only when the travelers used the app. Therefore, especially visits to neighboring, walkable attraction areas might be missed. Overall, considering these points, 0.3 appears to be reasonable. The number of persons based on GPS data with and without the threshold is shown in Fig. 8. In particular, we note that for weekdays the estimate with μ appears to be more realistic as clearly tourists tend to populate the touristic places during the day hours.

Overall, based on this analysis and the aforementioned research of Kawakami et al. [28], we concluded that $\hat{P}$ is a better estimate than $\check{P}$ for the number of tourists and that 0.3 tourist areas visited per day appears to be a suitable threshold to extract tourists from the longer-term GPS tracking data.

We conclude this section by noting the small number of observations we have in Fig. 9 for the GPS data, implying that if one wants to use the GPS data as a basis for tourist number estimation, additional information for appropriate scaling is required, which is our topic of discussion in the next section.

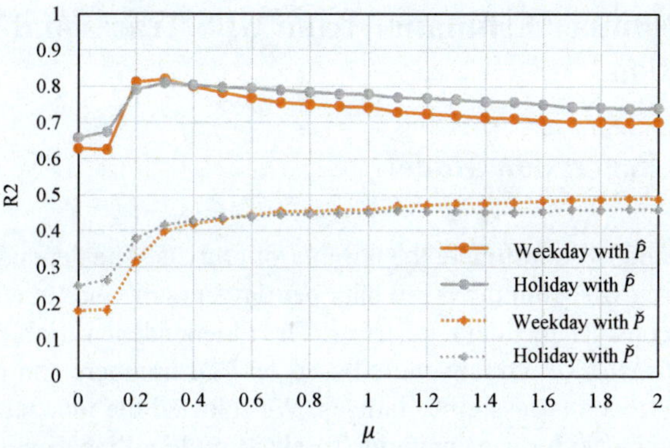

Fig. 7 Sensitivity analysis result with $\hat{P}$ and $\check{P}$

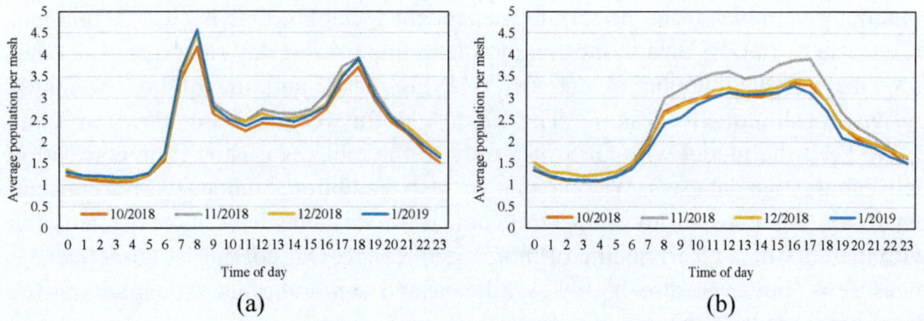

Fig. 8 Number of persons recorded in tourism areas with $\mu = 0$ based on GPS records **a** on weekdays (left) **b** on holidays (right)

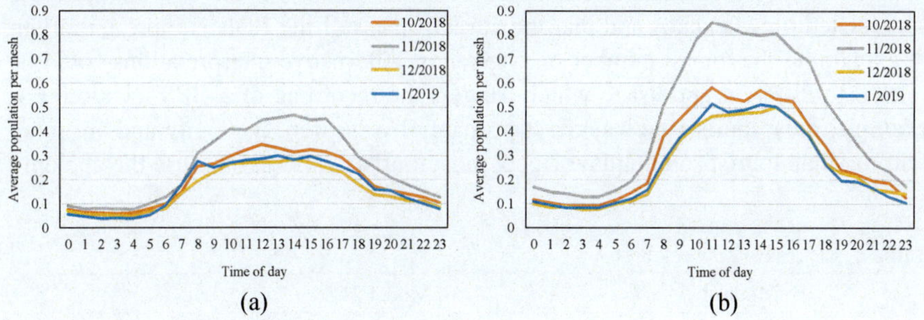

Fig. 9 Number of persons recorded in tourism areas with $\mu = 0$ based on GPS records **a** on weekdays (left) **b** on holidays (right)

5 Tourist Number Estimation from GPS Traces and POI Information

5.1 Linear Regression Model

We first conducted LR to estimate the number of tourists. The dependent variable is the number of tourists from the mesh data per day, month, and time of day in each mesh, i.e., we consider this data as "true." The independent variables are the GPS records, the estimates of attractiveness based on POI numbers and ratings as well as the aforementioned accessibility indices. We selected the independent variables following the stepwise forward method. To avoid multicollinearity, we did not use variables with absolute correlation coefficients exceeding 0.4 simultaneously.

As a consequence, our preferred models all only have two significant uncorrelated remaining variables from the set of independent variables. One is $g_{i,m,t}$ the number of tourists from GPS data in mesh *i*, month *m*, and time of day *t* per day. The other is $s_{i,t}$ the weighted number of stations (WNS) in mesh *i* and time of day *t* weighted by type mesh attractiveness w_j^a. The LR results for weekdays are shown in Table 2. We keep the model with GPS data only in the table as it shows the correlation between the two data sets. We note that we tested additional models with a constant but found this constant to be insignificant. Both variables have the expected sign with the GPS data clearly being of more significance. Our composite attractiveness measure is, however, also highly significant and can contribute to explaining the differences between the two data sets.

We also show the LR results for holidays in Table 3. The results were mainly the same as on weekdays, but β of GPS data was smaller, and β of WNS was larger than the result on weekdays. The result suggests hence that the two data sets are less related on weekends and that the attractiveness of the POIs is more important in explaining the tourist number on weekends. Alternative interpretations could be related to different app usage which triggers the recording of a GPS location. For example, most tourists on weekdays probably live nearer Kyoto city and are more familiar than tourists on holidays. If so, most tourists on weekdays use the app only

Table 2 LR results on weekdays

	Model 1				Model 2			
	B	S.E	B	T	B	S.E	β	t
$g_{i,m,t}$ [10^3]	1.29	3.49	0.911	368[b]	1.22	3.50	0.865	348[b]
$s_{i,t}$ [10^{-2}]	–	–	–	–	1.64	2.91×10^{-4}	0.140	56.4[b]
RMSE	424				402			
AIC	414,479.3				414,477.4			
N	27,744				27,744			

B: Non-standardized coefficient, S.E.: standardized error, β: standardized coefficient, [a]: $p < 0.05$, [b]: $p < 0.01$

around the station just to know when trains arrive at and leave the station because they have a clear understanding of the transportation system. On the other hand, tourists on holidays utilize the app frequently to search for the best way to their next destination because they are not familiar with transportation system in Kyoto city. In this case, the amount of GPS data could converge on weekdays, and be dispersed on holidays, which clearly explains our obtained result.

The scatter plot, whose x-axis is the estimated population and the y-axis is the mesh population for Model 2, is shown in Fig. 10a, and the fit of Model 4 is shown in Fig. 10b. As can be seen, the results are satisfactory with correlation coefficients above 0.91.

Table 3 LR results on holidays

	Model 3				Model 4			
	B	S.E	B	T	B	S.E	β	T
$g_{i,m,t}$ [10^3]	0.979	2.76	0.905	355[b]	0.892	2.50	0.824	356[b]
$s_{i,t}$[10^{-2}]	–	–	–	–	3.54	3.35×10^{-4}	0.237	103[b]
RMSE	556				474			
AIC	429,466				429,465			
N	27,744				27,744			

B: Non-standardized coefficient, S.E.: standardized error, β: standardized coefficient, [a]: $p < 0.05$, [b]: $p < 0.01$

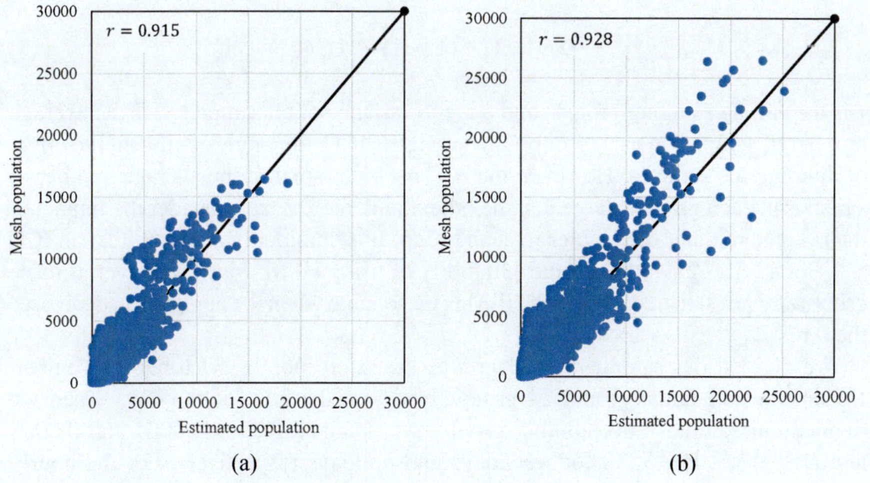

Fig. 10 LR results **a** on weekdays (left) **b** on holidays (right)

5.2 Hierarchical Linear Models

Tourism behavior in Kyoto changes with the seasons. In particular red leaves in autumn and winter scenery with snow in January or February tend to attract tourists to different sites. In addition to this, the time distribution of GPS data can be different from that of mesh data because GPS data were collected when the app was used. Considering these, we test if the coefficient values vary by month and time of day using HLM.

HLM is a way of considering the fixed and random effects within groups of the whole sample. Consideration of random effects means that group-specific variables are estimated, whereas the assumption of fixed effects means a "global" variable for the whole data set. In HLM, the variances of each coefficient among the group are considered if the coefficient has a random effect. The coefficient of each group is assumed to follow the normal distribution with the fixed effects as average and the variance. All coefficients including constant values can have a random effect, so that we must decide for which coefficient it is more suitable to estimate a random effect when applying the HLM. We test different specifications and select the best model based on terms of minimal AIC [38].

Criteria for the appropriateness of using an HLM approach are the intra-class correlation coefficient (ICC) and the design effect (DE). ICC follows Eq. (2), and DE Eq. (3) [38].

$$ICC = \frac{\tau}{\sigma} \tag{2}$$

$$DE = 1 + (k - 1) \times ICC \tag{3}$$

τ is the variance among groups, and σ is the variance of all samples. k is the average number of samples in each group. The larger the ICC, the more significant the effect of dividing the samples. However, the ICC becomes small if the average number of samples in each group is large. On the other hand, the DE can consider the impact of the average number of samples in each group. It is commonly accepted that an ICC > 0.1 or a DE > 2.0 indicates the suitability of using HLM. Moreover, even if these criteria are not satisfied, applying HLM is reasonable when the application decreases the AIC [38].

We divided our samples into 4 groups based on months (October, November, December, and January) and 24 groups based on the hour of the day. When we divided our samples into monthly groups, we obtained ICC = 0.0004 and 0.0005 and DE = 3.5, DE = 30 for weekdays and holidays respectively. For the hourly models, the ICC values were 0.0172 and 0.011 as well as 20 and 14 respectively. Considering these results, our adoption of the HLM is reasonable.

The monthly results are shown in Table 4. The AIC became the smallest when the coefficient of GPS data had a random effect, and the coefficient of WNS data did not have a random effect. The comparison of the RMSE of the LR and HLM models is shown in Fig. 11. In this figure, for a fairer comparison, we divided RMSE by the average population per month because a higher average population is associated with a bigger improvement in RMSE. Based on this, we can know for each month the improvement of using the HLM. RMSE values slightly decreased compared to the LR results, but the difference in RMSE is not large. Considering these results, GPS data represents the effect of the month well. The range improvement is slightly more significant on weekdays than on holidays. The fact that only GPS data has a random effect suggests that the inconsistency of mesh data and GPS data is more significant on weekdays than on holidays but that this effect depends on the month. An explanation is that mesh data on weekdays includes more non-tourists than on holiday and that the size of this effect has a seasonal dependence.

The hourly results are shown in Table 5. Also here we find that adding random effects for the GPS records but not for $s_{i,t}$ is the preferred model.

Table 4 HLM results considering the month's effect

	Model 5(Weekdays)				Model 6(Holidays)			
	B	S.E	β	t	B	S.E	B	t
$g_{i,m,t}$ [10^3]	1.27	58.0	0.892	21.9[b]	0.906	16.3	0.831	55.5 [b]
$s_{i,t}$ [10^{-2}]	1.58	2.86×10^{-4}	0.138	55.8[b]	3.49	3.45×10^{-4}	0.240	101[b]
Variance of coefficient								
$g_{i,m,t}$ [10^3]	13.3				1.04			
RMSE	393				472			
AIC	410,249				420,398			
N	27,744				27,744			

B: Non-standardized coefficient, S.E.: standardized error, β: standardized coefficient, [a]: $p < 0.05$, [b]: $p < 0.01$

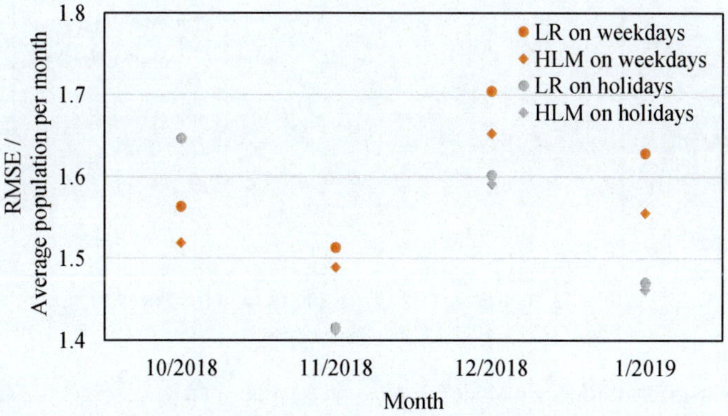

Fig. 11 Comparison of RMSE in Models 2, 4, 5, and 6 for the months Oct 2018–Jan 2019

Table 5 HLM results considering time of day effects

	Model 7 (Weekdays)				Model 8 (Holidays)			
	B	S.E	B	t	B	S.E	B	t
$g_{i,m,t}$ [10^3]	1.19	27.5	0.840	43.4[b]	0.963	36.7	0.883	26.2[b]
$s_{i,t}$ [10^{-2}]	1.63	2.84×10^{-4}	0.143	57.5[b]	3.32	3.15×10^{-4}	0.228	105[b]
Variance of coefficient								
$g_{i,m,t}$ [10^3]	17.5				32.1			
RMSE	390				429			
AIC	409,941				415,172			
N	27,744				27,744			

B: Non-standardized coefficient, S.E.: standardized error, β: standardized coefficient, [a] $p < 0.05$, [b] $p < 0.01$

The RMSE values decreased more significantly compared to the LR results for both weekdays and holidays. The comparison of the RMSE of the LR and HLM is shown in Fig. 12.

It can be observed that the difference in RMSE in the early morning was more significant than for other times of the day. The reason could be that many tourists used the app at that time of day. Further, comparing weekdays and holidays, the range of improvement on holidays is bigger than on weekdays possibly for the same reason. As additional evidence, we find that on weekdays the improvement is larger in earlier times of the day (predominantly 6–8 am.) than on holidays (8–10 am.) as presumably there is a larger proportion of non-tourists in the early morning weekday data. These effects are reflected in Fig. 13 in comparison to Fig. 10.

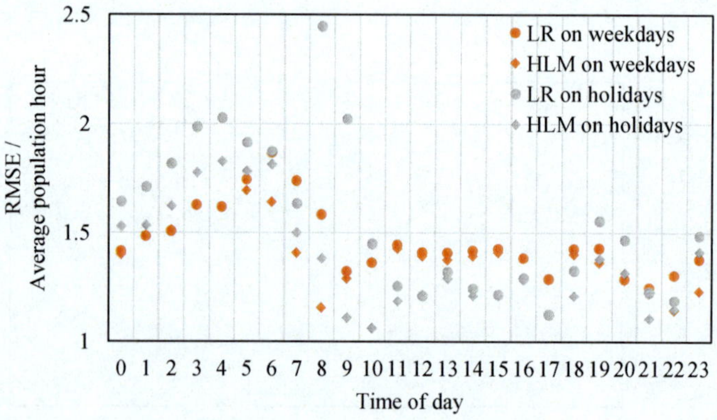

Fig. 12 Comparison of RMSE in Models 2, 4, 7, and 8 by time of day

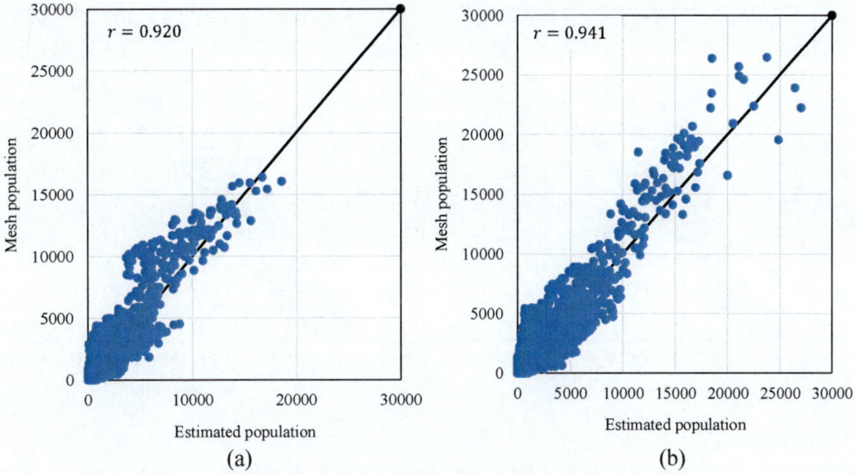

Fig. 13 HLM results **a** on weekdays (left) **b** on holidays (right)

5.3 Estimation of the Number of Tourists Within the Touristic Areas

In previous subsections, we estimated the LR based on the standard 1 km^2 meshes that are provided by the data provider. We now apply the preferred models found in previous subsections to the comparison between the GPS data (plus attractiveness and accessibility) models and the mesh data by considering the actual tourist areas shown in Fig. 2. Hence both data sets are adjusted to fit the revised areas.

The results of using the HLM models are shown in Fig. 14. The figure shows the comparison between the "estimated population" based on the HLM model and the "distributed mesh population" for each area, month and time of day. There are hence 3,552 (= $37 areas \times 4\ months \times 24h$) points in each graph. The distributed mesh population is obtained by a simple estimate of multiplying the population of each mesh that overlaps with the target area with a "GPS overlapping ratio". This ratio is defined as the percentage of the GPS data that are within the target area part of the mesh compared to all GPS records found in this mesh. The "GPS overlapping ratio" is used instead of simply taking the area overlap itself to correct for cases where, for example, half of the mesh area is mountainous and not accessible. In that case most GPS records will be found in the target area part of the mesh and hence the ratio will be near one so that also all the mesh population will be assumed to be in the target area.

Due to various assumptions discussed in this and previous sections, we acknowledge hence that both "distributed" and "estimated" values could be different from the actual ones. However, our matching shown in Fig. 14 gives us some confidence that our values are not too far from the ground truth. In general, we observe a good model fit for the two methods that need to overcome very different data limitations.

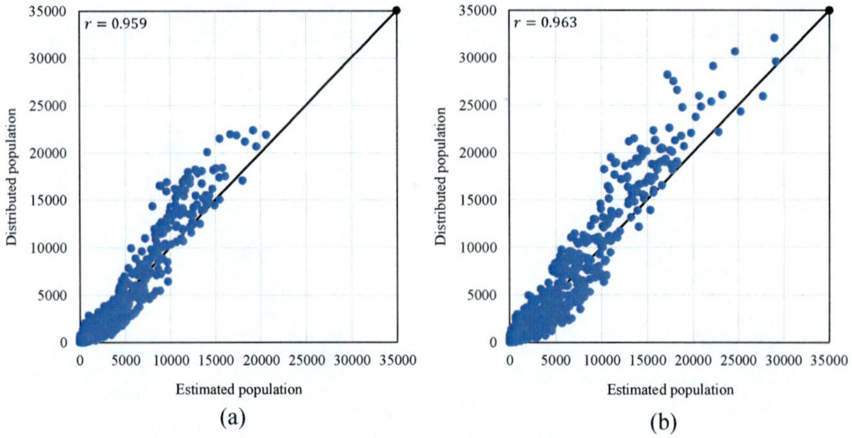

Fig. 14 Comparing the estimated and the distributed populations **a** on weekdays **b** on holidays

Among the areas with worse fit, that is a larger RMSE, are Areas 25 and 29 (refer to Fig. 2 and Table 1). These areas include stations with interchange between different train lines. Our estimated values might hence miss some tourists who only traverse this area but do not stay there for longer term. We further observe some errors in Areas 2, 35 and 37. These are far from the public transportation services and have a low number of tourists. Random under-sampling in these areas as well as systemic under-sampling of tourists coming by car to these areas–who are hence less likely captured with the GPS data from our travel planning app—might contribute to these errors.

6 Conclusion

Our study discussed the problem of estimating the number of tourists in specific areas and times given that most commonly available data sets are inadequate for this purpose. We suggest that, this is not only a common problem for many cities but also an often under-researched area by the transportation research community. Alternative methods to estimate the tourist population in specific areas at specific times are based on counting, tickets sales, hotel bookings, etc. However, in this paper, we showed that a range of map data in conjunction with relatively "small big data" could be a potentially useful alternative.

In the case of Kyoto City, as well as other cities, it might be possible to obtain visitor versus residential data, but clearly not all visitors are tourists. We first discuss how the mobile spatial statistics might be adjusted accordingly and then how an adjusted set of GPS data plus POI and accessibility data can be used to estimate the number of tourists. The mobile spatial statistics of non-residents, in general, appear to give a good estimate of tourist numbers, however, the estimate unsurprisingly appears to be better on holidays than on weekdays.

From the GPS tracking data, we found that taking a threshold of 0.3 tourist attraction visits per day is a reasonable threshold to distinguish tourists. As afore discussed, the value might seem low, but important to remember is that our sampling frequency is quite low such that a fairly large number of attraction visits are likely to be missed and that the data is recorded over a longer period such that also days without tourist activities could be included. One reason why the sampling frequency is quite low is that the GPS data is sampled mostly only when travelled used the app. Also, the utilization frequency of the app is for many tourists low, among others, because the attraction areas in Kyoto city are easily walked to and from. Therefore, if applied to other GPS tracking data sets, the threshold might have to be revised.

Our linear regression models matching the mobile statistics with GPS records plus additional information from the POIs and accessibility information generally, showed a good fit as illustrated by the R^2 values as well as the plot of the two estimates. The standardized coefficients of GPS data and the weighted number of stations differed between weekdays and holidays. This might suggest that the tendency to use the app which triggers the recording of the GPS locations is different on weekdays and holidays. The difference might also be due to different touring patterns on weekdays with a different weight also attached to accessibility. Yet another interpretation is again related to the different proportion of non-tourists in the data on weekdays.

The results improved further when using HLM. Dividing the samples by month also improved the model fit, though the increase was mostly insignificant. Instead, by dividing the samples based on the time of day, the model accuracy was improved more significantly. In both models, variable from GPS data have a random effect so that GPS data can have a bias based on month or time of day, and we could establish the model considering this bias. The range of improvement in the early morning, from 6 a.m. to 8 a.m. on weekdays and from 8 a.m. to 10 a.m. on holidays, was higher than that of other times during the day. These results also suggested different touring patterns on weekdays and holidays and different proportions of non-tourists in the data on weekdays and holidays.

We then applied the HLM model to estimate the tourist population within the touristic areas predefined by the Kyoto city government. Since there is no ground truth data for this estimation result, we compared the estimations by the two data sets. As a result, we gained some confidence that our estimation results are reasonable. Nevertheless, in future work, we certainly hope to obtain some observed data to obtain more evidence to affirm our present conclusions. In ongoing work, we are further utilizing the forementioned data to obtain additional information such as the stay duration in touristic areas and characteristics of tourist movements in the city.

Acknowledgements The first author was supported by JST SPRING, Grant Number JPMJSP2110. The data purchases were supported by JST SICORP projects, Grant Numbers JPMJSC1805 and JPMJSC20C4.

References

1. Imaizumi H (2020) Sustainable tourism for the resilience of vulnerable regions: Pro-poor tourism and over-tourism. J Econ 60(5・6):91–106. (in Japanese)
2. Kyoto city (2019) Kyoto sightseeing overall research, annual reports from 2001 to 2019. https://www.city.kyoto.lg.jp/menu2/category/22-6-0-0-0-0-0-0-0-0.html (in Japanese)
3. Schmöcker J-D (2021) Estimation of city tourism flows: challenges, new data and COVID. Transp Rev. Editorial 41(2):137–140
4. Bai Z, Wang J, Wang M, Gao M, Sun J (2018) Accuracy assessment of multi-source gridded population distribution datasets in China. Sustain 10(5)
5. Gao P, Wu T, Ge Y, Li Z (2022) Improving the accuracy of extant gridded population maps using multisource map fusion. GIScience & Remote Sensing 59(1):54–70
6. Bustos MFA, Hall O, Niedomysl T, Ernstson U (2020) A pixel level evaluation of five multitemporal global gridded population datasets: A case study in Sweden, 1990–2015. Popul Environ 42(2):255–277
7. Calka B, Bielecka E (2019) Reliability analysis of landScan gridded population data. The case study of Poland. ISPRS Int J Geo-Inf 8(5)
8. Chen R, Yan H, Liu F, Du W, Yang Y (2020) Multiple global population datasets: Differences and spatial distribution characteristics. ISPRS Int J Geo-Inf 9(11)
9. Mattos ACH, Mcradle G, Bertlolotto M (2020) Assessing the quality of gridded population data for quantifying the population living in deprived communities. arXiv preprint arXiv:2011.12923
10. Seike T, Mimaki H, Hara Y, Odawara T, Nagata T, Terada M (2011) Research on the applicability of mobile spatial statistics for enhanced urban planning. J City Plan Inst Jpn 46(3). (in Japanese)
11. Balakrishnan K (2020) A method for urban population density prediction at 30m resolution. Cartogr Geogr Inf Sci 47(3):193–213
12. Bakillah M, Liang S, Mobasheri A, Arsanjani JJ, Zipf A (2014) Fine-resolution population mapping using OpenStreetMap points-of-interest. Int J Geogr Inf Sci 28(9):1940–1963
13. Shimosaka M, Hayakawa Y, Tsubouchi K (2019) Spatiality preservable factored Poisson regression for large-scale fine-grained GPS-based population analysis. In: Proceedings of the AAAI conference on artificial intelligence. pp 1142–1149
14. Azar D, Graesser J, Engstrom R, Comenetz J, Leddy RM Jr, Schechtman NG, Andrews T (2010) Spatial refinement of census population distribution using remotely sensed estimates of impervious surfaces in Haiti. Int J Remote Sens 31(21):5635–5655
15. Douglass RW, Meyer DA, Ram M, Rideout D, Song D (2015) High resolution population estimates from telecommunications data. EPJ Data Science 4(4):1–3
16. Kikuchi M, Iwadate K, Hato E, Mogi W, Kato M (2018) Practical method to update master data of parson trip survey in metropolitan areas using the transportation big data. Proc Jpn Soc Civ Eng 74(5):667–676 (in Japanese)
17. Otake T, Kikuchi A (2019) Development of a simulator system for travel demand forecasting with data assimilation. Proc Jpn Soc Civ Eng 75(5):607–613 (in Japanese)
18. Khodabandelou, G., Gauthier, V., El-Yacoubi, M., Fiore, M. (2016). Population estimation from mobile network traffic metadata. In: 2016 IEEE 17th International Symposium on A World of Wireless, Mobile and Multimedia Networks (WoWMoM). pp 1–9
19. Cecaj A, Lippi M, Mamei M, Zambnelli F (2020) Forecasting crowd distribution in smart cities. In: IEEE international conference on sensing, communication and networking (SECON Workshops). pp 1–6
20. Bachir D, Gauthier V, El-Yacoubi M, Khodabandelou G (2017) Using mobile phone data analysis for the estimation of daily urban dynamics. In: IEEE 20th international conference on intelligent transportation systems (ITSC). pp 626–632
21. Aasa A, Kamenjuk P, Saluveer E, Šimbera J, Raun J (2021) Spatial interpolation of mobile positioning data for population statistics. J Locat Based Serv 15(4):239–260
22. Ahas R, Aasa A, Mark Ü, Pae T, Kull A (2007) Seasonal tourism spaces in Estonia: Case study with mobile positioning data. Tour Manage 28(3):898–910

23. Ahas R, Aasa A, Roose A, Mark Ü, Silm S (2008) Evaluating passive mobile positioning data for tourism surveys: An Estonian case study. Tour Manage 29(3):469–486
24. Ubukata Y, Sekimoto Y, Horanont T (2013) Availability as tourism statistical data of large scale and long term human mobility tracks by GPS: a study of Ishikawa Pref. Proc Jpn Soc Civ Eng 69(5):345–352 (in Japanese)
25. Kobayashi H, Zhang C, Schmöcker J-D, Nakao S, Yamada T (2021) Markovian analysis of tourist tours based on travel app data from Kyoto, Japan. In: Presented at 25th international conference of the Hong Kong society for transportation studies (HKSTS). December 12–14
26. Dantsuji T, Sugishita D, Fukuda D, Asano M (2017) Analysis of the properties of tourists' dwell time using Wi-Fi packet data a case study of the approach to Hase-Dera temple. J City Plan Inst Jpn 52(3). (in Japanese)
27. Nakanishi W, Kobayashi H, Tsuru T, Matsumoto T, Tanaka K, Suga Y, Kamiya D, Fukuda D (2018) Understanding travel pattern of tourists from Wi-Fi probe requests: a case study in Motobu Peninsula, Okinawa. Proc Jpn Soc Civ Eng 74(5):787–797 (in Japanese)
28. Kawakami R, Schmöcker J-D, Uno N, Nakamura T (2020) OD matrix estimation utilizing mobile spatial statistics with Kyoto tourism case study. Proc Jpn Soc Civ Eng 75(6):379–391 (in Japanese)
29. Gao Y, Schmöcker J-D (2022) Distinguishing different types of city tourists through clustering and recursive logit models applied to Wi-Fi data. Asian Transport Studies 8:100044
30. Takahashi K, Igarashi H (1990) Study on the recreation activity by recreation spot attractive index. Proc Jpn Soc Civ Eng 8:233–240 (in Japanese)
31. Kobayashi K, Sekihara Y (1991) Estimating the number of tourist visitors with destination-based surveys. Proc Jpn Soc Civ Eng 9:101–108 (in Japanese)
32. Mizokami S, Mogisugi H, Fujita M (1992) Modelling on the attraction of sightseeing area and excursion behavior. J City Plan Inst Jpn 27:517–522 (in Japanese)
33. Shen K, Schmöcker JD, Sun WZ, Qureshi AG (2022) Calibration of sightseeing tour choices considering multiple decision criteria with diminishing reward. Transportation. https://doi.org/10.1007/s11116-022-10296-7
34. Ishigami T, Kikuchi M, Inoue T, Iwadate K, Morio J, Ishii R (2017) Expectations and problems of traffic-related big data from a stand point of urban transport practical work. Jpn Soc Civ Eng 55. (in Japanese)
35. NTT Docomo Mobile spatial statistics. https://mobaku.jp/ (19 May 2022) (in Japanese)
36. Kyoto city. Fare table for Kyoto city subway. https://www.city.kyoto.lg.jp/kotsu/page/0000163782.html (in Japanese)
37. Kato H, Hashimoto T (2008) Mata-analysis on value of travel time savings in Japan. Jpn Soc Civ Eng 38. (in Japanese)
38. Simizu H (2017) Multilevel modelings for individual and group data. Nakanishiya Shuppan. (in Japanese)

Development of an Evaluation System for Virtual Ridepooling Stops: A Case Study

Dennis Harmann, Sefa Yilmaz-Niewerth, Riklas Häbel, Vanessa Vinke, Sarah Kögler, and Bernhard Friedrich

Abstract The majority of shared mobility services such as ridepooling use predefined locations to pick-up and drop-off passengers, so-called virtual stops. Compared to public transport stops, they do not require any infrastructural conditions. However, little to no attention is usually paid to the importance of the local environment when implementing virtual stops in an urban network. This paper presents an evaluation system to assess the feasibility of real-world virtual stops. We focus on essential criteria concerning admissibility, accessibility and the impact on traffic flow. To apply the proposed evaluation system, we used a case study of a mid-sized city in Germany with more than 4.300 virtual stops deployed by three different distribution methods. The results show that a vehicle may legally stop at 87% of these virtual stop locations and that 38% provide barrier-free access. In addition, at almost all stops (about 95%), the vehicle must remain on the roadway since there is no bay offered to leave the road. Regarding the distribution methods, the intersection approach shows benefits.

Keywords Ride access points · PUDO locations · Meeting points · Ridesharing · Shared mobility

1 Introduction

The location to access a transportation mode is one of the most essential elements of a trip. Especially for public transport systems, the stop positions must be chosen carefully. Moreover, they have to provide several essential infrastructure elements

---

D. Harmann (✉) · S. Yilmaz-Niewerth · R. Häbel · V. Vinke · S. Kögler · B. Friedrich
Technische Universität Braunschweig, Institute of Transportation and Urban Engineering, Hermann-Blenk-Str. 42, Braunschweig 38108, Germany
e-mail: d.harmann@tu-braunschweig.de

C. Antoniou et al. (eds.), *Proceedings of the 12th International Scientific Conference on Mobility and Transport*, Lecture Notes in Mobility,
https://doi.org/10.1007/978-981-19-8361-0_15

like a shelter, a bin, or seating possibilities, in terms of quality, cleanliness, and accessibility.

However, innovative and shared mobility systems like ridepooling also use specific stop locations to pick-up or drop-off (PUDO) passengers. As there are more and more cities offering new stop-based mobility services, the number of stops for these services will increase accordingly. Therefore, the study of these PUDO locations compared to other vehicle stops in public urban space is becoming increasingly important, especially in terms of infrastructural requirements.

According to Pettersson [15], there are three different variants of PUDO points for shared and on-demand mobility services. Besides the mentioned physical (bus) stops, there is the opportunity to operate in a taxi-like door-to-door service or to provide a stop-based service, using so-called virtual stops. However, it is proven by several research studies (i.e. [8, 17, 20]), that a stop-based system leads to a significant improvement for the service quality and efficiency compared to a door-to-door service. Though, this involves an additional walk for the customer to the vehicle's stop position [19]. The main difference of a virtual stop in comparison to a common public transport bus stop is that they are virtual, which implies that a virtual stop does not contain any additional physical infrastructure in a given road space and therefore can, theoretically, be located on any place of a roadside [16]. It is simply a digital location that can be identified with a mobile device including the related application of the provider. Consequently, they do not impact the cities infrastructure and do not show a specific design.

In comparison, public transport stops have a major impact on the cityscape. In addition to the necessary functional infrastructure elements, the design of the stops is becoming more and more important. Brovarone [2] concluded that a public transport stop can be developed as an active element that is embedded in its surroundings or serves as an eye-catcher. In general, five design principles for public transport stops could be highlighted: imageability, belonging, usability, transparency and accessibility. The aspect of accessibility is of particular importance for the attractiveness of a mode of transport with regard to demographic change [11]. As representing a public space, the requirements for public transport stops are defined by law. In Germany, for instance, every public transport stop must be designed barrier-free, according to the German guideline of the Passenger Transportation Act [13].

Apart from the conditions focusing on the user perspective, public transport bus stops must also take into account the requirements of vehicles at a stop location. Bus stop locations must provide sufficient space for a safe and comfortable arrival and departure to or from the stop, so aspects such as the width of the road and the required edge length must be considered when designing the bus stop [11].

In addition to public transport stops, stop areas reserved for taxis are another type of stop in cities. These taxi ranks are usually designed for parking several taxi vehicles. Therefore, planning such a taxi rank also means thinking about the number of taxis parking there. According to Cooper et al. [3], the main objective is to identify a location where taxis can wait for the next upcoming request, and where passengers can come and enter the vehicle comfortably, instead of hailing on street. Moreover, the parking taxis do not impact the traffic flow, due to their own reserved parking lane. On the other hand, they "occupy valuable space in central city locations" [3].

Concerning virtual stops for ridepooling, the literature mainly deals with simulative investigation or with the optimization of the service itself in terms of dispatching requests and vehicles or in terms of user experience. Engelhardt and Bogenberger [6] set up a simulation framework to analyze the impact of the boarding process and the stop's density and to optimize these virtual stops dynamically. To do so, they used a case study of Munich and created various scenarios using different sets of PUDO locations. As a result, the created model could increase the number of customers. However, virtual stops are only considered as digital points in an agent-based simulation setup and are located independently from the local environment or infrastructure. Similar to this work, Lotze et al. [12] investigated the difference between static and dynamic virtual stops. In their work, dynamic assigning of stops is based on current demand. They focus on the assessment of route length and travel time, and concluded, that dynamic stops lead to an improvement of both the criteria.

Bilali et al. [1] developed a model, which analyzed the service quality in terms of shareability. Besides constraints like detour time and waiting time they also considered the time to board and alight the vehicle. However, in their mathematical model, they do not consider the local conditions of PUDO locations within the shareability factor. In contrast to this, Hub et al. [10] investigated the user experience of an augmented reality-based prototype to visualize virtual stops concerning usability and intuitiveness. Due to the very specific sample of a total of 18 young people who were interviewed for this purpose, the diversity of society is not taken into account. This work concentrates on visualization and tries to ease the identification of virtual stops, with specific regard to autonomous vehicles. Czioska et al. [4, 5] points out that PUDO locations should be only located, where a safe boarding is ensured. In this context, they chose predefined and feasible areas like parking places, fuel stations, or cul-de-sacs. However, they investigated private and intercity ridesharing systems, which have other requirements in terms of routing and dispatching. Especially in urban areas, a high density of virtual stops is necessary to provide an efficient ridepooling service [5]. This is not possible by only focusing on these locations to implement virtual stops. In addition, Harmann et al. [9] conducted various expert interviews regarding PUDO points for ridepooling services. As a result, among others, unsuitable locations for virtual stops could be obtained: dead ends, railroad crossings, bicycle lanes, parking and stopping bans. However, the attractiveness of a future mobility service operating with automated vehicles highly depends on the users' experience at PUDO locations [14].

So far, however, there are no infrastructural requirements that a virtual stop must fulfill, even though it is located in a public space. In contrast, public transport stops must offer barrier-free access, among other things. Therefore, there is a need for clear and transparent criteria that a virtual stop should fulfill in order to be usable for both users and vehicles.

In this context, our aim with this paper is to propose and apply an evaluation system to assess the local conditions of virtual stops. For this purpose, we have implemented more than 4.300 virtual stops in the city of Braunschweig, Germany, using different distribution methods. Subsequently, we applied the evaluation system through empirical research to analyze the local environment of each stop.

2 Methodology

2.1 Evaluation Criteria

The evaluation system consists of four criteria, with the first criterion serving as initialization. Table 1 provides a comprehensive overview of the criteria considered. First, the distribution method for the stop location is given and which cardinal direction results from the center of the roadway. This information is not necessary for evaluating a PUDO location. However, it is needed for the unique identification of each location and for evaluation purposes such as filtering. For instance, when applying the streetlamp concept, it could be possible that a streetlamp is located on a central strip. Therefore, the value *middle* is noted for the direction criteria in this case. The information about one-way streets may be important for route planning.

All of the following evaluation criteria are considered in terms of whether the conditions are fulfilled or not. Consequently, the assessment values only distinguish between *True* and *False*.

First of all, it is of particular importance whether a vehicle's stop (of max. 3 minutes) at a certain place is allowed by law. Therefore, whether the local conditions provide for circumstances prohibiting a stop must be checked. In our case, we focus on German networks. Hence, we analyzed the PUDO locations according to the German road traffic act StVO [18]. According to this, stopping is not allowed in places with an absolute no-stopping zone or in the area of sharp curves.

In addition, stopping at taxi ranks is prohibited for shared mobility vehicles, so these optional stops are not eligible for virtual stops. A more detailed list of all prohibitions can be found in Table 1. Apart from these legal aspects, we have also designated PUDO locations on roads with a speed limit higher than 50 km/h as not permissible for a virtual stop for traffic safety reasons.

As far as the impact of virtual stops on flowing traffic is concerned, it is of crucial importance whether the vehicle has to stop on the roadway or next to it. To address this issue, the bay attribute is used. A bay can occur due to various local conditions at the stop. For example, a conventional bay is a driveway where the vehicle can briefly leave the roadway and stop next to it so that the stopping maneuver does not affect the traffic flow. Delivery zones and any type of bay can also be used as a stopping bay where a vehicle's stop is not prohibited. If there is no infrastructure for stopping at the edge of the roadway, the vehicle remains on the street. Thus, the impact on traffic flow depends on whether other vehicles can pass the stopped ridepooling vehicle. As a consequence, two infrastructural conditions may allow for passing. First is the option of a simple lane change, which is only possible if there are two or more lanes for that direction. Since the PUDO position is always on the

Table 1 Criteria and attributes considered within the evaluation system

Criterion	Attribute	Value	Description
Initialization	Distribution approach	Grid	Stop locations with a distance of 100 m
		Intersection	Stop locations at each intersection arm
		Streetlamp	Stop locations at each streetlamp
	Direction	North	Position of the stop according to cardinal direction (in relation to the center of the lane)
		South	
		West	
		East	
		Middle	Position in the middle on central strip
	One-Way	True	Stop is located in one-way street
		False	Stop is not located in one-way street
Admissibility	Permission	True	Stop is allowed according to StVO [18]
		False	Stop is not allowed according to StVO [18] or according to other conditions: – Absolute stopping prohibition – Cycling infrastructure on the roadway – Speed limit higher than 50 km/h – Railroad crossing – Entrance or exit lanes – Fire department access road – Taxi rank – In the running space of rail vehicles – Traffic roundabout – In unclear road sections – In the area of sharp curves
Impact on traffic flow	Bay	True	Vehicle can leave roadway for stop (bay, driveway, delivery/loading zone, etc.)
		False	Vehicle stops on roadway
	Pass-by (lane)	True	Other road users can pass the vehicle during the stop, as there is at least a second lane in the direction of travel
		False	There is no second lane in the direction of travel
	Pass-by (width of road)	True	If the roadway width is >= 4.75 m (according to the German guideline RASt06), it is possible for other vehicles to pass (if necessary using the lane for oncoming traffic) during the stop
		False	Passing of other vehicles during the stop is not possible
Accessibility	Barrier-free	True	Entry is possible directly from the curb
		False	Access is not possible without overcoming obstacles (bicycle parking facilities, ditch, railroad track) or no sidewalk available
	Public transport bus stop	True	Stop is located at a bus stop
		False	Stop is not located at a bus stop
	Green spaces	True	Stop is located at a green area
		False	Stop is not located at a green area
	Parking spaces	True	Stop is located at parking spaces on the roadside
		False	Stop is not located at parking spaces

right lane, the other one can be used for passing. However, depending on the traffic volume, this can also cause negative impacts as other vehicles have to wait behind the stopping vehicle. Another possibility for overtaking when there is only one lane relates to the width of the entire roadway. According to the German guideline RASt [7], a road width of more than 4.75 m is required for two vehicles to pass each other without conflict. Therefore, passing is possible if this width can be guaranteed at the virtual stop. However, it must be taken into account that the oncoming lane is used when passing. This can lead to risky situations.

Apart from traffic-related attributes, usability also needs to be considered. One of the essential aspects is accessibility since the service should be usable by everyone, regardless of possible mobility impairment. Therefore, providers should not only think about wheelchair users when it comes to accessibility. But also, elderly, disabled, visually impaired, and people with luggage or baby carriages should be provided with the opportunity to enter or leave a vehicle at a virtual stop. One aspect of particular importance is the height of the kerb-stone. Therefore, we consider whether it is possible to enter the vehicle directly from the curb for our evaluation system. Suppose access to the vehicle at the PUDO location is not possible without overcoming obstacles (bicycle parking facilities, ditch, railroad track) or there is no sidewalk. In that case, the attribute barrier-free is rated as *False*.

As far as the accessibility aspect is concerned, public transport bus stops provide a convenient place for entering and leaving a vehicle. Moreover, the infrastructural conditions at bus stops are better than anywhere on the roadside. Therefore, using a bus stop as a virtual stop is beneficial for shared mobility services. Consequently, it is important to know whether a virtual stop is located at a bus stop. However, depending on the service's operator (public transport or private company) the usage of public bus stops may not be allowed and must be adjusted with the local public transport company.

In some cases, virtual stops are located near green areas, such as lawns, bushes or trees. To filter such stops, we assigned to each stop position whether it is necessary to cross green areas or whether it is possible to reach the vehicle from asphalt ground.

Another important aspect is whether there are official parking spaces at the virtual stop. However, since it is not possible to monitor if parking spaces are available or not, the stopping position must be on the roadside. If a parking position is available at a particular virtual stop location, the vehicle may use it and leave the lane, if possible. However, this cannot be guaranteed. Therefore, virtual stops at public parking places should be considered with caution. In addition, if parking spaces are occupied by parked vehicles, accessibility to the ridepooling-car is limited.

2.2 Empirical Investigation

The application of the proposed evaluation system and assessing attributes for all possible virtual stop locations can only be made by practical empirical research. This means that each PUDO location must be verified individually.

In principle, we evaluated each stop independently and sequentially. First, we used open-source satellite imagery. Most criteria can be assessed in this way. Based on this data, most of the permissible aspects of a PUDO location can be easily identified. It is also possible to determine the direction of travel and whether the stop is on a one-way street or at a public bus stop.

Furthermore, in most cases it can be used to check if there is a bay at a particular location. In addition, the number of lanes per direction can be determined by analyzing satellite images. Using specific measuring tools provided by satellite imagery software allows getting the width of a road. Thus, whether the pass-by attribute is true or false can be analyzed. In addition to the road-specific attributes, it is also possible to evaluate the roadside area. According to the information from the satellite image, it can be determined if there are green spaces or parking spaces next to the roadway. Consequently, these attributes can also be evaluated using satellite imagery. However, sometimes there might be trees or bushes next to a parking lane. In this case, the parking attribute is assessed as *True* and green spaces as *False*. Hence, there is no barrier-free access to the vehicle, which leads to the corresponding assessment.

However, it is only possible to verify all attributes for a few locations using satellite imagery. Therefore, additional on-site observations are indispensable to investigate the missing attributes. The number of on-site observations required depends on the quality of the satellite images for a given area. However, traffic signs in particular, such as "absolute no parking", are usually not visible on satellite imagery. In addition, it is possible that trees or other obstacles on the image obscure the view of the road, so it is not clear whether there are parking spaces, for example. Therefore, these information must be determined by additional on-site observations. Hence, the procedure of applying the evaluation system can be very time-consuming.

It should also be mentioned that the environment determined from satellite images can deviate from the actual conditions. Depending on when the satellite image was taken, parking spaces may not be identified because, for example, they are vacant and not marked on the road. An on-site observation, e.g., in the evening, shows that there are usually many parked vehicles.

Another difficulty that must be considered is construction sites as they can lead to uncertainties. For example construction can affect traffic routing for a period of time, making it difficult to evaluate a PUDO location that may be present after construction is complete. Therefore, it may be advisable to note such locations and check the local conditions afterwards. In addition, regarding traffic regulations, there might be time restrictions in some cases, such as temporary parking restrictions. In these cases, the interim rule should be considered to be permanently in place.

3 Case Study

Our research work investigating virtual stops is based on the road network of Braunschweig, Germany. Therefore, we applied the proposed evaluation system for a case study within this network. The highlighted area in Fig. 1 shows the service area where

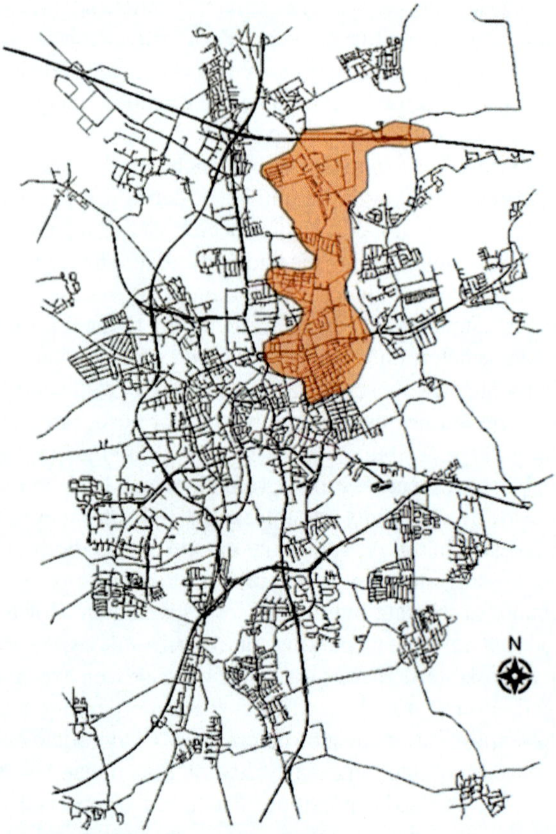

Fig. 1 Road network of Braunschweig with the highlighted service area

the virtual stops are implemented. This corridor connects the inner-city center and the northern part, Braunschweig's rural, outlying area. It consists of almost 1.000 edges with a total length of nearly 163 km, 425 intersections and covers an area of approximately 17 km^2. Moreover, there are six bus lines and two tram lines crossing this corridor, using around 65 public transport stops in this area.

To implement virtual stops into this area, we applied the three distribution methods proposed by Harmann et al. [9], to get three different sets of virtual stops: *Grid*, *Streetlamp*, *Intersection*. We chose a distance of 100 m for the grid approach. In one-way streets, the stop is located only on the right side, corresponding to the vehicle's entrance. According to the streetlamp method, we implemented a possible PUDO location in front of every lantern. All street crossings are considered for the intersection approach, and virtual stops are set into each intersection arm. The distribution approaches are applied to the whole service area, independently from any local restrictions or conditions. Exemplary, a segment of the service area including virtual stop locations according to the distribution approaches is shown in Fig. 2.

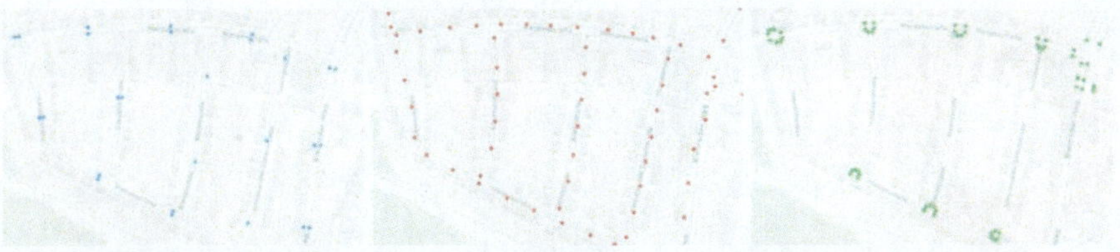

(a) Grid with a distance of $100m$. (b) Streetlamps. (c) Intersection.

Fig. 2 Virtual stops after applying the approaches by Harmann et al. [9]

Table 2 Quantity of virtual stops, per distribution method

Distribution method	Number of virtual stops
Grid	1036
Streetlamps	1816
Intersection	1484
$\sum$	4336

After applying the distribution methods, we obtained more than 4.300 virtual stops, all stored to a GeoJSON-File. Table 2 gives an overview of the number of stops created by each approach. Due to the high density of streetlamps in the urban area of Braunschweig, this approach leads to the most significant number of stops.

4 Results

The results are based on the case study's empirical investigation of all virtual stops by applying the proposed evaluation system. In the end, around 80% of the PUDO locations needed additional on-site observations to collect the missing data, which two persons conducted over several days. Each stop position was inspected once by one of these persons. In particular, the identification of no-stopping zones caused this high percentage of local examination, which could not be determined using satellite imagery.

For the presentation of results all virtual stops that meet the attributes are considered first. Figure 3 shows the calculated share of virtual stop locations per attribute rated as *True*.

Regarding the permissibility of a virtual stop, a stopped vehicle is allowed at the majority of locations (Fig. 3a). In 13% of all virtual stops, stopping is not legal due to restrictions, i.e. absolute parking prohibitions. This result is essential for further evaluation since a legal stop position is a basis for establishing a virtual stop. In our case study, 3.787 out of 4.336 can be used as a legal stop location. It can also be seen that the distribution methods used lead to a majority of legal stopping positions, although they are initially applied independently of the network structure.

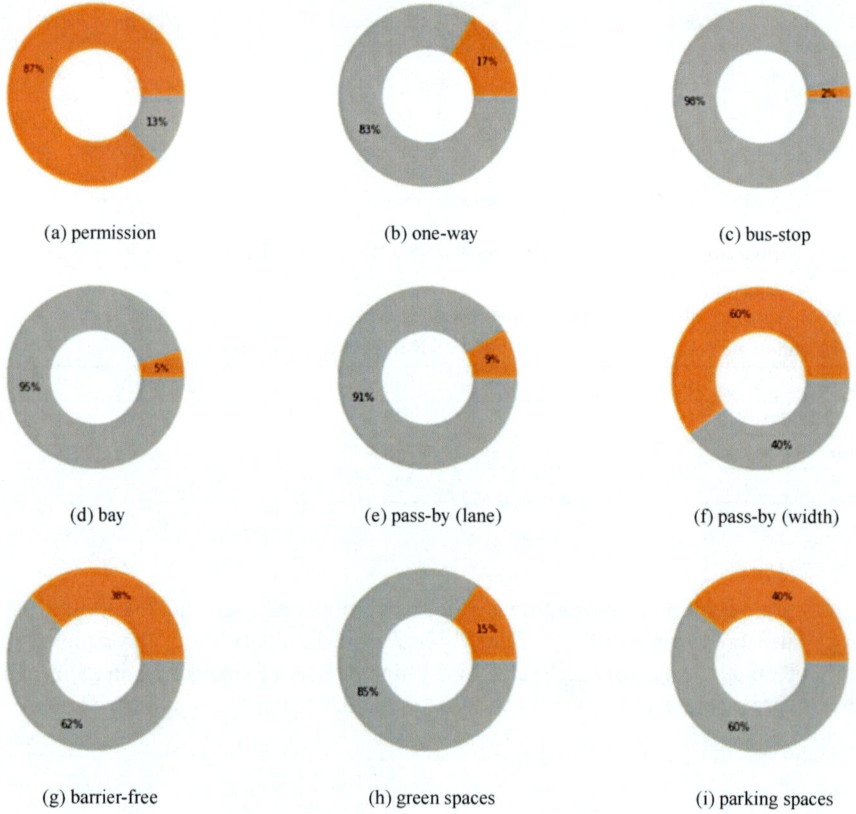

Fig. 3 Share of virtual stops according to attributes. (*orange = True; gray = False*)

The share of virtual stops in one-way streets is about 17% (Fig. 3b). The density of PUDO locations is influenced by the proportion of one-way streets since virtual stops are only required on the right side of the road. One-way streets are especially prevalent in dense urban and residential areas.

Regarding the virtual stops located directly at a bus stop (Fig. 3b), the applied distribution methods resulted in a share of 2%. However, public transport (bus) stops provide a comfortable infrastructure for customer boarding and alighting and should be included for the operation of a ridepooling service.

The three associated attributes show different results for the criteria that focus on the potential impact of virtual stops on moving traffic. Only 5% of all stops provide a bay (Fig. 3c), enabling the vehicle to leave the road for the PUDO event. Consequently, the majority of stops need to be done on the lane. For this case, the pass-by attributes were investigated. The results of the empirical study show that in only 9% of the cases, a second lane is available to pass the stopping vehicle (Fig. 3d). Due to urban and residential characteristics in the service area, this result is not surprising. Since major roads usually consist of two or more lanes in each direction, the proportion of virtual stops on this type of road is marginal.

However, overtaking a stopping vehicle is possible in many cases due to the road width. The results of this attribute show that at 60% of the stops, the road is wide enough to overtake a stopping vehicle on the roadway (Fig. 3e). It should be noted that all PUDO locations with more than one lane in each direction are included in this percentage. Nevertheless, passing a stopping vehicle is possible at more than 50% of virtual stops. Since this mostly require the use of the oncoming lane, these movements can lead to critical situations in terms of traffic safety. On the other hand, around 50% of stops lead to an obstacle for traffic flow due to a stopping vehicle on the roadway. In these cases, overtaking is not possible because of the road width.

The accessibility criteria are essential for the usability of a stop location. The results show, that more than 1.600 stop positions in the service area offer barrier-free access to the vehicle (Fig. 3f). However, it must be considered, that the barrier-free attribute only consists of the vehicle stopping directly at the roadside. Other infrastructural accessibility elements like benches or shelters cannot be guaranteed at a virtual stop.

The share of accessible virtual stops depends primarily on local conditions such as green spaces or parking spaces at a stop. Looking at the results of these attributes, it is noticeable that more than half of the virtual stop positions are located either at green spaces or parking spaces (Fig. 3g; h). Both of these can lead to restrictions on vehicle access. In particular, if parking spaces are occupied by parked vehicles, the PUDO event can be problematic, which is the case for 40% of all virtual stops in the service area.

For a more detailed view of the results, Fig. 4 gives an overview of the share of virtual stops per attribute, subdivided into the used distribution approaches. Since providers typically use only one method to implement virtual stops, an approach-specific evaluation is required to determine where each distribution method's advantages lie.

For most attributes, the results differ not significantly between approaches. For example, the lower percentage of legal stop positions within the intersection approach (permission) can be explained because it is not always allowed to stop directly in front of an intersection. However, this depends on the network. On the other hand, the lower rate for the grid approach for the one-way streets can be derived from the distribution method itself, which provides for a stop only on the right side of the street in the case of one-way streets.

The most conspicuous differences can be seen in the attributes pass-by, barrier-free and parking spaces, which show significant advantages using the intersection approach. Since roads widen before an intersection in many cases, the width of the road is sufficient to pass a stopped vehicle. This leads to about 70% of virtual stops realized by the intersection approach, providing the opportunity to pass the vehicle. Moreover, the intersection approach shows a promising result regarding the barrier-free attribute. Compared to the streetlamp and the grid approach, the share of PUDO locations that offer a stop position directly at the curb is almost twice as high, at around 56%. Again infrastructural conditions explain this result. Typically, there

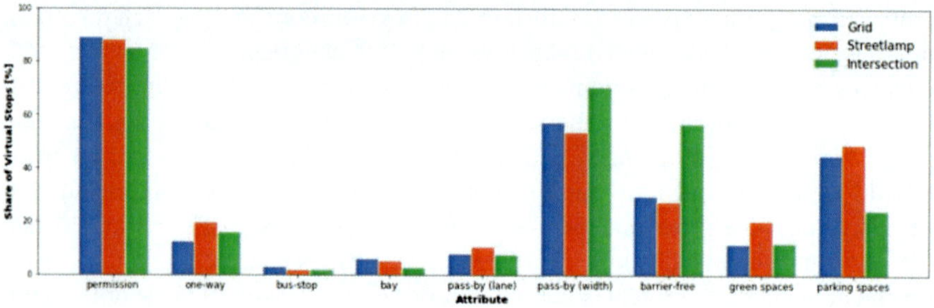

Fig. 4 Share of virtual stops per attribute and approach

are fewer trees or parking facilities near intersections due to better visibility. The resulting percentage shows this for green and parking areas. In this case, the grid and streetlamp approach indicates a higher share near one of these environmental elements. The probability of crossing an obstacle such as a green space or passing a parked vehicle is even lower for the intersection approach.

In addition to considering specific and unique attributes, the combination of attributes is critical to assessing the actual usability of a stopping location. First and foremost, locations must legally allow a vehicle to stop at a specific location. As mentioned above, most PUDO locations meet this requirement. However, other essential criteria must be considered determining the quality of stop locations. Therefore, we create several sub-datasets of virtual stops, based on the results of the evaluation system. Each dataset consists of virtual stops for which certain attributes are required. Consequently, it is possible, that a single stop position can be associated to more than one dataset, as it meets the requirements for all of them. The description and the used filters to create the datasets are listed below:

- **Dataset 1**: Legal and accessible
  - *(permission AND barrier-free == True) AND (green spaces AND parking-spaces == False).*
- **Dataset 2**: Legal and minimal impact on traffic flow
  - *(permission == True) AND (bay OR pass-by (lane) OR pass-by (width) == True).*
- **Dataset 3**: Legal, accessible and minimal impact on traffic flow
  - combination of Dataset 1 and Dataset 2
  - *(permission AND barrier-free == True) AND (green spaces AND parking-spaces == False) AND (bay OR pass-by (lane) OR pass-by (width) == True).*

Figure 5 shows the resulting number of stops that meet the attributes of each dataset. Dataset 2 includes all PUDO locations that are legally permissible while providing the ability to overtake the vehicle during the stop. Either by providing a

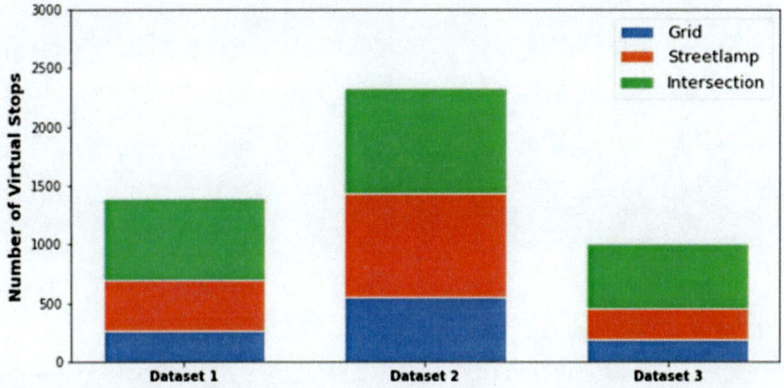

Fig. 5 Number of virtual stops according to dataset

bay or allowing the vehicle to pass due to sufficient requirements for passing on the road. Dataset 2 shows the highest percentage with 2.327 stops, about 53% of all virtual stops. It is also noticeable that most of the stop locations in this dataset are based on the intersection (897 stops) and streetlamp (881 stops) approach.

About Dataset 1, the number of stops that offer a legal and accessible location is less than 1.500, out of more than 4.300 stops in total. This is because Dataset 1 considers all stopping positions that meet the barrier-free attribute and offer a location where vehicles can legally stop. In this case, the intersection approach represents half of the PUDO locations.

A combination of both the pre-mentioned datasets results in Dataset 3. Hence, the filtering process presents the number of legal stopping positions providing barrier-free access and offering the possibility to pass the vehicle, either by providing a bay, a second lane or sufficient road width. Thus, this dataset can be used to filter out a user- and traffic-based optimal set of virtual stops. As the diagram shows, 1.004 feasible stops can be identified within the service area. However, more than half (54%) of these stops are implemented by applying the intersection approach. This results in 550 legal, accessible, and non-traffic impacting PUDO locations and leads to an average density of 296 m per stop considering the network length within the service area.

Figure 6 shows qualitatively the differences between the unevaluated set of virtual stops using the intersection approach with the resulting stop locations of Dataset 3 for the intersection approach. The number of stops is substantially lower, but the service area is still equipped with a comprehensive set of PUDO locations.

In comparison, the change in the streetlamp approach after the filtering process shows a different picture. As shown in Fig. 7, the number of virtual stops decreases drastically from 1.816 to 274 in the case of Dataset 3. Consequently, some gaps in the service area occur without any PUDO location. Since only a low percentage of all stops applied by the streetlamp method provides barrier-free access as well as options to pass-by, the result is a network with irregular distribution of virtual

(a) All Stops. (b) Remaining stops after filtering with Dataset 3.

Fig. 6 Virtual stops by applying the intersection approach

stops after filtering with the requirements of Dataset 3. In general, the grid approach performs worst in evaluating most attributes and can be evaluated as the least useful method for implementing virtual stops.

5 Discussion

In general, the results show that the distribution methods lead to a high number of legally admissible virtual stops in a network. On the other hand, the results also show that in practice, contrary to the theoretical assumption from Rissanen [16], not every roadside location can always be specified as a virtual stop.

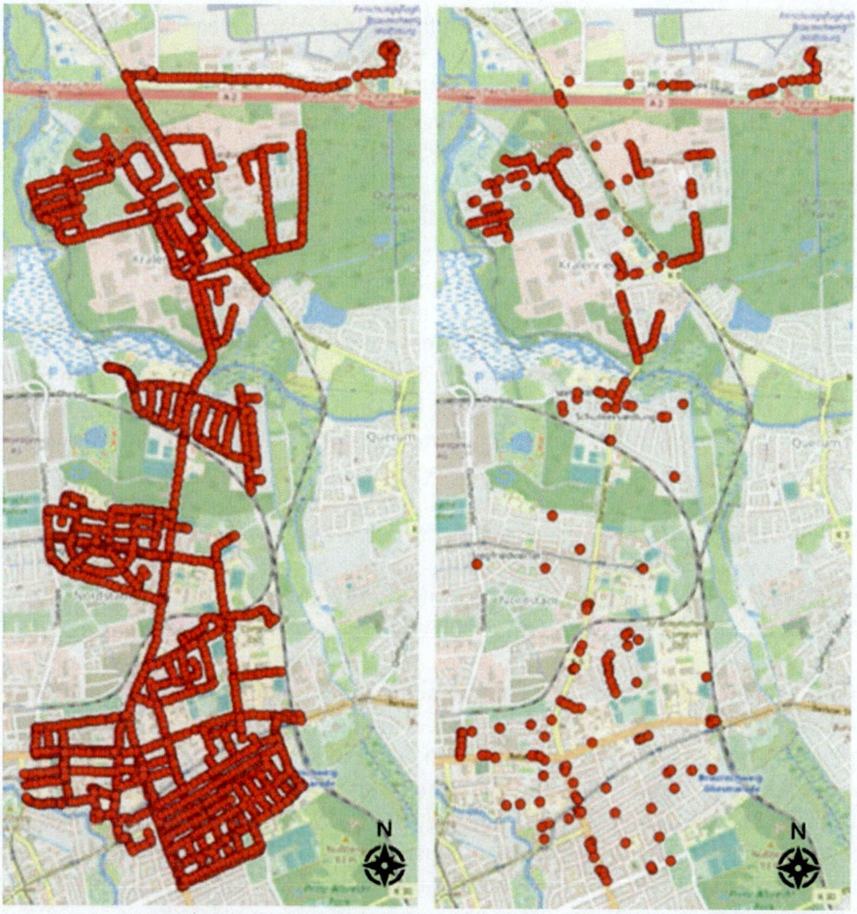

(a) All Stops. (b) Remaining stops after filtering with Dataset 3.

Fig. 7 Virtual stops by applying the streetlamp approach

Moreover, the admissibility cannot be the sole criterion. Even though the basic idea of virtual stops is to place them in a network as quickly and uncomplicated as possible, the locations still need to be carefully selected to ensure the general usability of the stop. Therefore, the proposed evaluation system provides a straightforward and comprehensive method to evaluate the feasibility of real-world virtual stops. Nevertheless, assessing all PUDO locations in a network with this evaluation method requires some effort. In particular, conducting on-site observations is time-consuming and should not be underestimated. However, most of the attributes can be assessed using satellite imagery.

Based on the case study's empirical investigation, we found that vehicles cannot leave the road at most locations due to the lack of bays and therefore have to stop on the lane. This finding emphasizes and confirms the need for research on the possible impact of virtual stops on traffic flow, which was also identified by Pernestål and Kristoffersson [14].

Concerning accessibility, the evaluation results show that about 38% of all virtual stops in the service area provide barrier-free access to the vehicle. On the other hand, this means that most stop locations cannot be used without overcoming an obstacle. In particular, regarding customers with impaired mobility and against the backdrop of demographic change, providers should give these customers the option to select a need for barrier-free PUDO location. Another option would be to generally include only accessible virtual stops in order to always ensure barrier-free boarding and alighting, regardless of the customers' impairments and needs. Therefore, the stops filtered by Dataset 1 and Dataset 3 can be used for a completely accessible service.

Zwick et al. [20] considered only public bus stops as virtual stops for their simulation of a stop-based ridepooling service in Munich. Thus, a convenient infrastructure is provided at each virtual stop as bus stops meet several requirements in terms of accessibility and quality. However, the primary advantage of a shared mobility service is the high density of stops, making the service more attractive and efficient. In terms of bus stop density in our case study, the coverage of public bus stops is significantly lower than when using the studied distribution methods to implement virtual stops, even when considering the highly filtered set of stops from Dataset 3.

Further research will mainly focus on the potential impact of virtual stops on traffic flow. For this purpose, a corresponding microscopic traffic flow simulation will be developed and analyzed. Moreover, the proposed evaluation system needs to be improved in terms of validity and transparency, for instance, by performing a more specific quantitative analysis of the resulting stop locations. Therefore, the results must be considered with a clear target function, for example, uniformly dense coverage of the service area. The requirements of the proposed datasets can be used as potential constraints. Another constraint could be the population size of an area, which was not considered in the method proposed in this paper. Additionally, the optimal combination of the distribution methods should be investigated, to merge the advantages of each approach. So far, the distribution approaches have been applied independently.

6 Conclusion

The majority of shared mobility services like ridepooling uses predefined and virtual stop locations to pick-up and drop-off passengers. Since these stop positions do not require any conventional infrastructural element, in theory they can be implemented in the street network independently from any local environment. In this paper, we present an evaluation system consisting of different essential attributes regarding the usability like permissibility and accessibility, to assess the feasibility of real-world virtual stops. We used a case study of the medium-sized city Braunschweig, Germany with more than 4.300 virtual stops, implemented by three different distribution methods, to apply the proposed evaluation system. The results indicate, that about

87% of all stop positions legally allow a vehicle's stop. However, in the majority of stops, the vehicle must stop on the road, since only about 5% of the PUDO locations offer a bay to leave the road during the dwell time. Concerning the accessibility, the evaluation results show a worrying result. Only about 38% of all virtual stops in the service area provide barrier-free access to the vehicle. Considering the distribution methods which were used to implement virtual stops into the service area, the intersection approach shows most advantages. More than half of the stops, which provide a legal stop location, where other vehicles have the option to pass the stopping vehicle and where barrier-free access is ensured, have been implemented by the intersection approach.

Acknowledgements This work was funded by the German Federal Ministry of Transport and Digital Infrastructure (BMVI) in the scope of the project ViVre (reference no. 01MM19014G).

References

1. Bilali A, Dandl F, Fastenrath U, Bogenberger K (2019) Impact of service quality factors on ride sharing in urban areas. In: 2019 6th international conference on models and technologies for intelligent transportation systems (MT-ITS). IEEE, pp 1–8
2. Brovarone EV (2021) Design as if bus stops mattered: exploring the potential role of public transport stops in the urban environment. Urban Design Int 26(1):82–96
3. Cooper J, Mundy R, Nelson J (2010) Taxi!: urban economies and the social and transport impacts of the taxicab. Ashgate Publishing, Ltd
4. Czioska P, Kutadinata R, Trifunović A, Winter S, Sester M, Friedrich B (2017) Real-world meeting points for shared demand-responsive transportation systems. Publ Transp 11(2):341–377
5. Czioska P, Mattfeld DC, Sester M (2017) GIS-based identification and assessment of suitable meeting point locations for ride-sharing. Transp Res Proc 22:314–324
6. Engelhardt R, Bogenberger K (2021) Benefits of flexible boarding locations in on-demand ride-pooling systems. In: 2021 7th international conference on models and technologies for intelligent transportation systems (MT-ITS). IEEE, pp 1–6
7. Forschungsgesellschaft für Straßen- und Verkehrswesen (2006) Richtlinien für die Anlage von Stadtstraßen (RASt 06). Ausgabe 2006
8. Fielbaum A, Bai X, Alonso-Mora J (2021) On-demand ridesharing with optimized pick-up and drop-off walking locations. Transp Res Part C: Emerg Technol 126:103061
9. Harmann D, Yilmaz-Niewerth S, Jacob C (2022) Methodological distribution of virtual stops for ridepooling. Transp Res Proc 62:442–449
10. Hub F, Wilbrink M, Kettwich C, Oehl M (2020) Designing ride access points for shared automated vehicles-an early stage prototype evaluation. International conference on human-computer interaction. Springer, Cham, pp 560–567
11. Kočárková D, Novotný V, Jíšová J (2019) Design of public transport stops and stations and its contribution to attractive and accessible public transport. In: 2019 smart city symposium prague (SCSP). IEEE, pp 1–7
12. Lotze C, Marszal P, Schröder M, Timme M (2021) Dynamic stop pooling for flexible and sustainable ride sharing. arXiv preprint arXiv:2108.00788
13. Personenbeförderungsgesetz in der Fassung der Bekanntmachung vom 8. August 1990 (BGBl. I S. 1690), das zuletzt durch Artikel 1 des Gesetzes vom 16. April 2021 (BGBl. I S. 822) geändert worden ist

14. Pernestål A, Kristoffersson I (2019) Effects of driverless vehicles-comparing simulations to get a broader picture. Eur J Transp Infrastruct Res 1(19):1–23
15. Pettersson F (2019) An international review of experiences from on-demand public transport services. K2 working paper
16. Rissanen K (2016) Kutsuplus-final report. HSL, Helsinki
17. Stiglic M, Agatz N, Savelsbergh M, Gradisar M (2015) The benefits of meeting points in ride-sharing systems. Transp Res Part B: Methodol 82:36–53
18. Straßenverkehrs-Ordnung (2021) Straßenverkehrs-Ordnung vom 6. März 2013 (BGBl. I S. 367), die zuletzt durch Artikel 13 des Gesetzes vom 12. Juli 2021 (BGBl. I S. 3091) geändert worden ist
19. Wang Z, Hyland MF, Bahk Y, Sarma NJ (2022) On optimizing shared-ride mobility services with walking legs. arXiv preprint arXiv:2201.12639
20. Zwick F, Kuehnel N, Moeckel R, Axhausen KW (2021) Agent-based simulation of city-wide autonomous ride-pooling and the impact on traffic noise. Transp Res Part D: Transp Environ 90:102673

Prediction of Signal Phase and Timing Information: Comparison of Machine Learning Algorithm Performance

Lena Elisa Schneegans, Josua Duensing, Kevin Heckmann, and Robert Hoyer

Abstract A significant amount of road traffic emissions are caused by traffic jams and stops in particular in front of (signalized) intersections. Reliable signal timing estimation methods could be the key to minimizing unnecessary braking and accelerating in front of traffic lights. In this paper, we present comparative experiments of machine learning algorithms to predict the switching times of traffic actuated signals considering classification as well as regression. Best results for the prediction of traffic actuated signals were obtained by methods based on decision trees. We came to the conclusion that extreme gradient boosting (XGBoost) shows the best performance for the two traffic signal systems under consideration. Here the most poorly performing methods were SARIMAX and neural networks (RNN, LSTM, GRU, and MLP).

1 Introduction and Background

The transportation sector is the second biggest emissions producer of greenhouse gas in Germany and accounts for 25% of all emissions from Germany [34]. For years, the annual NO2 limits have been exceeded in many major German cities: Major causes in cities here are traffic jams and stops in particular in front of (signalized) intersections. Reliable signal timing estimation methods could be the key to minimizing unnecessary braking and accelerating in front of traffic lights. By utilizing signal timing estimation methods, as used in Green Light Optimized Speed Advisory Applications (GLOSA), a driver can plan ahead and avoid unnecessary stops, thereby reducing the overall amount of emissions, see Fig. 1.

L. E. Schneegans (✉) · J. Duensing · K. Heckmann · R. Hoyer
Fachgebiet für Verkehrstechnik und Transportlogistik, Universität Kassel, Mönchebergstraße 7, 34125 Kassel, Germany
e-mail: lena.schneegans@uni-kassel.de

C. Antoniou et al. (eds.), *Proceedings of the 12th International Scientific Conference on Mobility and Transport*, Lecture Notes in Mobility,
https://doi.org/10.1007/978-981-19-8361-0_16

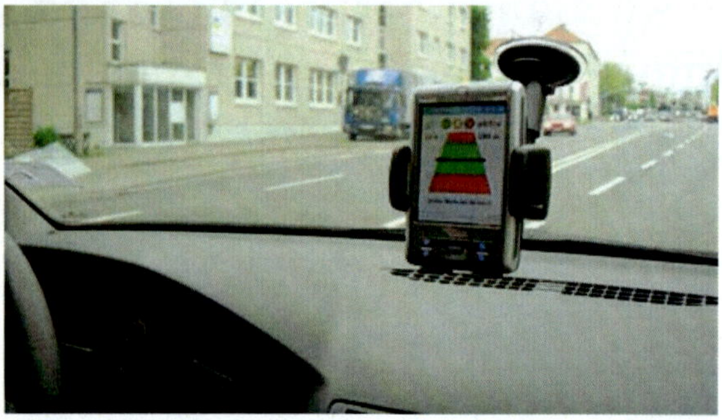

Fig. 1 Example of a residual red indication and speed recommendation of a GLOSA system from the AKTIV project [17]

Extensive research projects demonstrated the significant reduction of stops and emissions by the estimation of switching times. Concerning traffic signals and the optimization of approach speeds, Menig states that stops are almost completely avoidable [23].

Relevant articles also demonstrate the effectiveness of speed recommendations in case of unavoidable stops [1, 29].

The simulation experiments of Richter revealed that the influence of the driving behavior via speed recommendations resulted in up to 20% higher saturation traffic volumes during short green times and up to 8% higher saturation traffic volumes during long green times [30]. Fuel savings of 15% on average and 35% maximum were demonstrated. Nevertheless, there is a lack of switching times prediction procedures for these applications.

The demand for predictability of green starts and ends is increasing to make (smart) route planning and approach strategies to intersections more reliable and so environmentally friendly. Smart Routing is a method for urban navigation incorporating signalization information to avoid traffic jams and shorten travel times. Smart Routing in urban areas requires an exact time sequence of upcoming signal state changes of several (adaptive) traffic lights along the routes in question. For cooperative applications, e.g., in the field of autonomous driving, such Signal Phase and Timing information (switching times) predictions are even considered mandatory [31].

As described in [18] traffic lights are controlled in two different ways in Germany. Some traffic lights are controlled by fixed-time signal programs. Fixed-time controlled traffic lights are following a predefined pattern (order) of signals that is repeated in every single cycle. This makes signal timing estimation trivial. The second method for controlling traffic lights in Germany is by traffic actuated controllers. These take data from the area around the intersection into account when selecting the next stage of the traffic light. In addition to traffic detector data, a traffic light controller could also use information from a multitude of sources, e.g. public transport prioritization messages.

For these more complex and oftentimes custom-made programs signal timings are more difficult to estimate. This is exemplified by Figs. 2 and 3. Figure 2 shows the

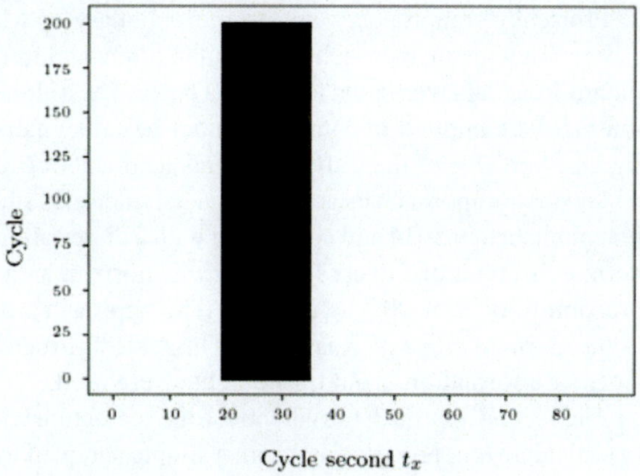

Fig. 2 Example of a fixed-time signal (green light is black)

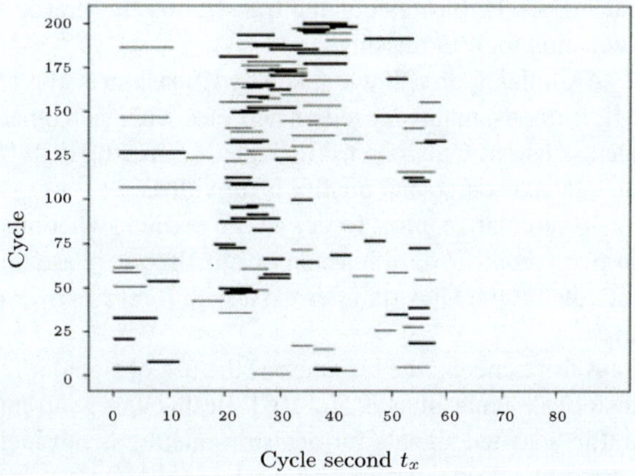

Fig. 3 Example of a stronger traffic-adaptable signalization, Druseltalstraße, SG 7 (green light is black)

phase timing of a fixed-time controlled signal while Fig. 3 shows the phase timing of a strongly traffic-adaptive signal. The green time of the fixed-time signal starts and ends at the same cycle second in each cycle (reference Fig. 2). In contrast, the start and end as well as the duration of the green time in Fig. 3 are adjusted to the current traffic situation within every cycle.

The estimation of traffic actuated signals cannot be done by simply collecting detector data and using the control program, due to the short horizon in time such a detection affects the control. To fulfill a horizon as demanded above, more sophisticated methods like machine learning (ML) methods are needed to predict future signal behavior.

To our knowledge, there are only a few authors who have used ML methods to predict the switching times of traffic lights. In addition, there are some statistical approaches, see for example [20, 21, 33]. However, the procedure of the latter is not considered suitable for practical use by [26].

Protschky's approaches are on the boundary of ML. He presents traffic signal estimation with an approach via Kalman filters and frequency distributions [26–28]. Floating car data were used as the data basis. The traffic light assistance and the smart routing were applied in Munich (model-based signalization) at approx. 600 traffic signals. For 71% of the traffic lights an accuracy of over 95% was reached.

Weisheit's approach uses a classical ML method: Support Vector Machines. It was first published in 2014 and completed with a dissertation in 2017 [36]. The prediction horizon is about one cycle long. For this horizon an accuracy to the second with a probability of over 90% is reached. The approach was tested at four intersections in the German cities of Kassel and Düsseldorf. Process data, e.g. saved signal and detector information, from traffic lights were used.

The second approach to switching time prediction with ML is presented by Genser in [12]. Here too, process data from a simple actuated traffic signal located in the city of Zurich is used, and as one of the first, Genser compared several ML methods with each other. Unfortunately, the transfer to other traffic signals with a more complex switching logic is missing.

A similar approach was taken by Kunashko et al. [22]. They also compare several ML methods, mainly based on neurons, with each other. The data model is comparable to Genser. Only one traffic signal is investigated. Unfortunately, no information on the data source and quality is provided.

To summarize, most forecasts can predict switching times with a tolerance of one to two seconds over a horizon of mostly one phase up to a maximum of one cycle into the future. Uncertainties exist with forecasts over the horizon of more than one cycle.

A difference in the literature is the use of signal information. In some cases, only historical signal images are used. In this case, current information that is used by traffic-actuated signals for decision-making is not included in the estimation. This contrasts with the use of current data from the last few seconds as in [22, 35] and [11]. So-called crowd data like floating car data is also used, see [26].

In this publication, current results on the switching time prediction of urban traffic-actuated signals using ML are presented. The results demonstrate an advancement of the dissertation by Weisheit [35]. The focus is on the research question: Which ML methods are suitable for predicting the switching times of traffic signal systems? The comparison considered here is more extensive than the one in [12], and [22] regarding adaptivity as well as the amount of considered signals and ML methods.

In the following sections, the methodology and data basis are first explained (Sect. 2). Then the results of these experiments are explained (Sect. 3) and finally summarized (Sect. 4).

2 Methodology and Data

2.1 Methodology

Initially, we selected ML methods for the experiments. The examined model types are:

1. K Nearest Neighbors (KNN) [6],
2. Support Vector Machines (SVM) (RBF ($_r$) and linear ($_l$) kernel) [25],
3. Decision Trees (DT) [4],
4. Random Forests (RF) [3],
5. AdaBoost (Ada) [7],
6. EXtreme-Gradient-Boosting (XGB) [5],
7. Multi Layer Perceptren (MLP) [15],
8. Recurrent Neural Networks (RNN) [37],
9. Long Short Term Memory neural networks (LSTM) [16],
10. Gated Recurrent Unit neural networks (GRU) [14],
11. Naive Bayes (NB) [2],
12. Seasonal AutoRegressive Integrated Moving Average with eXogenous regressors model (SARIMAX, SAR) [2].

We chose at least one representative method from each concept domain (neurons, decision trees, vector-based methods, statistics, and stochastics). For particularly popular or successful methods, several related methods were tested. If different implementations were available for classification and regression, both possibilities were tested (1–7 of the following list), since the search space of the targets can have one hundred and more possible classes, see Sect. 2.2. Results for the regressions were rounded to integers accordingly.

Only established implementations of standard ML-methods were used. For ML 1–5 as well as 7, 11 and 12 the package SKLearn was used; for 6 the corresponding package XGBoost; for the recurrent networks (8–10) torch and for SARIMAX statsmodels.

To prevent poor parameterization, the hyperparameters of all methods were tuned prior to testing and comparing their performances. For this bayes search was used [32]. Implementations in torch use 4 hidden layer with 30 neurons, Adam as optimizer, and cross entropy loss as training score. These experiments base upon four different data models/data sets (see Sect. 2.2). Root Mean Square Error (RMSE) (1) and accuracy (2) served as evaluation scores. Utilizing the number of estimations N, the true value y and the estimated value y'. The accuracy here thus represents the percentage of estimations that are accurate to the second. The significance test to compare the experiments is the Friedman Nemenyi post hoc test [8–10, 13].

$$\text{RMSE} = \frac{\sqrt{\sum_{n=1}^{N}(y'_n - y_n)^2}}{N} \tag{1}$$

$$\text{accuracy} = \frac{\text{number of correct estimations}}{\text{number of all estimations}} \quad (2)$$

2.2 Data Preprocessing

The urban traffic and roads authority of the German city of Kassel provided the data basis for the described experiments [19]. The data is available in OCIT format. It contains all input and output data required for control purposes. Data from September 2019 of two four-arm intersections was selected, cf. site plans in Figs. 10 and 11. Experiments were conducted for peak traffic volumes and off-peak hours during the daytime on Tuesdays, Wednesdays, and Thursdays. School vacations and holidays were excluded, as well as data on exceptional operating conditions.

Signal controls of these intersections are traffic-actuated. According to the staff of the responsible office in Kassel, they are representative for the traffic volume and traffic light controls in Kassel.

On the one hand, the intersection Druseltalstraße/Baunsbergstraße was selected. It is located on an access road to the city center to which traffic can approach unaffected by other junctions. This traffic light operates employing 22 signal groups (SG). Without a demand the main direction is signalized green. Switches only occur when a release is requested by other traffic streams.

On the other hand, the Katzensprung was selected. This junction is located in the city center of Kassel. This traffic light operates through 27 SG. The city center of Kassel has a high density of intersections. Only a few hundred meters spacing them. At the intersections, there are detectors. They detect motorized traffic demand, time gaps and pedestrian requests, and lanes with corresponding message notifications.

Among the examined SG, of the considered traffic lights, are strong and weak traffic dependent signal groups. The weaker the traffic dependency, the more similar the black and white pattern is to the pattern of a fixed-time controlled signal, see Fig. 2. Some SG are switched on demand and do not receive clearance every cycle, especially those for public transport.

The summary of the used features can be seen in Table 1. Target features of the predictions were the green starts (GS) and ends (GE) of all signal groups n. Specifically the count-downs (CD) in seconds until the next GS or GE of SG_i were used. As input features, the signal state information was used; first, in the fundamental representation (signal state (ST)) of the respective signal (0 stop, 1 clearance) for the current point in time; second, the duration since the last occurrence (time since (TS)) of these states in seconds [s]. In addition, two different input-data-sets of the signal groups were used: On the one hand, only the signal group also constituting the target, and on the other hand all signal groups available for the junction.

Table 2 depicts the executed experiments. For every SG a separate experiment was performed for the GS and GE, likewise for every method and intersection. The data models are kept as simple as possible. Therefore, the performance is low compared

Table 1 Summary of used features with assignment to data sets

Set	Feature (in/out)	Description
A_i	${}_{CD}SG_{GS,i}$	Countdown to next GS (int)
B_i	${}_{CD}SG_{GE,i}$	Countdown to next GE (int)
a_i	${}_{ST}SG_{GS,i}$	Signal state of GS (bool)
	${}_{ST}SG_{GE,i}$	Signal state of GE (bool)
	${}_{TS}SG_{GS,i}$	Time since last GS [s] (int)
	${}_{TS}SG_{GE,i}$	Time since last GE [s] (int)
b_n	${}_{ST}SG_{GS,i} \forall i \in n$	Signal state of GS (bool)
	${}_{ST}SG_{GE,i} \forall i \in n$	Signal state of GE (bool)
	${}_{TS}SG_{GS,i} \forall i \in n$	Time since last GS [s] (int)
	${}_{TS}SG_{GE,i} \forall i \in n$	Time since last GE [s] (int)

Table 2 Data sets for in- and output data as summary of experiments

target	input	repetitions for all data sets
A_i	a_i	- for all ML-methods of above listing
A_i	b_n	- likewise for DruseltalstraÃŸe
B_i	a_i	and Katzensprung
B_i	b_n	- targets $SG_i \forall i \in n$

to publications mentioned above. We aim for the relative performance of the methods not their absolute.

3 Results

Regarding the central question, of which ML method is best to use for signal timing prediction, there are several questions arising within.

- Should the problem be looked at as one for classifiers or regressions?
- What is the performance of the different data models?
- Can groups of users (individual motorized traffic, pedestrians and cyclists, public transport) be distinguished due to their impact on the prediction?
- Is there a difference between the results of the two here regarded traffic lights?
- Is there a difference in prediction quality for GS and GE?

The Tables 3, 4 and 5 show the obtained scores for exemplary signal groups of the traffic light Druseltalstraße which have been demonstrated in Sect. 2.2 with input a_i. Since approximately 50 signal groups were evaluated, not all results are shown. The evaluation was by comparison. Exploratory pretests with more complex data sets

Table 3 RMSE [s] for GS and GE with XGBoost

SG	4	11	18	24
GS	16.8	2025.4	11.6	8.1
GE	16.1	1654.9	5.1	9.7

Table 4 Accuracy [%] for all ML-methods for classifiers

SG	4GS	4GE	18GS	18GE	24GS	24GE
MLP	1.13	1.14	1.12	1.12	1.11	1.11
KNN	3.31	5.74	5.40	16.12	10.21	8.94
SVM_r	5.21	6.01	5.59	13.15	8.03	8.55
RNN	1.49	2.16	2.25	2.62	2.17	1.58
RF	5.63	7.07	6.77	28.09	16.36	10.93
XGB	8.47	9.57	10.62	44.01	22.28	20.88
Ada	4.10	5.09	5.05	6.81	5.76	5.86
GRU	5.67	6.60	6.03	12.43	8.61	8.90
DT	1.54	2.41	1.81	2.02	1.20	2.06
LSTM	1.05	1.05	1.41	1.11	1.59	2.30
SVM_l	4.10	5.09	5.05	6.81	5.76	5.86

Table 5 Accuracy [%] for all ML-methods for regressors

SG	4GS	4GE	18GS	18GE	24GS	24GE
NB	2.32	4.78	2.85	6.33	5.28	5.27
Ada	3.26	2.72	4.63	3.81	5.43	5.40
DT	4.73	5.21	5.48	14.61	8.50	6.67
SVM_l	3.63	3.98	4.32	42.62	9.83	5.05
MLP	2.01	1.10	2.58	1.11	1.36	2.63
KNN	3.35	2.93	3.05	12.84	5.44	6.06
SVM_r	3.16	4.36	4.21	24.82	10.48	6.89
RF	4.95	5.27	5.47	25.87	13.51	8.30
XGB	3.55	4.63	4.30	12.37	8.64	6.78
SAR	0	0	0	0.01	0	0

have shown that the scores are overall higher but conclusions drawn here sustain. Since they were sparser they are not shown here.

Table 3 shows the RMSE for the XGBoost classifier. Tables 4 and 5 show the accuracy for all considered methods. The evaluation scores RMSE and accuracy are equivalent to each other, with minimal differences in the ranking scores of Friedman-Nimenyi-post-hoc-test. SG 11 is omitted from Table 4 due to the poor results already seen in Table 3. This SG is according to technical documentation exclusively for public transport. Additional information about the arrival of public transport vehicles

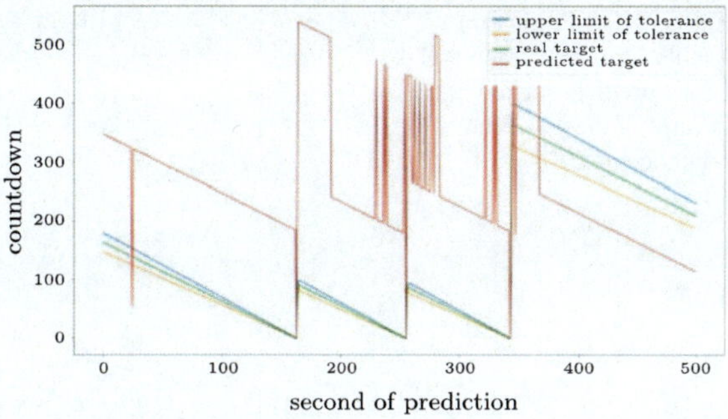

Fig. 4 XGB Prediction quality Druseltalstraße, SG 11 GS of prediction exemplary excerpt of the test data, tolerance of 10%

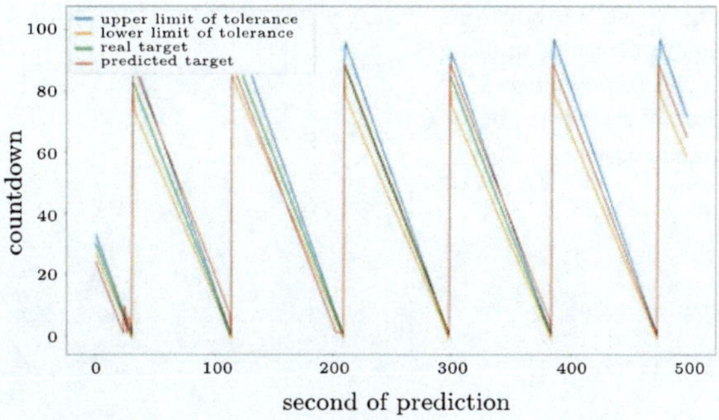

Fig. 5 XGB Prediction quality Druseltalstraße, SG 18 GE of prediction exemplary excerpt the test data, tolerance of 10%

in the data models is needed for such SGs. The information on the SG about itself does not allow conclusions about the arrival of these vehicles.

Table 4 displays the differences in quality between the predictions of the methods. The Friedman Nemenyi test confirms the differences seen in the table: The methods XGBoost and Random Forests offer significantly better results compared to all other considered methods. The significance level was 5%. Regarding these two winners, classifiers are performing better than regressors, see Tables 4 and 5. No difference was observed between the groupings of SG by user group. GS generally appear to be more challenging to predict than GE, which can be ascribed to the permission periods during rush hour. They lead to a reasonable constant green phase, but not always to a similarly consistent timing for GS. Figures 4, 5, 6 and 7 show excerpts of the predictions. They depict the target variable in green and its prediction in red. Furthermore, a tolerance funnel of 10% illustrates in blue and orange the time remaining until the signal switch, emphasizing user acceptance [24].

Figures 8 and 9 compare the confusion matrices for SG 24 (GE), which had a RMSE of 9.7 s and an accuracy of 21% at best, between XGBoost and RNN. The matrices map the predictions with the targeted values. The darker the colour, the higher the frequency. RNN prioritize certain timings of the countdown (see Fig. 8). The MLP regressor does similarly. XGB, on the other hand, detects all classes of the target. In case of an inaccurate hit, this method chooses neighboring classes, depicted as lighter lines (see Fig. 9).

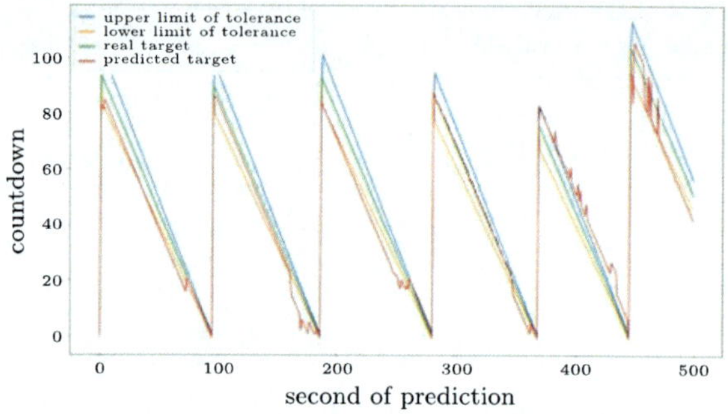

Fig. 6 XGB Prediction quality Druseltalstraße, SG 4 GE of prediction exemplary excerpt of the test data, tolerance of 10%

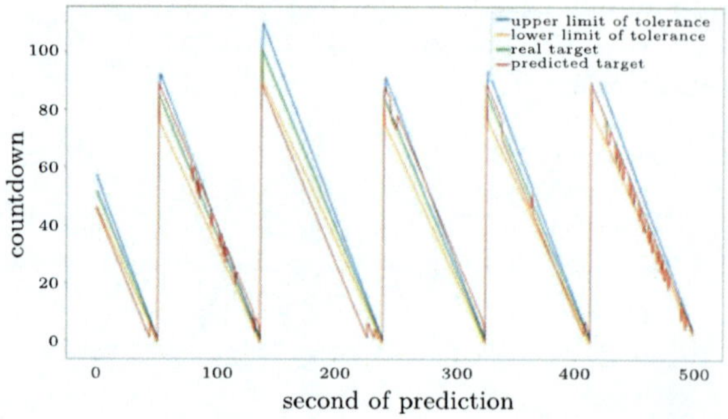

Fig. 7 XGB Prediction quality Druseltalstraße, SG 24 GS of prediction exemplary excerpt of the test data, tolerance of 10%

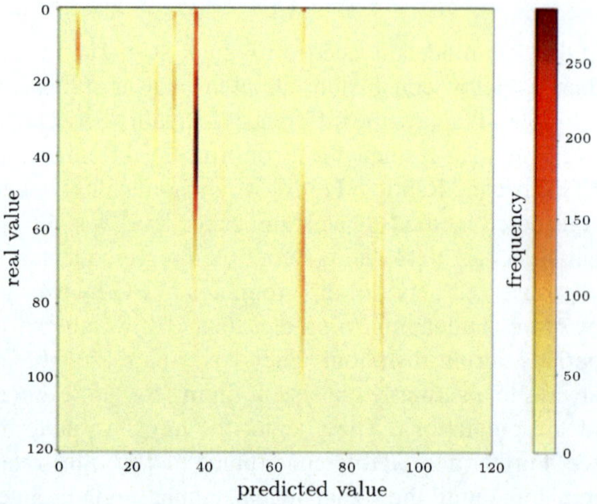

Fig. 8 Confusion matrix for RNN (LSA, Druseltalstraße, SG 24 GE)

The accuracy is obviously low, but regarding the RMSE deviations are only about a few seconds. Considering, that the accuracy needed by the user depends on his or her distance to the signal, slightly deviating estimations might be tolerable especially with an increasing forecast horizon.

The experiments on Katzensprung confirm the results presented here for the Druseltalstraße. The same applies to the results for the off-peak hours and inputs b_n. Between the user groups that can be assigned to SGs, no significant difference

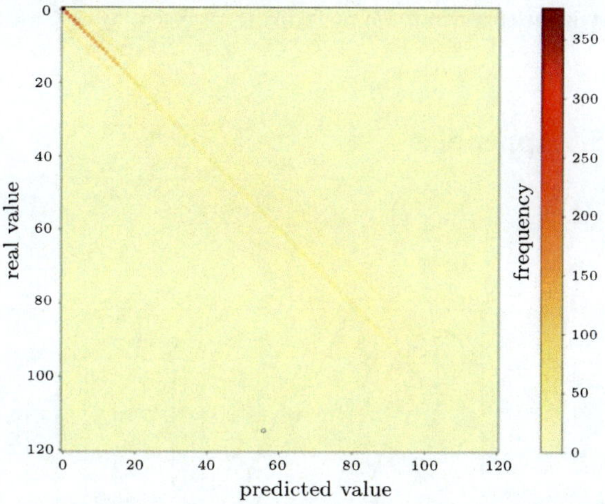

Fig. 9 Confusion matrix for XGB (LSA, Druseltalstraße, SG 24 GE)

was found. Regarding the data sets, the results with b_n were slightly better than those for a_i. However, no significant difference between a_i and b_n could be found. Additional experiments have shown, that a dedicated feature selection for every target based on b_n and further inputs e.g. detection information increase the estimation performance.

In all experiments, the decision tree-based classifiers XGBoost and random forest dominate among the best five ML methods ranked by Friedman-Nemenyi. The classifier XGBoost performs best in most cases. In cases where XGBoost does not lead to the best results, random forests does. In addition, SVM based on RBF and occasionally KNN can be found in the top 5, but never among the best 3. The best two methods were usually **significantly** better than the worse half. SARIMAX and all methods based on neural networks (RNN, LSTM, GRU, and MLP) belonged to the inferior half. SARIMAX ranked ordinarily last.

4 Conclusion

The experiments have shown by relative comparison, that the two decision tree based methods XGBoost and random forest are the best approaches for the considered traffic systems and data models. Of these XGBoost is significantly better. Neural networks (RNN, LSTM, GRU und MLP) perform worst alongside SARIMAX.

In further investigations, we would like to confirm the described results. For example, we want to test additional data models including detector data and communication with public transport. These models include another representation of the data as well as detector data and public transport messages of the junctions. Furthermore, we want to incorporate data from neighboring intersections.

Acknowledgements We would like to thank the city of Kassel for providing the data basis.

5 Appendix

See Figs. 10 and 11.

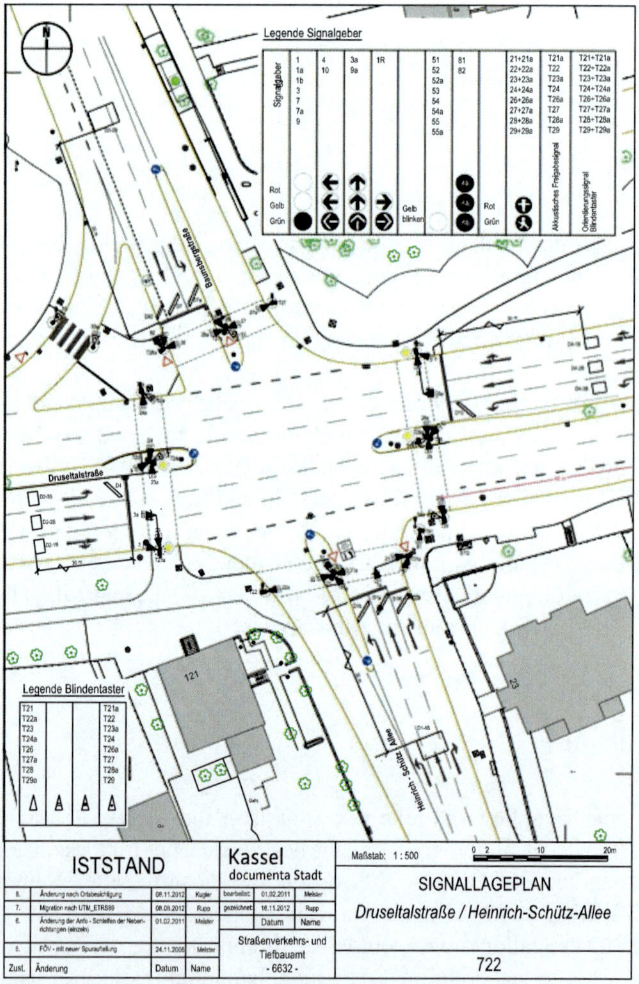

Fig. 10 Signallageplan Druseltalstraße/Baunsbergstraße (722) [19]

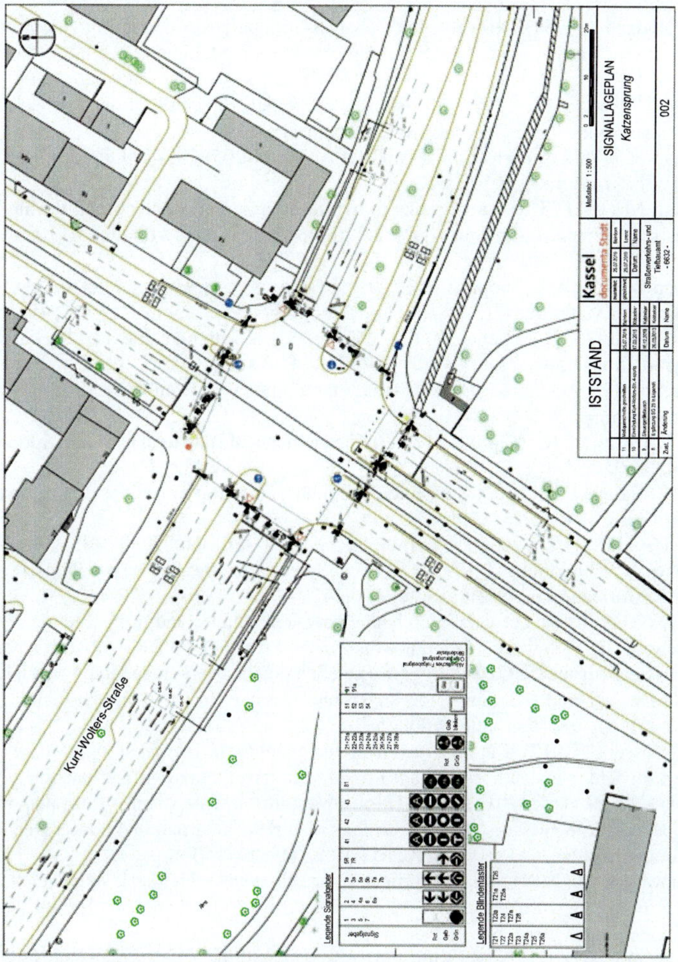

Fig. 11 Signallageplan Katzensprung (002) [19]

References

1. Asadi B, Vahidi A (2010) Predictive cruise control: utilizing upcoming traffic signal information for improving fuel economy and reducing trip time. IEEE Trans Control Syst Technol 19(3):707 714
2. Bishop CM (2009) Pattern recognition and machine learning. InforInformation science and statistics, Jordan M, Kleinberg J, Schölkopf B (eds), 8th edn. Springer (2009). ISBN: 978-0387-31073-2
3. Breiman (2001) Random forests. Mach Learn 45(1):5–32
4. Breiman L et al (2017) Classification and regression trees. Routledge
5. Chen T, Guestrin C (2016) XGBoost: a scalable tree boosting system. In: Proceedings of the 22nd ACM SIGKDD international conference on knowledge discovery and data mining, KDD '16. ACM, San Francisco, California (2016), pp 785–794. ISBN: 978-1-4503-4232- 2. https://doi.org/10.1145/2939672.2939785. http://doi.acm.org/10.1145/2939672.2939785

6. Fix E, Hodges J (1951) Discriminatory analysis, nonparametric discrimination: consistency properties. Technical report, USAF School of Aviation Medicine, Randolph Field (Texas), USA
7. Freund Y, Schapire RE (1997) Adecision-theoretic generalization of on-line learning and an application to boosting. J Comput Syst Sci 55(1):119–139
8. Friedman M (1939) A correction. J Am Stat Assoc 34(205):109–109. eprint: https://doi.org/10.1080/01621459.1939.10502372
9. Friedman M (1937) The use of ranks to avoid the assumption of normality implicit in the analysis of variance. J Am Stat Assoc 32(200):675–701. https://doi.org/10.1080/01621459.1939.10502372
10. Friedman M (1940) A comparison of alternative tests of significance for the problem of m rankings. Ann Math Stat 11(1):86–92. https://doi.org/10.1214/aoms/1177731944
11. Genser A et al (2020). Enhancement of SPaT messages with machine learning based time-to-green predictions. https://doi.org/10.3929/ETHZ-B-000458670
12. Genser A et al (2021) Time-to-green predictions: a framework to enhance SPaT messages using machine learning
13. Hastie T, Tibshirani R, Friedman J (2009) The elements of statistical learning. Springer. ISBN: 978-1-4899-0519-2
14. Heck JC, Salem FM (2017) Simplified minimal gated unit variations for recurrent neural networks. In: CoRR. arXiv:1701.03452
15. Hinton GE (1989) Connectionist learning procedures. Artif intell 40(1):185–234
16. Hochreiter S, Schmidhuber J (1997) LSTM can solve hard long time lag problems. In: Advances in neural information processing systems, pp 473–479
17. Hoyer R (2013) Effects of data interchange between vehicles and infrastructure at signalized intersections. In: 13th world conference on transport research (WCRT), Rio de Janeiro, Brazil
18. Hoyer R, Schneegans LE (2021) Prognose von Schaltzeiten verkehrsabhängiger Lichtsignalanlagen in Deutschland - Heraus-forderungen und Ansätze. In: at - Automa- tisierungstechnik 69.11 (2021), pp 931–939. https://doi.org/10.1515/auto-2021-0097
19. City of Kassel (2019) Traffic light technical documentation. Technical report. City of Kassel - urban traffic and roads authority (Stadt Kassel - Straënverkehrs-und Tiefbauamt)
20. Koukoumidis EI et al. (2011) SignalGuru: leveraging mobile phones for collaborative traffic signal schedule advisory. In: Proceedings of the 9th international conference on Mobile systems,applications, and services. ACM (2011), pp 127–140
21. Krumnow et al M (2014) Schaltzeitprognose verkehrsabhängiger Lichtsignalanlagen im Rahmen des Forschungsprojektes EFA 2014/2. In: Aspekte der Verkehrstelematik-ausgewählte Veröffentlichungen 2014, p 7
22. Kunashko A et al (2021) Application and comparison of machine learning algorithms in traffic signals prediction. In: 17th scientific and technical conference transport systems theory and practice, September 20–21
23. Menig C (2012) Optimierung von LSA-Fahrzeug Systemen durch Car-2-X-Kommunikation. Technical report, 15
24. Noll B (2018) Stadtverkehr der Zukunft- Stressfrei durch den Strassenverkehr. In: Strassenverkehrstechnik 62.3. Ed by Bonn-Bad Godesberg: Kirschbaum-Verlag
25. Platt JC (1999) Probabilistic outputs for support vector machines and comparisons to regularized likelihood methods. In: Advances in large margin classifiers. MIT Press, pp 61–74
26. Protschky V (2016) Schaltzeitprädiktion und Routenoptimierung für die Ampelassistenz in Smart Cities. PhD thesis. Ludwig-Maximilians-Universität München
27. Protschky V (2013) Verfahren zur Prognose des Verhaltens von verkehrsadaptiven Ampelverkehrsknoten mithilfe von Historiendaten. Ludwig-Maximilians-Universität München, Technical report
28. Protschky V et al (2014) Adaptive traffic and light prediction and via Kalman and filtering. In: IEEE intelligent vehicles symposium (IV). Dearborn, Michigan. Technical. IEEE
29. Raubitschek C (2011) Predictive driving strategies under urban conditions for reducing fuel consumption based on vehicle environment information. In: (2011) IEEE forum on integrated and sustainable transportation systems. IEEE, pp 13–19

30. Richter A (2005) Geschwindigkeitsvorgabe an Lichtsignalanlagen: technische Aspekte und volkswirtschaftlicher Nutzen. PhD thesis. Universität der Bundeswehr Hamburg
31. Santa C et al (2014) Potenziale kooperativer Lichtsignalsteuerung zur Steigerung der Verkehrseffizienz und-sicherheit. In: Straën-verkehrstechnik 10, p 676
32. Scikit-optimize-community. Scikit-optimize. Oct. 2020. https://scikitoptimize.github.io/stable/
33. Uebel S et al (2014) Energieeffizientes Fahren 2014 (EFA2014)-2. Projektphase Erhöhung der Reichweite von Elektrofahrzeugen, Technical report
34. Umweltbundesamt, (German Environment Agency). Entwicklung der energiebedingten Treibhausgasemissionen nach Quellgruppen. Sept. 2021. https://www.umweltbundesamt.de/daten/energie/energiebedingte-emissionen#energiebedingtetreibhausgas-emissionen
35. Weisheit T (2017) Ein Verfahren zur Prognose verkehrsabhängiger Schaltzeiten von Lichtsignalanlagen. PhD thesis. Universität Kassel
36. Weisheit T, Hoyer R (2014) Prediction of switching times of traffic actuated signal controls using support vector machines. In: Fischer-Wolfarth J, Meyer G (eds) Advanced microsystems for automotive applications 2014. Springer International Publishing, Cham, pp 121–129. 978-3-319-08087-1
37. Williams RJ, Hinton GE, Rumelhart DE (1986) Learning representations by back-propagating errors. Nature 323(6088):533–536. ISSN: 1476-4687